Green Energy and Technology

More information about this series at http://www.springer.com/series/8059

Walter Leal Filho · Dinesh Surroop
Editors

The Nexus: Energy, Environment and Climate Change

Springer

Editors
Walter Leal Filho
Manchester Metropolitan University
Hamburg University of Applied Sciences
Hamburg
Germany

Dinesh Surroop
Department of Chemical and Environmental Engineering
University of Mauritius
Réduit
Mauritius

ISSN 1865-3529 ISSN 1865-3537 (electronic)
Green Energy and Technology
ISBN 978-3-319-63611-5 ISBN 978-3-319-63612-2 (eBook)
https://doi.org/ 10.1007/978-3-319-63612-2

Library of Congress Control Number: 2017950023

Printed on acid-free paper

This Springer imprint is published by Springer Nature
The registered company is Springer International Publishing AG
The registered company address is: Gewerbestrasse 11, 6330 Cham, Switzerland

Contents

Introduction

Energy can transform lives. There is a direct and proven correlation between development and energy. People from developing countries have far greater prospects of lifting themselves out of poverty if they have access to affordable power. Energy usage affects all aspects of life, from employment and livelihood activities to education services. In fact, energy or rather clean energy is extremely important for structural transformation and poverty reduction.

The world is developing and progressing at an unprecedented pace and this has given rise to an increase in the energy consumption in order to meet all the demands and needs. The increase in energy consumption is not only ensuring human comfort or maintaining certain lifestyle, but it is also causing environmental problems such as climate change among others. Energy conversion in a sustainable manner and with a minimum effect on climate change and other strata of the environment are almost reaching crisis levels. Energy and environment are the two sides of the same coin called climate change.

With all the environmental impacts associated with the use of energy, there is an urgent need to come up with other alternatives or strategies, like alternative or clean energy. In this context, this book has been produced to provide the scientific community with latest research on the "Nexus Energy, Environment and Climate Change". The different chapters in this book provide the readers an overview of the recent research in the field of energy, environment and climate change. The studies in this book were presented at the first International Conference on energy, Environment and Climate Change. The book consists of three parts namely energy, environment and climate change.

The first part of the book focusses topics like energy system in the transport sector, modelling of electricity generation in the sugar industry, rural electrification using agricultural residues, stand-alone wind and photovoltaic system, restoration of closed landfill for installing renewable energy technologies, energy potential from waste tyres, thermochemical technologies for conversion of waste to biofuel and energy, energy production from municipal solid waste, replacing fossil fuels with PV systems and bioenergy conversion of organic fraction of municipal solid waste.

The second part of the book focuses on more environmental issues like process optimization of bioconversion of lignocellulosic to biofuel, co-digestion of lawn grass with cow dung and pig manure, application of different pretreatment techniques for enhanced biogas production, design and operation of a single-stage plug-flow digester using organic wastes, regeneration of absorbent in biogas purification and wastewater biorefineries which integrates water treatment and value recovery.

The third part of the book highlights climate change issues and provides topic like approach in climate change mitigation, the impact of a carbon tax on employment, calculation of the grid emission factor for the national electricity system, greenhouse gas emission reductions from solid waste management system under the Clean Development Mechanism and implementation of Eco-Driving as a Climate Change Mitigation Measure.

Walter Leal Filho & Dinesh Surroop

Part I
Energy

Decarbonising the Finnish Transport Sector by 2050—Electricity or Biofuels?

Klaus Skytte and Rasmus Bramstoft Pedersen

Abstract Finland has set ambitious long-term targets, which aim to reduce greenhouse gas emissions from the transport sector and the whole energy system by 2050. By utilising the energy system model STREAM, which includes the power, heat and transport sectors, this paper develops two alternative scenarios for the transport sector by 2050—one with a high percentage of electric vehicles (EV) and another with a high percentage of biofuels (BIO), and compares the scenario results with a known Carbon-Neutral Scenario (CNS) which is adopted from the Nordic Energy Technology Perspective (IEA in Nordic energy technology perspective—pathways to a carbon-neutral energy future, 2013a). The socio-economic value of the total system cost is computed and the system integration of the transport sector with the electricity and heating sectors is simulated with an hourly time resolution. This study finds that a Finnish transport sector with a high share of EV by 2050 leads to the lowest total annual system cost of the scenarios and yields a reduction by 2.3% compared to CNS. While the transport configuration in the BIO scenario achieves the highest total annual system cost which is 0.4% higher than CNS. The robustness of the results is tested through a sensitivity analysis which shows that the costs (investment and maintenance) of biodiesel cars and EV are the most sensitive parameters in the comparative analysis of the scenarios.

Keywords Energy system modelling · Transport · Electric vehicles · Biofuels STREAM model

---

K. Skytte (✉) · R. Bramstoft Pedersen
DTU Management Engineering, Technical University of Denmark, Produktionstorvet 426, 2800 Kongens Lyngby, Denmark
e-mail: klsk@dtu.dk

R. Bramstoft Pedersen
e-mail: rabpe@dtu.dk

W. Leal Filho and D. Surroop (eds.), *The Nexus: Energy, Environment and Climate Change*, Green Energy and Technology, https://doi.org/10.1007/978-3-319-63612-2_1

1 Introduction

In recent years, reduction of greenhouse gas (GHG) emissions from the power, heat and transport sectors have received worldwide attention. To mitigate climate changes, both national and international agreements on climate and energy related targets have been stated. Finland has set ambitious targets which aims to reduce the GHG emissions by 20% by 2020 and 80–95% by 2050 compared to the 1990 level (EC 2011; IEA 2013b). As an intermediate target, Finland has the ambition of integrating renewable energy sources (RES) in the energy system that accounts for 38% of the gross final energy consumption by 2020 (IEA 2013a).

In EU, a target share of at least 10% RES in the energy used in the transport sector by 2020 is stated (EC 2009). In 2011 the Finnish transport sector accounted for approximately 20% of the national GHG emissions (FMEE 2014; IEA 2013a). As stated in the Finnish national and climate strategy, emissions from the transport sector can be reduced by 15% by 2020 and up to 35–40% by 2030 compared to the 2005 level (FMEE 2014). Finland has, furthermore, a notable goal of implementing 20% renewables in the road transport by 2020 (IEA 2013a).

The Finnish power and heating system will experience a remarkable transition in the future, heading towards power and heat production based on carbon-neutral energy sources. Currently, the Finnish electricity generation portfolio is very diverse, and is primarily supplied by fossil fuels and nuclear. However, the future power generation portfolio in Finland is expected to rely on nuclear and forestry-related renewables (IEA 2013b). The share of electricity generated by nuclear power plants is expected to increase to the double by 2025, since the Finnish Parliament has approved new nuclear power plants to be constructed (IEA 2013b). Finland is the most forested country in Europe, hence there is a major potential of incorporating a large share of forestry-related renewables in the energy system. Biomass can be used in the power, heating and transport sector and will therefore play a prominent role in the future energy system (IEA 2013b).

Currently, the transport sector is highly dependent on oil-based fossil fuels, i.e. diesel and gasoline. Hence, a radical restructuring of the fuel use and vehicle stock is required in order to reduce the GHG emissions in the transport sector. This restructuring has to be carried out by introducing governmental initiatives in combination with improving technological development in the non-fossil technologies and enhancing infrastructure.

The transition towards a green transport sector by 2050 is related to stronger synergies with the power and heating sector (Skytte et al. 2017b) prompted by higher utilisation of electric and biofuel transportation. Hence, for analysing the transport sector, a complete analysis of the entire energy system, including the power, heat and transport system is required.

By using the simulation model STREAM (Münster et al. 2014), a holistic analysis of the transport scenarios and their system integration with the electricity and heating sectors will be conducted. Two scenarios for a fossil free Finnish transport sector by 2050 are developed. The first scenario (EVS) includes a high percentage of electric vehicles (EV) and the second scenario (BIOS) includes a high percentage of biofuel use in the transport sector.

Following the introduction and a review of previous studies, Sect. 2 describes the energy system model which is utilised to perform the simulation of the scenarios. Section 3 describes the 2050 scenarios developed in this paper. Section 4 presents and discusses the results obtained by the model simulations. Finally, a conclusion summing up all the findings is carried out in Sect. 5.

1.1 Previous Studies

Extensive studies considering a complete energy system analysis including the transport sector has been conducted. Nordic Energy Technology Perspective 2013 provides pathways to a Carbon-Neutral Energy Future by 2050 for the Nordic countries (IEA 2013a). The entire energy system in the Nordic countries is described by a Carbon-Neutral Scenario (CNS). In this paper, CNS represents the reference scenario and it is used to carry out a comparative evaluation of the two alternative transport scenarios.

Research on benefits for the power system and the customers have been carried out within the field of integrating a larger share of EV in the energy and transport system (Skytte et al. 2017a). The concept of vehicles-to-grid (V2G) has been explained, and the potential benefits of implementing V2G in the energy system have been investigated (Kempton and Tomic 2005). The influence of the power system investments when integrating a larger share of the variable renewable energy (VRE) source wind in combination with plugin electric vehicles, and heat storages has been studied (Kiviluoma and Meibom 2010). The results pointed to the findings that EV can store the electricity for later use and thus can increase the flexibility of the power system. Furthermore, the optimal configuration of an integrated power and transport system has been identified by utilising a linear optimisation model (Juul and Meibom 2011).

Comprehensive studies have been conducted considering a larger implementation of bioenergy and biofuels in the energy and transport system (e.g. Skytte et al. 2006). Börjesson et al. (2013) presented a review of studies which investigate biofuels future in the transport sector by utilising energy economic modelling. The study is comprehensive and finds that the future market share of biofuels in the transport sector varies significantly in the review studies. They find that biofuels play a key role in the medium term, while most of the reviewed studies favour hydrogen or electricity in the longer term.

2 The STREAM Model

The different transport scenarios are implemented in the energy system model Sustainable Technology Research and Energy Analysis Model (STREAM), which is a simulation tool that includes the complete energy system. The model provides an overview of the energy flow across all sectors, from the primary fuel through conversion technologies to the end-users. Furthermore, STREAM allows simulating different scenarios in order to perform comparative analyses of their solutions, considering the electricity, heat and transport sectors simultaneously (Connolly et al. 2010; Skytte and Christensen 1999).

The modelling tool consists of two sub-models, i.e. the energy flow model and the duration curve model, as illustrated in Fig. 1, both of which relies on a bottom-up approach. Even though STREAM is not an optimisation model, the operation of the energy system is optimised under the given conditions.

The model working flow and interactions of the two sub-models in STREAM is as follows (Münster et al. 2014; Skytte et al. 2017a)

- The Energy Flow Model provides an overview of demand and supply in a given year. Projection of demands for energy services is specified and consequently, the final energy consumption is calculated, according to the defined inputs for supply. Finally, an overview of GHG emissions, energy resources and fuel conversion is created.
- The Duration Curve Model computes the system balancing on an hourly time resolution. It includes correlations in the electricity and district heating systems: uncontrollable versus dispatchable electricity, combined heat and power, district heating generation, storage facilities, flexible and non-flexible demand, etc. The results computed in the Duration Curve Model is later used as input is to the Energy Flow Model in an iterative procedure.

In STREAM, the transport system is modelled as follows. The transport system is divided into four independent sectors: passenger, freight, agriculture and fishery.

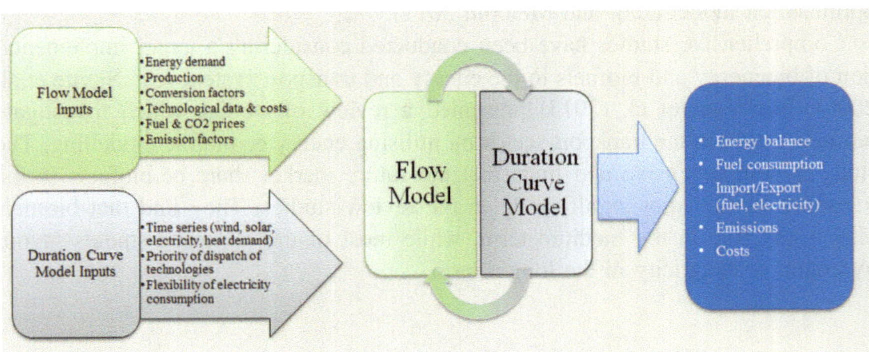

Fig. 1 Linkage between flow model and duration curve model in STREAM

The current transport work (passenger-kilometre, tonne-kilometre and energy consumption in agriculture and fishery) is specified. The projection for the demand in the year of simulation is computed by defining metrics for expected economic growth and specific energy intensity factors. Driven by the computed demand, the transportation sectors are described along with their synergies with the energy system. Furthermore, the transport work is allocated by vehicle type (e.g. car, bus, train, plane, bike, etc.), the utilisation degree (or stocking density), the composition of fuels for transportation means for both passenger and freight transport along with agriculture and fishery is defined according to the purpose of the scenario. Associated costs and emissions related to fuel production and consumption as well as vehicle acquisition and maintenance are accounted for. Finally, modelling of flexible fuel production and charging of EV are facilitated by the Duration Curve model.

2.1 *Main Data Assumptions*

To perform the quantitative assessment, metrics for the electricity and heat sector along with specific prices are adopted from the anticipative Carbon-Neutral Scenario (CNS) outlined in the NETP 2013 project (IEA 2013a). CNS push beyond the IEA's global two Degree Scenario (2DS) (IEA 2012), towards a near carbon-neutral energy system by 2050. CNS takes into account interactions between countries while hold promise of achieving the national energy policy targets. Furthermore, CNS representing a pathway which can potentially reduce the aggregated emissions by 85% by 2050 compared to 1990 levels.

Data for generation technologies and costs for vehicles (DEA 2013) are adopted from publicly available catalogues (DEA 2012, 2013).

Fossil and biofuel prices along with the CO_2 price are parameters that potentially have a sustainable influence on the final results. The price levels by 2050 of these parameters are presented in Table 1.

The price projections for fossil fuels, i.e. hard coal, crude oil and natural gas is adopted from NETP 2013 (IEA 2013a). The projections for biofuel prices are

Table 1 Fuel and CO_2 prices by 2050 which are implemented in STREAM

Fuel prices		
Natural gas	6.02	€/GJ
Nuclear–uranium	4.00	€/GJ
Biomass (straw, woodwaste)	9.10	€/GJ
Biomass (energy crops)	9.78	€/GJ
Biomass (manure)	0.00	€/GJ
Imported electricity	31.13	€/GJ
Coal	1.58	€/GJ
Oil	16.40	€/GJ
CO_2 price	120.30	€/t CO_2

adopted from EA Energy Analyses (EA 2013) and are an output from the Global Assessment Model (GCAM). The CO_2 price reflects the marginal abatement costs in the electricity system and shows a steep increment in price towards 2050 aided by the ambitious and strict energy policy targets in the 2DS.

3 Description of the 2050 Scenarios

In this study, two alternative transport scenarios towards a fossil free transport sector will be investigated, namely, EVS and BIOS, and compared with CNS. In NETP 2013, CNS contains a very diverse fuel mix in the transport sector, also including different biofuels. However, in this paper, the preferred biofuels are biodiesel and UPGR biogas.

The transport sector is currently almost independent of the power and heat generation, however, in the future their synergies will be stronger. Thus restructuring of the fuel use in the transport sector has an influence at the entire energy system. Heading towards a transport sector with a higher share of EV implies increased electricity consumption. Furthermore if EV relies on the concept of G2V, demand-side flexibility is provided to the system.

In this paper the power, district heat production and transport sector are implemented based on the following assumptions:

- The electricity system is designed in order to ensure consistencies in the final energy production between the two alternative transport scenarios and CNS, except for the electricity production by wind turbines. In EVS onshore wind is increased in order to cover the increased electricity demand, while in BIOS offshore wind production is reduced in order to meet the demand.
- In the district heating and industry sectors, the percentages allocation of the heating supply mix as in CNS are adopted in the STRAM model, thus inconsistencies in the actual production may appear.
- The transport sector is designed according to the purpose of the alternative scenarios, i.e. EVS or BIOS.

3.1 Resources

The chain from primary resources to the end consumers and thus the relation between the availability of resources and the service provided to the consumers is essential to investigate in order to evaluate the degree of self-sufficiency. By applying STREAM which covers the complete Finish energy system, this analysis is performed.

Bioenergy, in particular, plays an important role in the future energy system and as stated in NETP 2013, the Nordic region becomes a net importer of bioenergy in

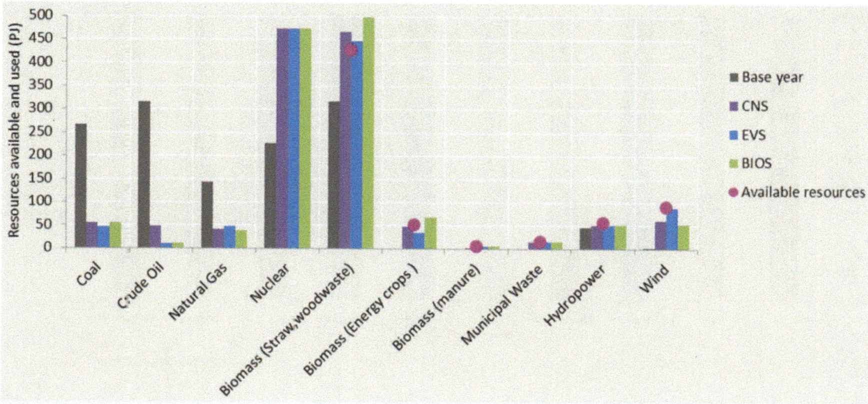

Fig. 2 Resource use compared to available resources for the scenarios

CNS 2050 and imports 13% of its supply. Considering the import of bioenergy to Finland in CNS, an energy content of almost 40 PJ is imported by 2050. Based on this assumption, import of bioenergy resources are allowed in the alternative transport scenarios, however they are compared with the national techno-economical available resources in Finland, which are estimated by VTT Finland and used in the NETP 2013 project.

Figure 2 presents the total use of resources in the base year (use in 2012) and in CNS, EVS and BIOS by 2050 compared to the techno-economic potentials.

Figure 2 shows the shift from the current fossil-fuel based energy system to an energy system which mainly relies on carbon-neutral energy sources. The contribution of each fuel-type varies between the scenarios, however the use of bioenergy, wind and nuclear in the 2050-scenarios accounts for a significant amount compared to the base year (2012).

In CNS, the Nordic region is assumed to become a net importer of bioenergy, i.e. straw and wood waste, by 2050. Furthermore, the use of fossil fuels, i.e. coal, oil and natural gas are still used to a small extent by 2050.

In EVS, the use of wind energy is just below the techno-economical potential. Furthermore, bioenergy resources are used in all sectors, and Finland imports a small amount of straw and wood waste. Moreover, the Finnish energy system is still dependent on fossil-resources by 2050.

BIOS is characterised by a large share of biofuel driven transportation which consequently causes a higher consumption of different types of bioenergy. Both straw and wood and energy crops exceed the techno-economical national available resources. According to the calculations, almost 70 PJ of straw and wood waste and 17 PJ of energy crops need therefore to be imported by 2050.

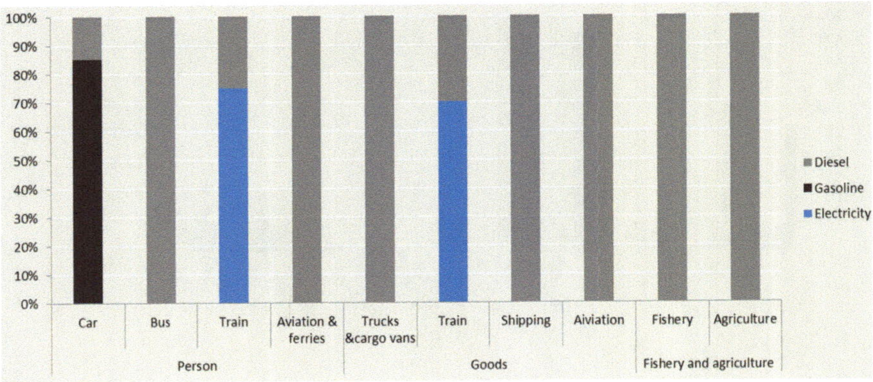

Fig. 3 Fuel use in Finland (2012). Measured as fuel consumption in percent person transportation work (pkm) or percent freight transportation work (tkm)

3.2 Description of the Transport Systems

Currently, the Finnish transport sector relies on fossil fuels as illustrated in Fig. 3 which shows the allocation of fuel consumption in 2012 by vehicle types.

Evidently, the fuel consumption is homogeneous and is dominated by oil-based fuels, i.e. diesel and gasoline, which accounts 96% of the total final consumption in the transport sector. Diesel is the dominating fuel used for aviation, shipping, busses and cargo vans transportation, whereas gasoline is commonly used by cars where it accounts for 85% of the total fuel consumption in this sector. Electricity is used for train transportation, however the use of electricity in the others sectors are limited to a minor share of the total energy use.

In order to meet the ambitious 2050-goal, radical restructuring of the fuel use and vehicle stock is required in the transport sector. Hence, the 2050-scenarios allocation of the fuel consumption is very diverse compared to the current situation. A careful description of the transport system for CNS, EVS and BIOS will be addressed in the following.

The projection for passenger and freight work by 2050 is based on CNS projections for all three transport scenarios. These projections are computed based on historical trends for transport work and economic growth along with assumptions for future transportation demand, transport work, efficiency improvements and technology switch.

The three transport scenarios vary in the composition of the fuel mix. The composition of the fuel mix in CNS is adopted from NETP 2013 and is shown in Fig. 4.

In CNS, 32% of the fuel use for cars is electricity whereas only 4 and 8% of the fuel use is diesel and gasoline respectively. The rest of the fuel mix for cars will be diverse with 29% share of bio-ethanol, 14% biodiesel, 4% UPGR biogas and 9% hydrogen (H_2). Additionally, the composition in the other transport categories has

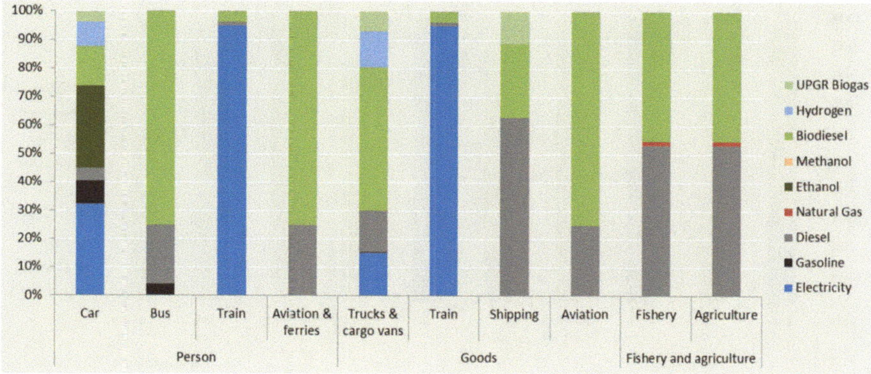

Fig. 4 Fuel use in the transport sector in CNS by 2050. Measured as fuel consumption in percent person transportation work (pkm) or percent freight transportation work (tkm)

much higher shares of electricity and biofuels. Within the biofuels, biodiesel is used as the main substitute for traditional diesel.

In EVS a high share of EV is integrated. Figure 5 shows the fuel mix in the transport sector chosen for EVS.

The electrification of the transport sector applied in EVS is illustrated in Fig. 5, and shows that approximately 90% of the cars by 2050 are EV and all trains uses electricity. The bus, cargo vans and short to medium distance trucks sectors become more electrified in the future, however the remaining fuel used in these sectors are biofuels, mainly biodiesel and upgraded biogas. In the shipping and aviation sector, biodiesel is the preferred fuel.

In BIOS, high utilisation of biofuels is the way to a fossil free transport sector. In the development of the scenario, the same shares of EV in the fuel mix as in CNS are adopted. Therefore, biofuels cover the remaining share of transport demand,

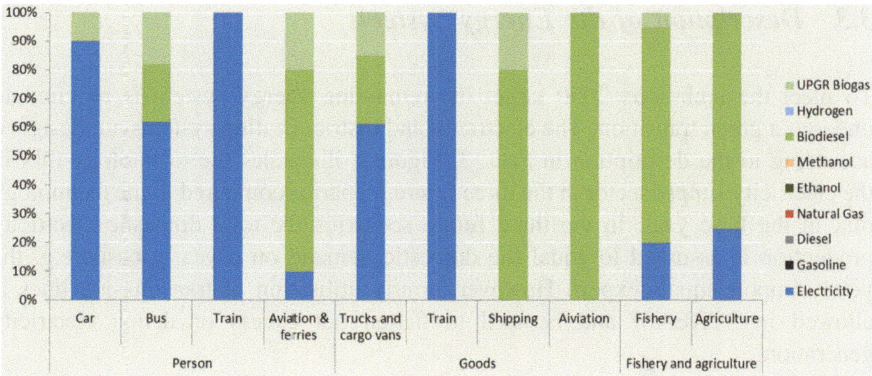

Fig. 5 Fuel use in EVS by 2050. Measured as fuel consumption in percent person transportation work (pkm) or percent freight transportation work (tkm)

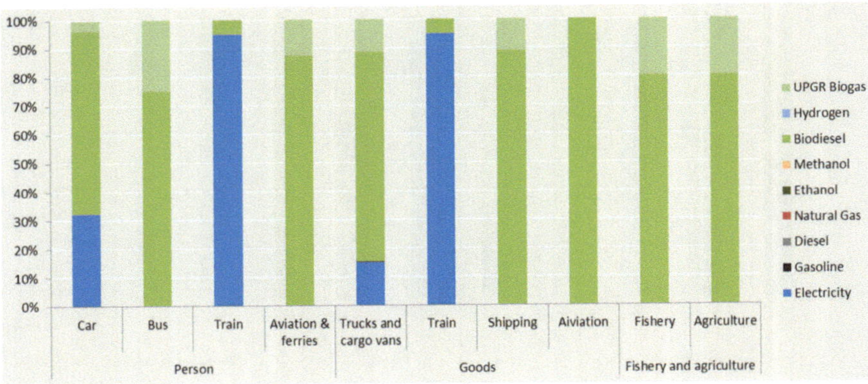

Fig. 6 Fuel use in BIOS by 2050. Measured as fuel consumption in percent person transportation work (pkm) or percent freight transportation work (tkm)

where biodiesel and upgraded (UPGR) biogas is the preferred fuel choices. Biodiesel chosen to be the key fuel in the future and UPGR biogas is used to the limit of the available resources.

The composition of the fuel use in BIOS is illustrated in Fig. 6.

As illustrated in Fig. 6, the fuel mix in BIOS is more homogeneous compared to CNS. Biodiesel is diesel-equivalent and thus it is heavily used in the future transport sector. Biodiesel used for car transportation is 64 and 75% for busses. Considering the transportation of goods 89% of shipping and 100% aviation uses biodiesel as fuel. When considering the share of UPGR biogas in the passenger transport sector, 4% of all cars and 25% of all busses uses UPGR biogas. Furthermore, 11% of cargo vans and short to medium distance trucks and shipping transport use UPGR biogas by 2050.

3.3 Description of the Energy Systems

To meet the ambitious 2050 target the remaining energy system is required to undergo a green transition. The electricity and district heating systems are designed according to the description in Sect. 3. Figure 7 illustrates the technology mix in the electricity supply sector in the three future scenarios compared to the technology mix in the base year. In the three future scenarios the total domestic electricity production is assumed to equal the domestic demand on a yearly basis, e.g. the yearly import equals export. However, hourly utilisation of transmission lines is allowed in STREAM and is used to flatten out excess or deficit electricity generation.

The Finnish electricity generation is, in the base year (2012), dominated by coal and nuclear power plants, which accounts for 22 and 25% of the total electricity

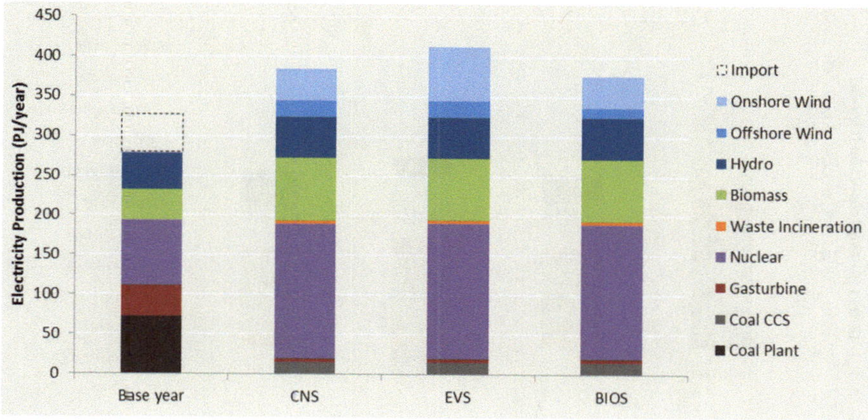

Fig. 7 Technology mix in the electricity in three 2050 scenarios compared to the base year

generation, respectively. Hydro, gas turbines and biomass based power generation have shares of 14, 12 and 12%, respectively, and therefore are important technologies at present. It is notable that 14% of the total electricity consumed in Finland is imported from the neighbouring countries.

According to CNS, a considerable reduction in the use of coal and gas will be replaced by carbon-neutral generation technologies. In Finland, nuclear power will be the major supplier of electricity by 2050, and Finland is expected to more than double the power generation from nuclear, i.e. from 2.7 GW to 6.4 GW by 2050 (IEA 2013a). Renewables will contribute to the electricity generation, where particularly the production from wind and biomass is expected to increase the contribution in the future. By 2050, electricity generated from wind turbines is 17 TWh and electricity produced using biomass as fuel is 21.8 TWh. Electricity production from hydro power plants is almost at the same level as today, while waste incineration utilises the full techno-economical potential to produce both heat and electricity, and therefore contributes with 1.1 TWh to the electricity production.

Figure 7 shows how the metrics for the electricity production in CNS is adopted in the two alternative transport scenarios, with onshore and offshore wind generation as variable parameters to match the annual demand. A higher electricity production is found in EVS as a consequence of the higher penetration of EV and heat pumps for the purpose of individual heating. The electrification of the transport sector and consequently higher demand for electricity leads to 7% higher electricity production from the electricity generation technologies in EVS compared to CNS. In order to meet the higher demand, this implies an increment in the electricity generated by onshore wind power 7.8 TWh, which means a total wind production at 24.8 TWh in EVS by 2050. In BIOS, the lower annual electricity demand compared to CNS is met by reduced offshore wind power production, as illustrated in Fig. 7.

The CHP-biomass plants are running in back pressure mode and thus produce both electricity and heat. In STREAM, electricity production is modelled according

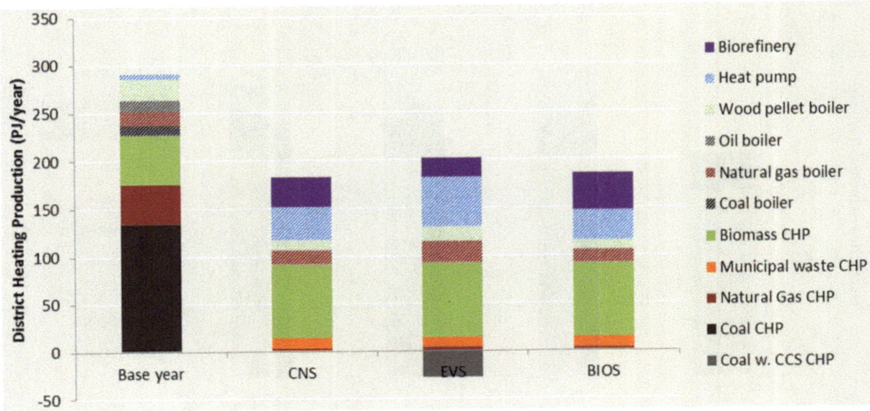

Fig. 8 Technology mix in district heating in three 2050 scenarios compared to the base year

to the electricity demand. Only as a consequence of this modelling structure, the heat production, which is associated with the electricity production appear due to the back pressure mode.

In 2050 the synergies between the transport, power and heating sectors are more coupled. Figure 8 compares the heating systems for the three future scenarios (CNS, EVS and BIOS) with the current situation.

The results regarding the district heating sector obtained by the model simulation is represented in Fig. 8. The technologies used to produce heat for district heating by 2050 includes CHP plants, district heating boilers and excess heat from biorefinery processes. From the investigation of the results, evidently, the heat production from the CHP plants and the refinery processes are not enough to meet the heat demand without boiler usage for all the scenarios.

EVS has the largest heat production from district heating boilers at 89.6 PJ while the values are 59.8 and 54.4 in CNS and BIOS, respectively. These production levels are allocated with the identical input of the percentage of the district heating boilers, namely 25% natural gas boiler, 58% heat pumps and 17% wood pellets boilers in the STREAM model.

Heat pumps are introduced and allow utilising excess wind energy for producing heat. The highest utilisation of heat pumps appears in EVS where 52 PJ heat is produced. This result shows that EV together with heat pumps allows a larger integration of VRE, e.g. wind energy.

Excess heat from biorefinery production can replace traditional heat supply. In BIOS, the highest contribution of heat from biorefinery is found. The heat production from biorefinery is 39 PJ in BIOS while 31 PJ and 20 PJ are produced in CNS and EVS, respectively.

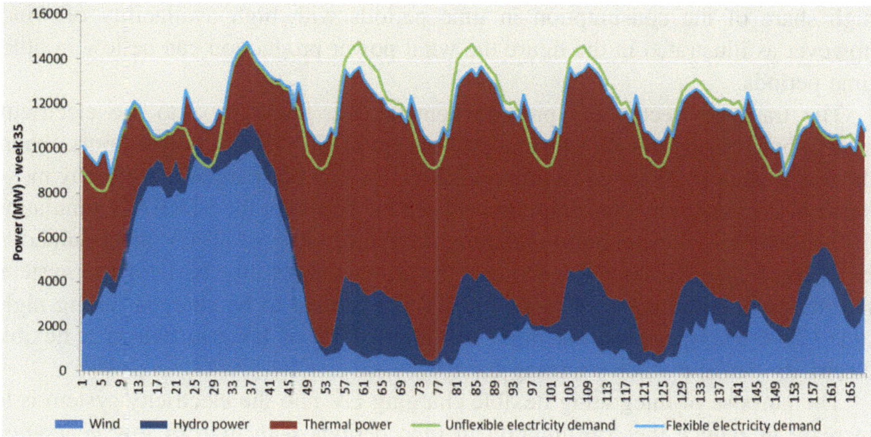

Fig. 9 Electricity generation and consumption in EVS with and without flexible demand in week 35 in the year 2050. Model simulation

4 Flexibility and System Costs

In general, the electricity generation portfolio by 2050 primarily relies on dispatchable power plants, i.e. nuclear, coal, biomass and hydro. The shares of non-dispatchable VRE generation, i.e. wind power, vary between the 2050-scenarios and are 21.6% in EVS, 15.9% in CNS and 14% in BIOS. The increased implementation of VRE, which has a stochastic electricity production profile, requires a flexible energy system in order to meet the associated challenges. In Finland, the integration of wind power is facilitated by the flexibility that hydro power reservoirs can provide and the transmission capacity to the surrounding countries.[1] However, the implementation of a large share of EV can provide additionally demand-side flexibility into the energy system.

4.1 Flexibility Added to the System

Figure 9 illustrates the electricity generation and consumption in EVS. The production is represented by the different electricity generation technologies in order to meet the required electricity demand. In the figure, the demand is represented by both the domestic electricity demand (green) and the electricity consumption including flexibility (cyan).

In the figure, the different electricity generation technologies are represented by different colour codes. Evidently, the wind power production in Finland can cover a

[1]This analysis investigate Finland as operating in an island mode.

high share of the consumption in time periods with high availability of wind, however as illustrated in the figure the wind power production can be low in other time periods.

The transport sector can provide demand-side flexibility into the electricity system by charging the EV utilising the concept of G2V. During night hours, where the traditional electricity consumption is low, EV can be charged and thereby move a part of the electricity consumption between the hours. In the STREAM simulation of the electricity consumption including flexibility, 40% of all EV are assumed to be charged in a flexible way in order to provide flexibility best for the system. Furthermore, additionally, 40% of the EV is assumed to be charged during night time (23–6 h). Hence, during night maximum 80% of EV can deliver a flexible electricity demand when charging the EV.

The purpose of integrating flexible charging EV into the electricity system is to increase or reduce the consumption in time periods with high or low generation from VRE technologies. In Fig. 9, the model increases the electricity demand in time periods with high shares of wind production and lower the electricity demand in hours with less VRE production. With increased demand-side flexibility, the system integration of a high penetration of VRE can be facilitated in a cost efficient way that keeps the cost for backup capacity down.

To keep the cost for backup capacity down, the model uses the flexible demand to minimise the thermal capacity i.e. the difference between the demand and the VRE production (red area) for each hour. During day time hours, which is characterised by high traditional electricity consumption, the demand flexibility from the EV reduces the demand in order to have less production from thermal power plant and thus less need for thermal generation capacity.

Since the charging of EV is more flexible during night hours, unusual demand peaks appear which indicates the demand shift from daytime to the night hours.

In BIOS, the contribution of demand-side flexibility added to the system is limited, due to the lower amount of EVs. The result of the hourly electricity generation and consumption from the model simulation of BIOS is illustrated in Fig. 10.

4.2 Systems Costs

The STREAM model computes the socio-economic cost of the entire energy system. Figure 11 presents the total annual system costs obtained by the model simulations. To identify the drivers for discrepancies in the system costs between the scenarios, the total annual cost for the entire energy system is disaggregated into: Cost of Energy Savings, Capital Cost, O&M Cost, Fuel Cost and CO_2 Cost.

The STREAM runs find the total annual system cost in CNS to 30,358 mill. €. By using CNS as a reference, the model computes the total annual system cost in EVS to be 29,659 mill. €, which corresponds to a reduction at 2.3%. BIOS is

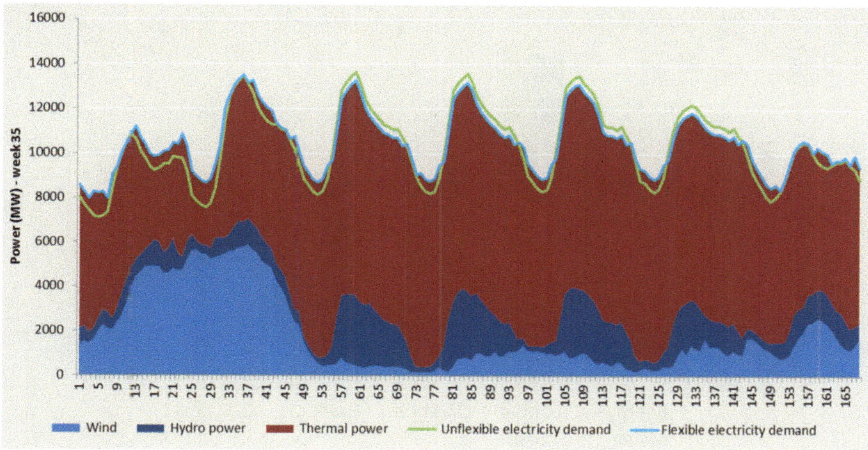

Fig. 10 Electricity generation and consumption in BIOS with and without flexible demand in week 35 in the year 2050

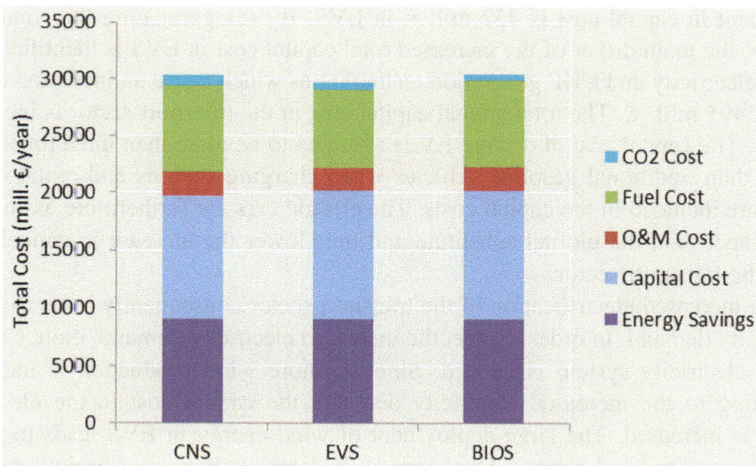

Fig. 11 Total annual system cost in the three scenarios by 2050

determined to be the most expensive with a value of 30,480 mill. € corresponding to an increment of 0.4% compared to CNS.

The quantification of the total annual system costs allows for a comparative analysis of the differences in the total system costs between the alternative transport scenarios and CNS, which is illustrated in Fig. 12.

In the following section, the main drivers for the differences in the system cost will be identified and further evaluated. EVS is the cheapest scenario, due to low fuel and CO_2 costs, however the capital costs are higher than in CNS. The total

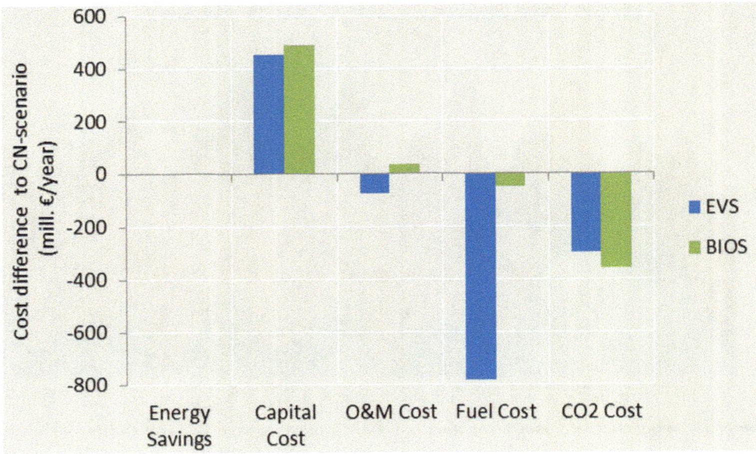

Fig. 12 Cost difference in total annual system cost by 2050 between the two alternative transport scenarios and CNS

increment in capital cost is 452 mill € in EVS. By disaggregating this value into sectors, the main driver of the increased total capital cost in EVS is identified to be in the electricity and CHP generation technologies which cause an increased capital cost at 495 mill. €. The total annual capital cost in the transport sector is increased by 5%. The capital cost of driving EV is assumed to be more than three times more costly than traditional gasoline vehicles when charging stations and rental of batteries are included in the capital costs. The electric cars are furthermore assumed to be cheaper than the biofuel substitute and thus lower the increase in capital costs from the transport sector.

The increased electrification of the transport sector consequently leads to higher electricity demand. In order to meet the increased electricity demand, more capacity in the electricity system is needed. Since onshore wind production is increased according to the increased electricity demand, the capital cost in the electricity sector is increased. The large deployment of wind energy in EVS leads to higher capital cost for wind energy. Furthermore, it demands increased backup capacity such as natural gas turbines. This subsequently implies increased capacity of natural gas turbines and thus increases the capital costs in EVS compared to CNS.

The results in Fig. 12 show a significant reduction in the costs related to the fuel in EVS. The main driver of reduced fuel cost in EVS compared to CNS is the costs related to the fuel used in the refinery processes. The model finds 81% higher fuel costs in the refinery processes in CNS compared to EVS. In total, this implies a large saving in fuel costs, even though the model finds approximately 50% higher fuel cost for district heating production technologies in EVS compared to CNS, due to higher demand for district heating boilers.

The model results indicate that BIOS is the scenario that has the highest total annual system cost. The increment in the capital cost compared to CNS is 489 mill.

€, corresponding to an increase at 10%, appear mainly due to high investment costs in biofuel transportation. The capital cost for bioenergy transportation is increased by approximately 60% compared to CNS. The CO_2 cost is reduced in BIOS compared to CNS. In CNS, fossil fuels are still used in the transport sector and thereby are related to a CO_2 cost.

4.2.1 Clarify Sensitivities of Main Assumptions

The robustness of the results obtained in the quantitative assessment will be tested by separately varying four parameters, i.e. prices of bioenergy, cost (investment and maintenance) of cars both for EV and biodiesel, and the CO_2 price, and evaluate the effects on the total annual system cost.

STREAM is a simulation tool and therefore changes in prices will not influence the composition of technologies in the sectors (Ravn and Skytte 2000). By varying the input prices, the total annual system cost will change with a constant slope. The percentage change caused by changing the parameters 1% is illustrated in Fig. 13.

The sensitivity analysis shows that the total annual system cost is most sensitive to changes in the prices of bioenergy resources for all three scenarios. The sensitivity to the costs (investment and maintenance) of the car types varies significantly

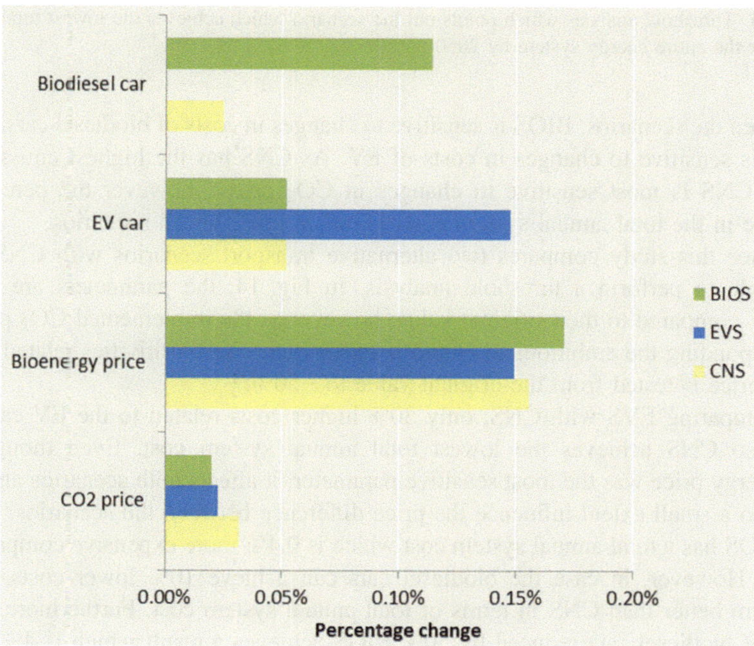

Fig. 13 Sensitivity analysis for the three 2050 scenario. The figure shows the percentage change caused by changing the parameters 1%

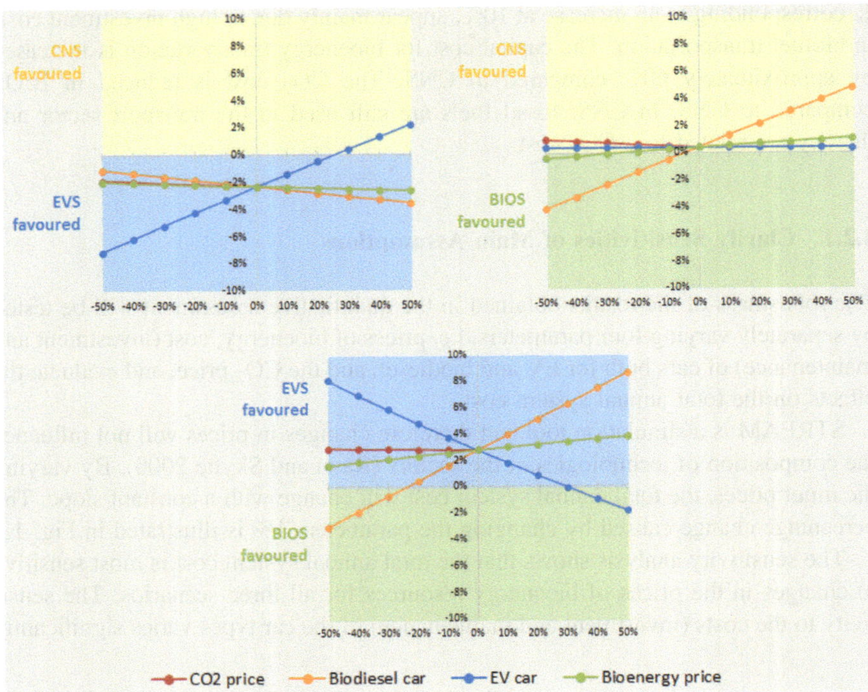

Fig. 14 Threshold analysis which points out the scenario which achieves the lowest total annual cost for the entire energy system by 2050

between the scenarios. BIOS is sensitive to changes in costs of biodiesel cars, while EVS is sensitive to changes in costs of EV. As CNS has the highest emission of CO_2, CNS is most sensitive to changes in CO_2 prices, however the percentage change in the total annual system costs is rather small for all scenarios.

Since this study compares two alternative transport scenarios with CNS, it is valuable to perform a threshold analysis. In Fig. 14, the parameters are varied ±50% compared to their original value, however as the implemented CO_2 price is corresponding the ambitious 2DS and is rather high, the sensitivities related to the CO_2 price is tested from the original value to −50%.

Comparing EVS with CNS, only 30% higher costs related to the EV causes a shift so CNS achieves the lowest total annual system cost. Even though the bioenergy price was the most sensitive parameter, it affects both scenarios and thus only to a small extent influence the price difference between the scenarios.

BIOS has a total annual system cost which is 0.4% more expensive compared to CNS. However, in case the biodiesel cars can achieve 10% lower costs, BIOS perform better than CNS in terms of total annual system cost. Furthermore, if the cost of biodiesel cars reduced by 50% BIOS achieves a result which is 4% better than CNS. High utilisation of bioenergy appear in both scenarios, however BIOS may perform better in case the bioenergy prices reduced by 30% and more.

Comparing the two alternative transport scenarios, it can be observed that the costs of cars is the parameter which can make BIOS better performing compared to EVS. In case either the cost of biodiesel cars reduces by 30% or the cost of EV is more than 30%, BIOS can achieve lower total annual system costs compared to EVS.

5 Conclusion

This study investigated the long-term role of EV or biofuels in facilitating the decarbonisation of the Finnish transport sector by 2050. By utilising a holistic energy system perspective, the transport sector was elucidated while taken into account future synergies between the power and heating sectors. The energy system model, STREAM, computed the socio-economic cost of the system and simulated the system integration of the transport sector with the electricity and heating sectors.

This paper compared a known scenario (CNS) adopted from Nordic Energy Technology Perspective 2013 with two alternative transport scenarios—either with a high percentage of EV (EVS) or with a high percentage of biofuel use (BIOS) in the transport sector.

The result shows that a Finnish transport sector with a high share of EV by 2050 leads to the cheapest cost solution under the given assumptions and yields a reduction in the total annual system cost by 2.3% compared to CNS. While the transport configuration in BIOS achieves the highest total annual system cost which is 0.4% higher than CNS.

Even though large bioenergy resources appear in Finland, the use of bioenergy resources scenarios exceeds their available domestic resources. The market price of bioenergy resources depends on the utilisation worldwide. Through a sensitivity analysis, the robustness of the scenario results was tested by changing bioenergy prices. Each of the scenarios was highly sensitive to changes in bioenergy prices, however introducing identical bioenergy prices in all scenarios, the results did not lead to different best performing solutions.

Biodiesel cars and EV are key technologies in the two alternative transport scenarios. It was shown that the scenario results were sensitive to changes in the costs (investment and maintenance) of these vehicle types and therefore show the importance of the future development of these two vehicle technologies. By reducing the costs of biodiesel cars by around 30% while keeping other input parameter constant, BIOS will be the scenario which achieves the lowest total annual system cost by 2050. Furthermore, by increasing the costs of EV by approximately 30%, EVS will no longer achieve the lowest total annual system cost.

References

Börjesson, M., Grahn, M., & Ahlgren, E. O. (2013). *Transport biofuel futures in energy-economic modeling—A review*. Report no 2013:10, f3 The Swedish Knowledge Centre for Renewable Transportation Fuels, Sweden.

Connolly, D., Lund, H., Mathiesen, B. V., & Leahy, M. (2010). A review of computer tools for analysing the integration of renewable energy into various energy systems. *Applied Energy, 87,* 1059–1082.

Danish Energy Agency and COWI. (2013). *Alternative drivmidler i transportsektoren 2.1.* May 2013.

Danish Energy Agency and Energinet.dk. (2012, May). *Technology data for energy plants—Individual heating and energy transport.*

EA Energy Analyses. (2013). For Danish Energy Agency. In C. Bang, A. Vitina, J. Sterling Gregg, H. H. Lindboe (Eds.), *Analysis of biomass prices—Future danish prices for straw, wood chips and wood pellets.* Published by: Ea Energy Analyses, Copenhagen, Denmark. Available online at: http://www.ea-energianalyse.dk/reports/1280_analysis_of_biomass_prices.pdf.

European Commission. (2009). *Directive 2009/28/EC of the European Parliament and of the Council of 23 April 2009 on the promotion of the use of energy from renewable sources and amending and subsequently repealing Directives 2001/77/EC and 2003/30/. Official Journal of the European Union, L 140/16-L 140/61.*

European Commission. (2011). *A Roadmap for moving to a competitive low carbon economy in 2050. Communication from the Commission to the European Parliament, the Council, the European Economic and Social Committee and the Committee of the Regions.*

Finish Ministry of Employment and the Economy. (2014). *Energy and climate roadmap 2050, report of the Parliamentary Committee on Energy and Climate Issues on October 16, 2014.*

International Energy Agency. (2012). *Energy technology perspectives 2012—Pathways to a clean energy system.*

International Energy Agency and Nordic Energy Research. (2013a). *Nordic energy technology perspective—Pathways to a carbon neutral energy future.*

International Energy Agency. (2013b). *Energy policies of IEA Countries, Finland, 2013 review.*

Juul, N., & Meibom, P. (2011). Optimal configuration of an integrated power and transport system. *Energy, 36,* 3523–3530.

Kempton, W., & Tomic, J. (2005). Vehicle-to-grid power fundamentals: Calculating capacity and net revenue. *Journal of Power Sources, 144,* 268–279.

Kiviluoma, J., & Meibom, P. (2010). Influence of wind power, plug-in electric vehicles, and heat storages on power system investments. *Energy, 35,* 1244–1255.

Münster, M., Larsen, H., & Iversen, J. (2014). DTU Risø and EA energy analysis, STREAM (sustainable technology research and energy analysis model) manual. August 2014. http://www.streammodel.org.

Ravn, HF., & Skytte, K. (2000). Uncertainty in energy-economic modelling of the electrical power sector. *Annals of Operations Research, 97*(1/4), 213–229.

Skytte, K., & Skjerk Christensen, P. (1999). An energy system simulator for long term planning. *Revista Mexicana de Fisica, 45*(5), 532–538.

Skytte, K., Meibom, P., & Henriksen, T. C. (2006). Electricity from biomass in European Union—With or without biomass import. *Biomass and Bioenergy, 30,* 385–392.

Skytte, K., Pizarro, A., & Karlsson, KB. (2017a) Use of electric vehicles or hydrogen in the Danish transport sector in 2050?. *Wiley Interdisciplinary Reviews: Energy and Environment, 6* (1), e233.

Skytte, K. Olsen, OJ., Soysal, ER., & Sneum, DM. (2017b) Barriers for district heating as a source of flexibility for the electricity system. *The Journal of Energy Markets, 10*(2), 1–19.

Modelling Electricity Generation from Sugarcane Production System Using Systems Dynamics

Mutanga Shingirirai Savious and Marne De Vries

Abstract Current global energy systems have proven unsustainable amid effects of the cumulative greenhouse emissions and climate change. The drive towards a low carbon future has precipitated the consideration of alternative energy sources. Among these sugar cane, grown widely in African countries, is known to be one of the most productive species in terms of its conversion of solar energy to chemical potential energy. However the supply of feedstock is limited to the harvest or crop season. More-so the sugarcane industry is faced with a plethora of threats and challenges. This paper seeks to broaden the understanding of the complexity in bio-electricity generation through a systems dynamics model. The model provides certain considerations for optimization of the energy value in sugarcane production systems. Among these is the use of trash as additive feedstock, and improvement in feedstock productions through enhanced sugarcane production systems. Apart from illustrating some of the policy considerations on land use change, sugarcane production, and improved technological efficiency the paper provides the effect on emission avoidance.

Keywords Energy · Bio-electricity · Sugarcane · Systems dynamics

1 Introduction

Current global energy systems have proven unsustainable amid effects of the cumulative greenhouse emissions (Jacob and Hilaire 2015; McGlade and Ekins 2015) and climate change (UNFCCC 1997). A recent surge in the area of renewable energy technologies in response to climate change (Cacho et al. 2003; Goldthau 2011; Oxfam 2010) fluctuations in international oil prices (UNIDO 2008),

M.S. Savious · M. De Vries
Department of Industrial Engineering, University of Pretoria, Pretoria, South Africa

M.S. Savious (✉)
Human Science Research Council, Pretoria, South Africa
e-mail: smutanga@hsrc.ac.za

W. Leal Filho and D. Surroop (eds.), *The Nexus: Energy, Environment and Climate Change*, Green Energy and Technology, https://doi.org/10.1007/978-3-319-63612-2_2

dependence on imported fossil fuels (Ramjeawon 2008) and the need for energy security (Cherp and Jewell 2014) have been witnessed in the last decade. Among the various renewable energy technology options are bio-derived fuels' in particular sugar cane, grown widely in African countries, is known to be one of the most productive species in terms of its conversion of solar energy to chemical potential energy. Similar to other renewable energy technologies sugarcane production systems generate electricity in a variable manner. The supply of feedstock is limited to the harvest or crop season. More-so the sugar industry has faced a plethora of threats and challenges which have hindered their growth across the Southern African region. One of the threats is the decline in sugar prices which affected the sugar industry in small developing islands, such as Mauritius. Competing priorities for land and water resources (Howells et al. 2013) and inefficient production plants (Mbohwa and Fukuda 2003) which have stalled the potential of electricity generation from bagasse. Despite these the opportunity of harnessing energy co-products from the sugar industry, such as ethanol and electricity is becoming increasingly attractive. This paper seeks to broaden the understanding the complexity in bio-electricity generation. Central is the drive towards optimization of the energy value in sugarcane production systems using a systems dynamic model. The demonstration model is used to:

- Determine the potential electricity and threshold of bagasse/trash as an energy source in Mauritius.
- Predict the environmental benefits from optimizing electricity value of sugarcane production systems.

The remaining content of the article is structured as follows. Section 2 defines the key terms applied in this paper and presents systems thinking as an approach to unpack the complexity in bio-electric generation. Section 3 defines the constructional components (four sub models) of the system dynamics demonstration model for Mauritius. Section 4 presents model results and a discussion; Sect. 5 is the conclusion with suggestions for future work.

2 Complexity Science: Sugarcane Production Systems

a. Defining salient terminology

Complexity theory: This can be defined as means of simplifying seemingly complex systems (Manson 2001). There is no single identifiable theory, instead a number of theories concerned with complexity system gather under the general umbrella of complexity research.

Energy and power: The three are often considered synonymous. Despite the fact that they are interrelated they are not the same. While energy is the ability to do work, power is its measurement, which calculates the time by which the energy has been used. In other words:

- Energy is a measure of how much fuel is contained within something, or used by something over a specific period of time.

Energy = power * time
kWh = Kw * h
Where: KWh is the energy
kW is the power
h is the time in hours

- Power is the **rate** at which energy is generated or used.
  kW is a unit of power.
  KWh is a unit of energy:

$$\text{Power} = \frac{\text{energy}}{\text{time}}$$

$$\text{kW} = \frac{\text{kWh}}{\text{h}}$$

Systems: *A system* can be defined as a 'complex whole of related parts' (Cabrera et al. 2008). It is the summation of different parts or entities related to each other, that constitute the observed whole. Considering a system implies cognisance that an observed phenomenon is an outcome of underlying complex interrelationships (Amigun et al. 2011). Similarly, alternative forms of energy such as biofuel encompasses a highly heterogeneous set of socio-technical systems implying there are underlying complex interrelationships too.

Systems analysis: System analysis is a structured way of analysing complex interrelationships that are problematic or simply of interest to mankind (Kirkwood 1998).

Systems thinking: At the heart of systems thinking is the recognition that factors behind the problematic situations are interdependent, that causal effect between these factors is often two-way, and that the impact of action is neither instantaneous nor linear. It is a formal, abstract, and structured cognitive endeavor on thinking about systems in general (Cabrera et al. 2008). Systems thinking make explicit causal-effect assumptions between related variables in a system, enabling independent assessment and improvement of mental models behind particular thinking (Datta 2008).

Systems dynamics is a methodology based on systems thinking. Systems dynamics provide the means to capture complex relationships and feedback effects within a set of interrelated activities and processes (Vennix 1999). The field developed initially from the work of Jay W. Forrester of the MIT Sloan School of Management with the establishment of the MIT System Dynamics Group (Forrester 1991). Systems dynamics (SD) therefore is a framework of modeling and simulation that can be used to understand the complex adaptive processes, also operating as a tool for experimenting with scenarios and policies for using bio-fuels as the successor of the fossil fuel regime in a virtual environment.

b. Unpacking the complexity in sugarcane production systems

Land has been identified as one of the major constraint for energy production in Mauritius. The total land and elasticity of arable land available determines the land available for sugarcane production. The cause-and-effect relationships are demonstrated in Fig. 1. The conversion rate from sugarcane land to other land use can be increased by the reduced sugarcane market price given that farmers will be reluctant to invest their time and money on an unlucrative and unviable farming practice. This can thus increase abandoned land and other land uses such as land for infrastructure development depicted as tourism land. The causal loop on Fig. 1 illustrates Policy interventions, such as provision of incentives to farmers, derocking farming practise coupled with increased desired land for sugarcane production. This shows the possibility of retaining or increasing the conversion rate from other land use to agriculture thus increasing land for sugarcane production and reduced abandoned land.

The land use for sugarcane is a key factor determining the total yield for sugarcane production. There are other factors that influence production among which includes water availability which can influence the yield rate and subsequently the total yield production. Policy intervention in the form of improved mechanisation,

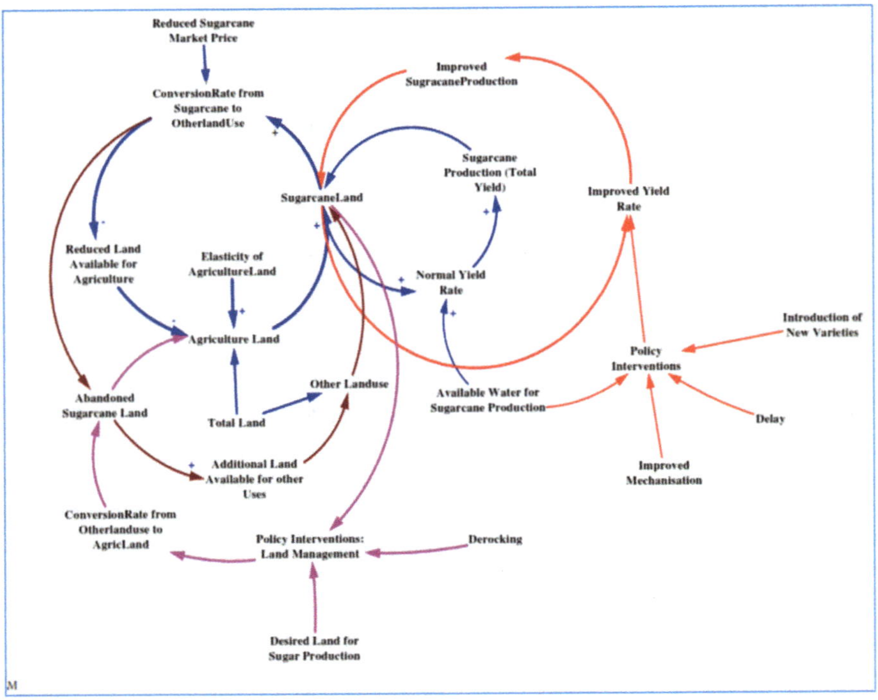

Fig. 1 Causal loop diagram for land and sugarcane production

introduction of new crop varieties can improve the yield rate subsequently improving the sugarcane production.

Illustrated in Fig. 2, high sugarcane yield with green harvesting technologies imply good supply of sugarcane processing waste such as trash and offcuts. In addition, increased cane processing lead to the production of more sugar and other by products. such as Bagasse. Increased production of bagasse positively affects available stock for electricity generation and an increase in power generation. Bagasse availability can attract capital investment in electricity generation, hence increased power generation. There are other exogenous factors in electricity production among which includes technology costs and maintenance costs. Supply of bagasse and trash feedstock is limited to the crop season. The more the feedstock to meet the desired feedstock for electricity generation the more the chances to preserve the feedstock inform of briquettes. The higher the briquette production the more the electricity generation off crop season period. Apart from feedstock supply, exogenous factors for electricity production may include among others moisture content.

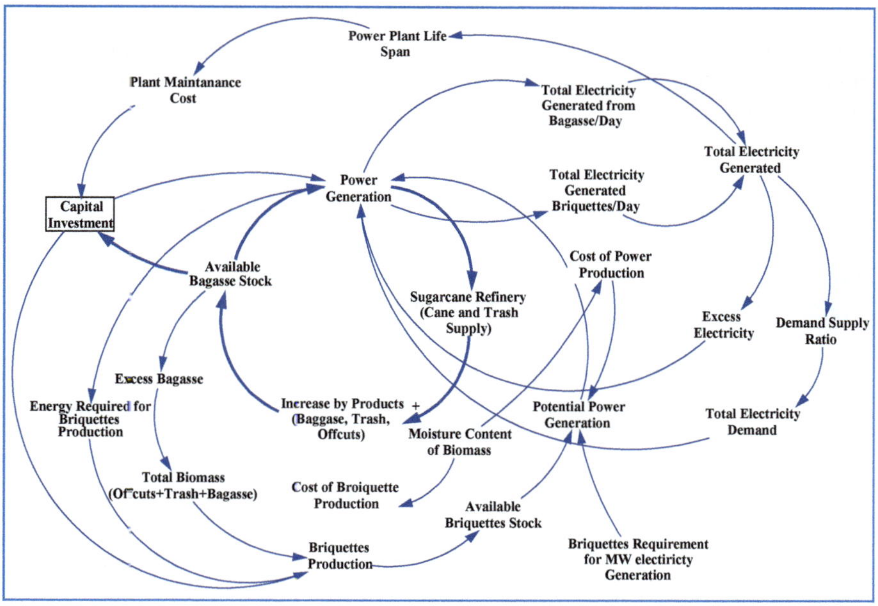

Fig. 2 Complexity in electricity production systems

3 Model Development

Based on the above premise system dynamics models were developed to optimise electricity value of the sugarcane production system in Mauritius. Figure 3 presents the models assumptions and boundaries. The next section provides the key assumption and the constructional elements of the model.

A. The Main Assumptions and Constraints of the Model

The model assumes a homogeneous landscape when simulating the land use change dynamics. The spatial variations on the landscape are not taken into consideration hence the production of sugarcane is influenced by the area under cultivation. The land available for sugarcane production is controlled by the total area under cultivation. A percentage of arable land is used for sugarcane production, however the changes in land use or total arable land might vary with increase in other crops. Since this study was undertaken for the entire Island, the threshold of sugarcane area has been based on the highest area under sugarcane production of 78,000 ha.

This study extracted the electricity production process requirements and parameter estimates from a previous study by Ramjeawon (2008) on life cycle assessment of sugarcane production systems in Mauritius (Fig. 3).

Based on the input data, electricity production process requirements and constraints, the following section describes the constructional components of the demonstration model.

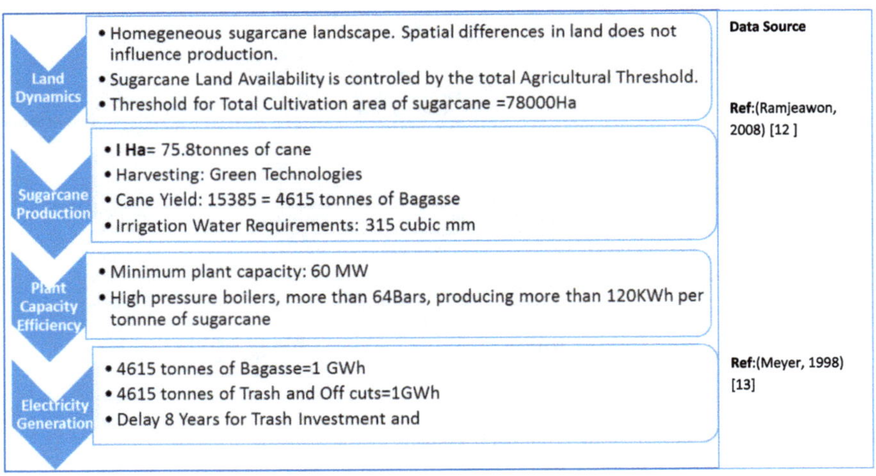

Fig. 3 The main assumptions and constraints of the model (Meyer 1998; Ramjeawon 2008)

B. Constructional Components of the Demonstration Model

I. Land and Biomass Production Sub Model

Essentially this sub model takes recognition of the land use changes which have been identified as one of the major constraining factor in Mauritius's sugarcane and bio-fuel production sector. Most importantly the sub model captures the competing priorities on land resources and the implications thereof. The changing land-use patterns are related to the total cane production and the subsequent bagasse output as feedstock for energy production. The rate of conversion of one land use to the other is also taken into account. While other studies have used land use parameters (Musango et al. 2012) they did not evaluate the impact of changing land use on production potential. This sub model also incorporates other resource constraints, such as water and climate, which have already been captured in previous studies on the CLEWS framework (Howells et al. 2013; Welsch et al. 2014). The sub model allows for investigating policy implications, such as the reformation of the sugarcane sector in Mauritius (Mutanga 2013). Population dynamics have also been taken into account since they influence settlement, food crops, and livestock requirements.

II. Harvesting and Electricity Feedstock Supply Sub Model

This sub model primarily focuses on building scenarios for preservation of sugarcane waste, in particular bagasse and trash. The sub-model infuses the green technology options that can be undertaken to ensure better utilization of sugarcane waste. While the known by-products of sugarcane production systems include ethanol, steam and electricity, this study focuses on electricity generation, since the potential of ethanol has already been captured in previous studies (AIEA 2011).

III. Electricity Generation and Technology Sub Model

The sub model takes into account technology efficiency as a critical factor for optimum electricity production (Mbohwa 2009). The effect of cogeneration is modelled in this sub model. The sub model builds scenarios comparing bagasse feedstock based plants including trash and offcuts feedstock, taking into account other competing priorities for these feedstocks.

IV. Bio-fuel Environmental and Profitability Sub Model

The sub model takes into account the potential environmental benefits. Essentially it models the emission avoidance based on the total annual sugarcane based electricity generation. In particular predictions focus on carbon dioxide and Sulphur dioxide avoidance based on the total annual electricity generation potential. The four sub models were constructed with the aid of systems dynamics software (STELLA). The next section report on a number of scenarios that were generated to guide decision-making.

4 Selected Model Results

(a) Modelling the effects of land use change dynamics on the current and future potential of cogeneration
From the Systems Dynamics model two scenarios were generated. Figure 4 illustrates the business-as-usual model compared to the alternative scenario. The results indicate that the business as usual (BAU) scenario is a gloomy picture for the sugar industry of Mauritius, characterised by continued decline in land for sugarcane production.

The depicted decline illustrated on Fig. 4(a) is likely to continue if no intervention policy measures in the sugar industry are put in place. Rapid population growth and growth of other sectors, such as tourism, result in conversion of agriculture land to other land uses. Projections show a decline to less than 55,000 ha by 2035 if no or little intervention measures are put in place. However policy interventions in the form of de-rocking (Deepchand 2005) incentives to farmers need to be intensified to help protect the sugarcane farmlands on the island. Although land used for other developmental priorities, such as infrastructure development cannot be reclaimed for sugarcane production, there is room for optimising the available land resources. As illustrated on the alternative scenario [see Fig. 4, Alternative Scenario (AS)] abandoned land can be reduced to almost 300 ha by 2025. Further research is key to ensure development of optimum land use strategies for the island to contain the threatened sugarcane land from other competing priorities. These efforts will reduce the increasing abandoned sugarcane land irrespective of the drop in sugar market price and other competing priorities, which have been identified to be an indirect factor in reducing the sugarcane land.

Land use change dynamics and sugarcane production: From the Systems Dynamics model two sugarcane production scenarios were generated. Figure 5 illustrates the business-as-usual model compared to the alternative scenario. The BAU on Fig. 5, illustrates a continued decline in sugarcane production to as little as just above 55,000 tonnes by 2035. This is not only a threat to the sugar industry but also the energy sector which has been relying largely on contribution

(a) Business as Usual (BAU) **(b) Alternative Scenario (AS)**

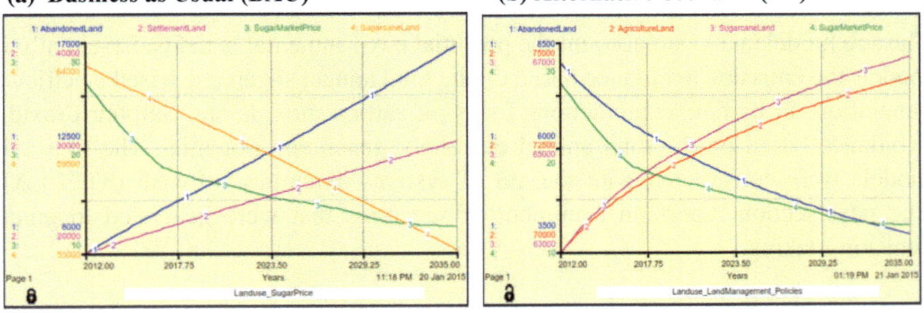

Fig. 4 Land use change dynamics

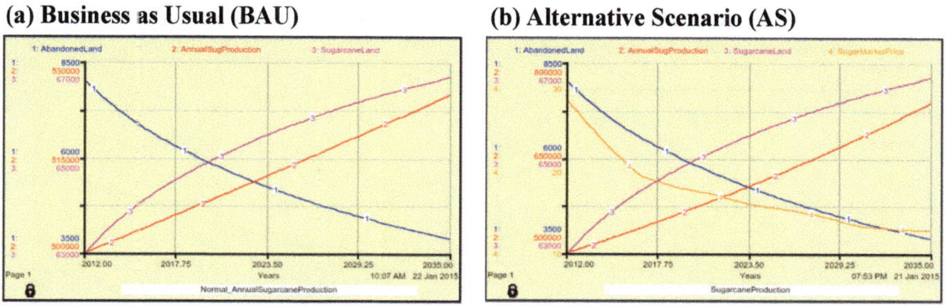

Fig. 5 Land use change dynamics and sugarcane production

of bagasse derived energy production mainly from the Independent Power Producers within the island. However the alternative scenario with intervention measure indicates that the highest recorded sugarcane production can still be achieved by 2025 through policy interventions. While production of sugarcane is not dependent on land only, other variables may also improve yields, such as climate factors, improved mechanisation, change of crop varieties and improved fertilisation. Monitoring and forecasting crop growth can aid in ensuring optimum yield production in the sugar industry.

(b) Determine the potential threshold of sugar systems as an energy source.

From the Systems Dynamics model two feedstock supply scenarios were generated. Figure 6 illustrates the business-as-usual model compared to the alternative scenario. The current generation is mainly based on bagasse as a feedstock. The business as usual scenario illustrated in Fig. 6 anticipates a continuation of this trend. Assuming the current feedstock supply is retained, the business as usual shows an increase in bagasse derived electricity to above 400 MW per annum by 2035. However this scenario is also dependent on the constant supply of sufficient feedstock (bagasse).

As indicated in Fig. 6, the alternative scenario projects that Bagasse and Trash derived electricity can contribute close to 500 GWh per annum by 2035 with strong policy interventions. This entails investment in trash and offcuts processing for

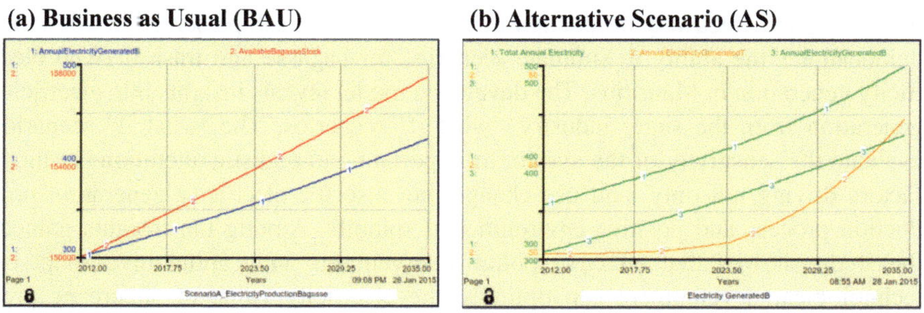

Fig. 6 Bagasse and trash electricity generation potential

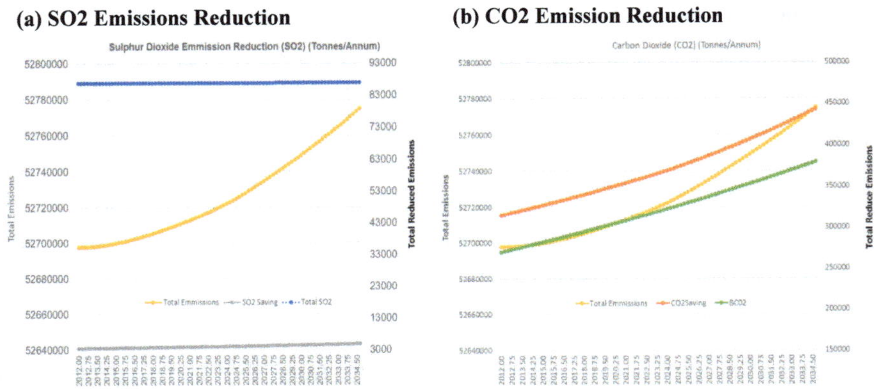

Fig. 7 Emission avoidance from sugarcane derived electricity production

optimum electricity from sugarcane production systems. The contribution of trash and offcuts as feedstock is only anticipated to take off 3 years from the time of this writing for this has been tried and tested in countries such as Brazil. Optimisation of electricity production can also be attained through improved technology. Some of the technological considerations could include but not limited to: improving steam conditions (Mbohwa and Fzweie 2002) making use of lower grade vapour for hearting purposes, improving boiler efficiencies, as well as replacement of steam driven mill drivers with electric DC motors (Mbohwa 2009).

Investment in bio-derived electricity generation can achieve more than 400,000 CO_2 tonnes and over 83,000 SO_2 tonnes of avoided emissions by 2025, as illustrated on Fig. 7(b) and (a) respectively. This illustrates the environmental benefits that can be accrued from optimising energy produced from the sugar based industrial systems.

5 Conclusion

The simple system dynamics model of land use change, sugarcane production, harvesting and electricity production from bagasse and trash presented in this study demonstrates the ability to simulate scenarios for bagasse and trash derived electricity generation in Mauritius. The developed model reveals insights into electricity generation from the sugar industry, 'what if' scenarios. The 'what if' scenarios evaluate the sensitivity of the system to important and realistic alterations in those factors driving not only land use change, but also the electricity generation production process and positive environmental spinoffs. Among the insights gained, the study showed that effective policy interventions and capital investment on technological development can optimise the electricity value of sugarcane production systems throughout the simulation period.

This work provides a good foundation for further research of the energy systems of not only small developing islands but most of the developing countries. Mauritius, while more advanced than most African countries in terms of sugar based electricity generation, is not alone in its trials to ensure energy security and drive towards a low carbon future and reduced importation and use of fossil fuels. The systems dynamics approach presented here provides a basis for further analysis of electricity generation and the various conversion pathways from sugarcane production to electricity production taking into account some of the limitations highlighted in this model.

Acknowledgements The authors gratefully acknowledge the contribution of the Southern African Young Scientist Programme under the auspice of the International Institute for systems Analysis (IIASA) and, National Research foundation of South Africa (NRF), Department of Science and Technology (DST) and the University of Free State. Special mention to mentors of the programme, Charles Mbohwa, Holger Rogner, Manfred Strubbeger, Dillip Kumar & Harold Annegun for their support. The authors thank the university of Mauritius, Central Electricity Board (CEB) of Mauritius and Independent power producers such as Ominican and Terragen, for the valuable information on energy in Mauritius.

References

AIEA. (2011). *Seeking CLEWS-climate, land, energy and water strategies—A pilot case study in Mauritius*. Vienna: AIEA.

Amigun, B., Musango, J. K., & Stafford, W. (2011). Biofuels and sustainability in Africa. *Renewable and Sustainable Energy Reviews, 15*(2), 1360–1372.

Cabrera, D., Colosi, L., & Lobdell, C. (2008). Systems thinking. *Evaluation and Program Planning, 31*(3), 299–310.

Cacho, O. J., Hean, R. L., & Wise, R. M. (2003). Carbon-accounting methods and reforestation incentives. *Australian Journal of Agricultural and Resource Economics, 47*(2), 153–179.

Cherp, A., & Jewell, J. (2014). The concept of energy security: Beyond the four As. *Energy Policy, 75*, 415–421.

Datta, L. E. (2008). Systems thinking: An evaluation practitioner's perspective. *Evaluation and Program Planning, 31*(3), 321–322.

Deepchand, K. (2005). *Parliamentarian forum on energy legislation and sustainable development*. Cape Town:South Africa.

Forrester, J. W. (1991). System dynamics and the lessons of 35 years. *A Systems Based Approach to Policymaking, 3*(2), 1–35.

Goldthau, A. (2011). Governing global energy: Existing approaches and discourses. *Current Opinion in Environmental Sustainability, 3*(4), 213–217.

Howells, M., Hermann, S., Welsch, M., Bazilian, M., Segerström, R., Alfstad, T., et al. (2013). Integrated analysis of climate change, land-use, energy and water strategies. *Nature Climate Change, 3*(7), 621–626.

Jacob, M., & Hilaire, J. (2015). Unburnable fossil-fuel reserves. *Nature Climate Change, 517,* 150–153.

Kirkwood, C. W. (1998). *System dynamics methods: A quick introduction*. Arizona: Arizona State University.

Manson, S. M. (2001). Simplifying complexity: A review of complexity theory. *Geoforum, 32*(3), 405–414.

Mbohwa, C. (2009). The energy and environmental impacts of a coal and bagasse-fired power plant in the sugar industry. *South Africa Sugar Technology Association, 82,* 214–224.
Mbohwa, C., & Fukuda, S. (2003). Electricity from bagasse in Zimbabwe. *Biomass and Bioenergy, 25*(2), 197–207.
Mbohwa, C., & Fzweie. (2002). Bagasse energy cogeneration in Zimbabwe: The technology, possible improvements and setting the right environment. *Harare* 1–8.
McGlade, C., & Ekins, P. (2015). The geographical distribution of fossil fuels unused when limiting global warming to 2 °C. *Nature, 517*(7533), 187–190.
Meyer, E. (1998). A model to estimate combine harvester and infield transport performance and costs. *Proceedings of the Annual Congress of the South African Sugar Technologists' Association* (vol. 72, pp. 58–60). South Africa.
Musango, J. K., Brent, A. C., Amigun, B., Pretorius, L., & Müller, H. (2012). A system dynamics approach to technology sustainability assessment: The case of biodiesel developments in South Africa. *Technovation, 32*(11), 639–651.
Mutanga, S.S, Pophiwa, N., & Simelane, T. (2013). Cities as green economy drivers: Making a case for green cities in South Africa. In S.S. Mutanga, T. Simelane, & N. Phophiwa (Eds.), *Africa in a changing global environment. Perspectives of climate change adaptation and mitigation strategies in Africa.* Pretoria: AISA.
Oxfam. (2010). Now more than ever: Climate talks that work for those who need them most. London. In Oxfam Godwell Nhamo (2011 Eds.), Green Economy and Climate Mitigation. Topics of Relevance to Africa. AISA South Africa.
Ramjeawon, T. (2008). Life cycle assessment of electricity generation from bagasse in Mauritius. *Journal of Cleaner Production, 16*(16), 1727–1734.
UNFCCC (1997). *Kyoto Protocol.* Kyoto: UNFCCC.
UNIDO. (2008). Scaling up renewable energy in Africa. Accessed February 20, 2012. http://www.unido.org/fileadmin/user_media/Services/Energy_and_Climate_Change/Renewable_Energy/Pub, Available from: http://www.unido.org/fileadmin/user_media/Services/Energy_and_Climate_Change/Renewable_Energy/Publications/ScalingUpweb.pdf.
Vennix, J. A. M. (1999). Group model-building: Tackling messy problems. *System Dynamics Review, 15*(4), 379–401.
Welsch, M., Hermann, S., Howells, M., Rogner, H. H., Young, C., Ramma, I., et al. (2014). Adding value with CLEWS—Modelling the energy system and its interdependencies for Mauritius. *Applied Energy, 113,* 1434–1445.

Author Biographies

Dr. Shingirirai Savious Mutanga is a well-established researcher in field of applied systems analysis, GIS and Remote Sensing. Recently he completed his doctoral studies in Industrial Systems Engineering with the University of Pretoria. Currently working as a research specialist under the Science and Technology programme of the Africa Institute of South Africa a division within the Human Science Research Council. He hold an M.Sc. in Geo-Information Science and Earth observation for environmental modeling and management, from a consortium of four universities namely Southampton (UK), Lund (Sweden), Warsaw Poland and ITC (Netherlands); and a Hons in Geography and Environmental Science from Midlands University, Zimbabwe. His work experience spans from academic, government, private and to research think tanks. Among his recent publications is the edited book Africa in a changing global environment, peer reviewed journal articles, policy briefs, monographs.

Dr. Marné De Vries started her career at a management consultancy company and worked as a consultant for seven-and-a-half years. During these years she worked as a Business and Systems Analyst where she was involved in various projects. In 2003 she joined the University of Pretoria. She completed her Ph.D. in 2012 within Enterprise Engineering, titled: *A process reuse identification framework using an alignment model.* She lectures on Information Systems Design and Enterprise Architecture.

Techno-Economic Viability of Husk Powered Systems for Rural Electrification in Uganda: Part I: Sustainability and Power Potential Aspects

Yonah K. Turinayo and Mackay A.E. Okure

Abstract Rural electrification by biomass gasification is a fairly new technology in East Africa. A Husk Power System (HPS) is one of such technologies utilizing agricultural residues as feedstock for power generation. Since biomass is considered carbon neutral, replacement of fossil fuel-based electricity by HPS derived electricity is one way to mitigate climate change through greenhouse gas reduction. Uganda is one among the countries in Africa having limited modern energy services such as electricity, though endowed with abundant biomass resources. Limited technical capacity to efficiently utilize biomass resources has been one of the major constraints of modern energy accessibility. In the more advanced developing countries like India however, efficient and environment-friendly technologies, such as HPS, have been developed and disseminated. In this study, a pilot HPS was imported from India and installed in Tiribogo village, Muduma Sub-county, Mpigi district in 2012 to supply electricity to rural communities through an isolated grid. But before extending such modern energy service to other parts of the country, there was a need to assess its technical performance, power potential of the fuel source and its sustainability in the selected agricultural rural areas. Coffee husks and rice husks were considered as fuel sources of a 34 kW_e HPS. The plant was operating for 6 h per day at a specific fuel consumption rate of 0.798 and 1.235 kg/kWh and overall conversion efficiency of 28.6 and 23.4% for coffee husks and rice husks, respectively. It was found that 36,355 tons of coffee husks and 12,591 tons of rice husks, equivalent to 21.1 and 4.6 MW_e basing on HPS performance, were generated annually from coffee and rice processing villages; this could sustain a total of 615 and 136 HPSs, respectively. However, it was more difficult to operate and sustain HPSs in villages connected to grid electricity and villages generating less than 480 tons of rice husks per year.

Y.K. Turinayo (✉)
National Forestry Resources Research Institute (NaFORRI), National Agricultural Research Organisation (NARO), P.O. Box 1752, Kampala, Uganda
e-mail: t2rinayonah@gmail.com

M.A.E. Okure
College of Engineering, Design, Art and Technology (CEDAT), Makerere University, P.O. Box 7062, Kampala, Uganda

W. Leal Filho and D. Surroop (eds.), *The Nexus: Energy, Environment and Climate Change*, Green Energy and Technology, https://doi.org/10.1007/978-3-319-63612-2_3

Keywords Agricultural residues · Husk power system · Rural · Electricity Coffee husks · Rice husks

1 Introduction

Due to the increase in global energy demand, diminishing fuel reserves, and increasing environmental concerns, the need to find alternative energy sources is crucial. Uganda is endowed with renewable resources which can, to a large extent, meet its energy demand if used sustainably. Within its national development plan and Renewable Energy Policy, the Government of Uganda has outlined the need to develop a wide range of energy sources which include: increased biopower, hydropower, gas- and oil-fired power plants, geothermal as well as grid extension and rural electrification (GOU 2010; MEMD 2007). However, the country ranks among the lowest in Africa in terms of National electricity access, estimated at 14%, (EU-AITF 2014). Basing on recent estimates by Uganda Bureau of Statistics (UBOS 2014), Uganda has a total installed electricity capacity of 827.5 MW [12.0% Thermal, 83.6% Hydroelectric, and 4.4% Biomass (bagasse)].

Biomass constitutes over 90% of Uganda's energy and is estimated at 460 million tons of standing stock, with a sustainable supply of 50 million tons (Energypedia 2015). However, it has continued to be used in its traditional form, largely as fuelwood and crop residues, using less efficient energy technologies (Okello et al. 2013). Overdependence on fuelwood has contributed to deforestation whereby 28.5% of forest area has significantly reduced between 2005 and 2010, leading to climate change and loss of biodiversity (UBOS 2014). Biomass consumption is increasing at about 6% per year with a potential for a shortage in future (Farley et al. 2011). Use of agro-wastes and efficient biomass energy technologies for energy production could, therefore, reduce on the rate of deforestation, hence promoting environmental and energy conservation.

Uganda being a predominantly agricultural country within the tropics motivates the exploration of the utilization of biomass energy resources. Indeed, growing of crops including coffee and rice is widespread. The processing of such crops generates waste biomass in form of husks that are ready for use as a feedstock for electricity generation if converted to producer gas (Buragohain et al. 2010). Nevertheless, there must be a sustainable supply of the biomass feedstock for gasification to be a reliable technology for rural electrification (Scalia et al. 2012). In this regard, agriculture in the rural areas generates a substantial amount of agro-waste (about 4.5 million tons/year) that could potentially produce over 400 MW_e (MEMD 2007, 2012; MFPED 2014). Estimates of 16 and 95 MW_e are expected to come from rice and coffee husks, respectively (MEMD 2007). Given this energy potential, the technology of making that energy available to Ugandan consumers has been a challenge. This could be due to inability of biomass energy technologies to compete (in terms of cost, energy conversion efficiency and resource availability) with fossil fuels, solar, hydro and other energy technologies.

2 Husk Power System (HPS) and Gasification Technology

Among the new emerging biomass energy technologies, an HPS has surfaced as a model in the renewable energy sector and has proved to be an attractive option for rural electrification (Singh 2010). Husk Power Systems are biomass gasification power plants (BGPPs) utilizing agricultural residues as feedstock, gasification as a technology and a syngas engine-generator combination as a prime mover to supply rural households with electricity. Figure 1 shows a typical setup.

As outlined by Hasler and Nussbaumer (1999), a downdraft fixed bed gasifier, coupled to an internal combustion engine, is the most appropriate technology for small-scale electricity generation (lower than 1 MW_e). In this type of a gasifier, the gas produced in the reactor during gasification is forced to pass through a high-temperature throat, producing a low tar content syngas. The syngas can, to some extent, replace engine oil in diesel engines, or be used in gasoline engines, where a generator coupled to the engine produces electricity. However, the gas is first cleaned and cooled using wet gas scrubbers before being used as a fuel in an engine. Biomass gasifier-engine sets usually exist in capacities varying from 10 to 500 kW_e and, operate on either dual-fuel producer gas engines or 100% producer gas engines. In the former, diesel engines are customized to run on a mixture of diesel and producer gas where, usually, 60–85% diesel is substituted for producer gas. In the later, 100% producer gas engines operate on 100% producer gas, similar to HPS in the present work.

Data on the performance of gas turbines and engines running on fossil fuels is available, but little information has been found on biomass-derived fuel gas engines, owing to the unknown costs of modification, maintenance and engine life (Bridgwater 1995). Although both dual-fuel diesel and gasoline engines, operating on low heating value gases, may be considered as fully developed, combination of a biomass gasifier and engine is not fully established (Bridgwater et al. 2002). Therefore, there is a need for further study on development and performance of integrated biomass gasifiers and engines, to generate electricity.

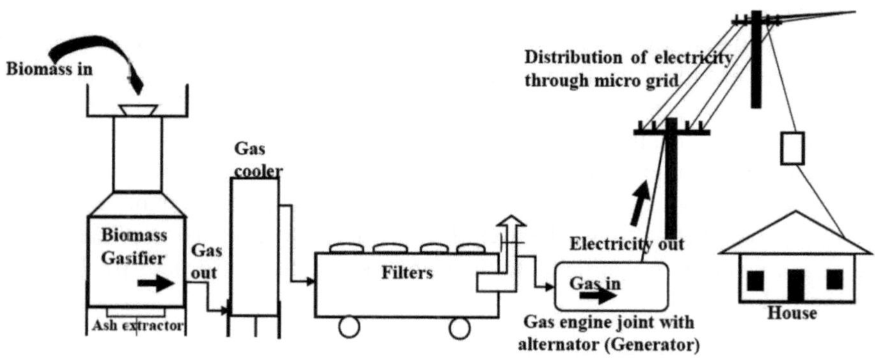

Fig. 1 A generic model of electricity generation and distribution by HPS

Several studies on the development of small-scale distributed BGPPs running on wood biomass have been conducted in Uganda (Buchholz and Da Silva 2010) and elsewhere in the world (Bridgwater 1995; Caputo et al. 2005; Fracaro et al. 2011). However, little has been done on BGPPs utilizing agricultural residues as fuel source, such as HPSs, in Uganda. Basing on feedstock availability, agricultural residues are not available throughout the year, and the quantities available depend upon agro-climatic conditions such as harvesting time, rain fall distribution, soil characteristics, storage related characteristics and the storage facility (Hyuha et al. 2005; Tripathi et al. 1998). On this note, all agricultural residues are not available in all parts of the country. It is with this background that estimation of the rice and coffee husks availability in particular areas of Uganda could be made. This paper presents findings on a study to evaluate the technical viability of HPSs in mainly agricultural villages of Uganda to support its adoption for rural electrification. The focus of this study was on assessment of fuel feedstock (coffee and rice husk) availability, its power potential and sustainability for HPS in supplying off-grid electricity to the rural communities in Uganda. Performance characteristics, considering fuel conversion efficiency, net electrical power output and biomass consumption of HPS were also determined (Caputo et al. 2005; Scalia et al. 2012).

3 Methods

A technical feasibility study to assess the sustainability of HPS and power potential of crop residues produced in rice and coffee processing rural areas of Uganda was done using a pilot HPS located in Mpigi District as a model. The electrical power output of HPS, biomass fuel feeding rate into the system, lower heating value (LHV) of rice and coffee husks, overall electrical energy conversion efficiency of the system, crop residual ratio and the total amount of waste husks produced by rice and coffee mills in particular locations, were important parameters to consider. The electrical power output of HPS was measured using a top view power analyser (Top View HT PQA 824) connected to 3 phase lines of power output from the generator as shown in Fig. 2.

Biomass fuel feeding rate was estimated by measuring each quantity fed into the gasifier and recording the time it takes for biomass to gasify to completeness (i.e. when gas is just not enough to run the engine). Considering an HPS as a black box having a transfer function between the input biomass flow rate $\dot{M}$ (kg/h) and the net electrical power output P (kW) (Caputo et al. 2005), the plant was modelled following Eq. 1. More specifically, P is directly proportional to the biomass amount $\dot{M}$, the biomass LHV (kJ/kg), the overall electrical energy conversion efficiency η of HPS, and inversely proportional to the plant operating time (seconds per hour), as shown in Eq. 1.

Fig. 2 The electrical power output recording using top view power analyser (HT PQA 824)

$$P = \frac{\dot{M} \times \eta \times \text{LHV}}{3600} \quad (1)$$

Hence, the overall electrical energy conversion efficiency (η) of HPS was estimated (Eq. 2).

$$\eta = \frac{P \times 3600}{\dot{M} \times \text{LHV}} \quad (2)$$

The LHV of rice and coffee husks were assumed to be 12.3 MJ/kg (Olupot et al. 2015) and 15.97 MJ/kg (Collins 2010), respectively. Surveys to estimate amount of agricultural residues generated from rice production rural areas in Pallisa and Butaleja, and coffee production areas in Masaka, Lwengo, Ssembabule, Bukomansimbi, Mubende and Mityana Districts (Fig. 3) were carried out using a checklist. Study sites in the respective districts were strategically selected in collaboration with the District Production and Agricultural Officers, and local community leaders. The target population involved rice or coffee milling machine operators, managers, owners, and farmers at village level. Data on machine processing pattern (hours of milling per day/month/year), amount of husks generated from a given quantity of crop processed (crop residual ratio, CR) and machine milling rate was sourced from operation records (availed by the machine operator/manager/owner). In case mill operation records were not availed, raw data was collected through checklists, questionnaires, direct measurements, observations and interviews. Each milling machine in a selected site was studied carefully to predict its processing pattern (hours of milling per day, month and year), quantity of

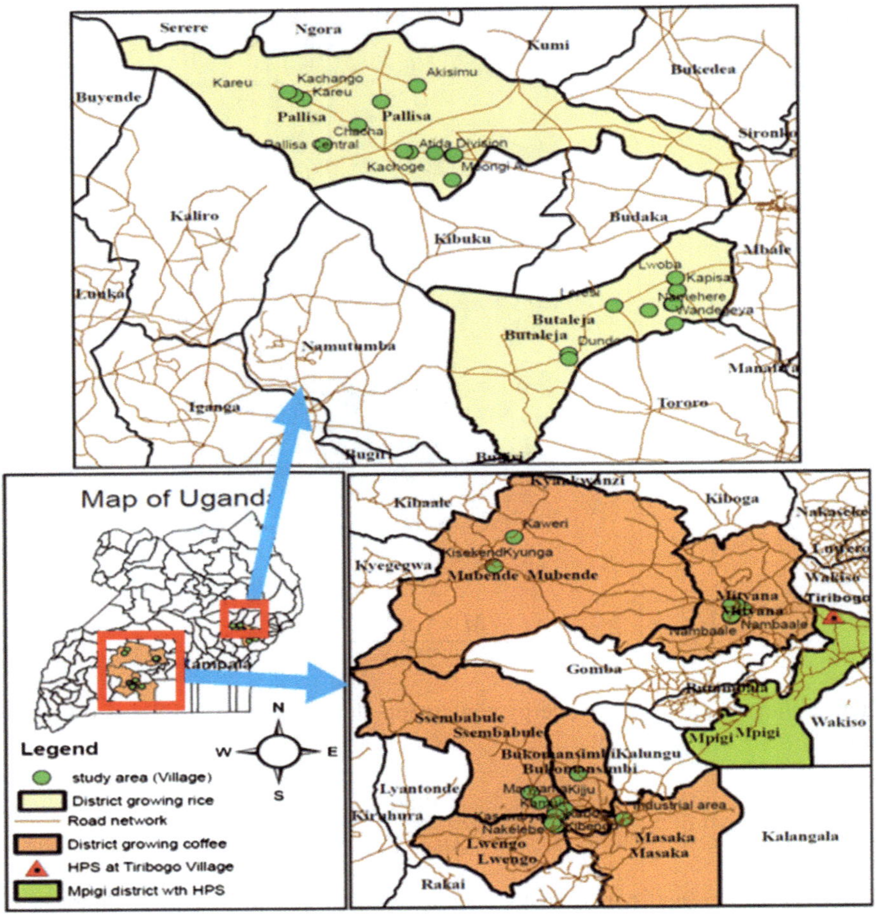

Fig. 3 Map of Uganda showing the study areas

husks generated per given quantity of crop processed (CR) and the amount of crop processed in a given period.

Assuming 30 days in a month and 365 days in a year, the total amount of husks generated (W) in a given location (i) per a given period (day, month or year) was estimated basing on crop milling rate, $\dot{R}$ (kg/h/month/year or ton/h/month/year), number of hours (H) of machine operation (in a day, month or year)—CR, and the number of rice or coffee milling machines (N) operating in a particular location (Eq. 3).

$$W_i = \dot{R}_i \times H_i \times \mathrm{CR} \times N_i \tag{3}$$

Given the fuel consumption rate ($\dot{M}$, kg or tons/h) of HPS and number of hours (h) the HPS operates per day, the amount of husks which can sustain the system throughout the year (π) was thus obtained using Eq. 4.

$$\pi = \dot{M} \times h \times 365 \tag{4}$$

The number of HPSs (n) that can be sustained in a given location (i) was as well estimated following Eq. 5.

$$n = \frac{W_i}{\pi} \tag{5}$$

Thermal energy potential (E_i, kW) of husks generated annually from respective rice and coffee processing areas was furthermore estimated (Gómez 2010) following Eq. 6; whereas the total electric energy potential (ψ, kW) was determined as the product of the thermal energy potential and the overall electrical energy conversion efficiency of HPS (Eq. 7).

$$E_i = \frac{\left(\sum_{\text{January}}^{\text{December}} W_i\right) \times \text{LHV} \times 10^6}{(3600 \times 24 \times 365)} \tag{6}$$

$$\psi_i = E_i \times \eta \tag{7}$$

4 Limitation

A number of limitations were noted and these include limited information sharing, low levels of local participation in research exercise and language barrier. Most of the mill machine owners in villages were not willing to share the information regarding their crop production. Simply because they believed that this study was a new strategy for the government to source information for imposing high taxes on their businesses. This might have led to giving erroneous information hence affecting final results. In some areas, local farmers were not willing to co-operate in the research exercise. They had been biased by the previous researchers who always collect information from them and they don't go back to implement. Therefore, access to data on rice and coffee production and waste-husk generation was limited. This could have resulted into the inaccurate estimation of quantities of waste-husk especially in areas where processing was evident. The language was also a barrier in some areas. Though the researcher used a translator, it wasn't accurate enough and this could have led to some data not being captured.

5 Results

5.1 Performance Characteristics of the HPS

Figure 4 shows the efficiency, the net electrical power output and the biomass (coffee and rice husk) consumption of the HPS.

The consumption of rice husks was about 1.5 times higher than that of coffee husks. The difference was significant ($p = 0.015$), though electrical power output did not significantly change ($p = 0.113$) with consumption of either rice or coffee husks (Fig. 4). The net electrical power output of HPS run on either rice husks or coffee husks as fuel, remained 34 kW. Its overall efficiency was different and found to be 23.4 and 28.6% when running on rice and coffee husks, respectively. As shown in Fig. 5, it was found that the consumption of coffee husks is less than rice husks at any given operating plant efficiency.

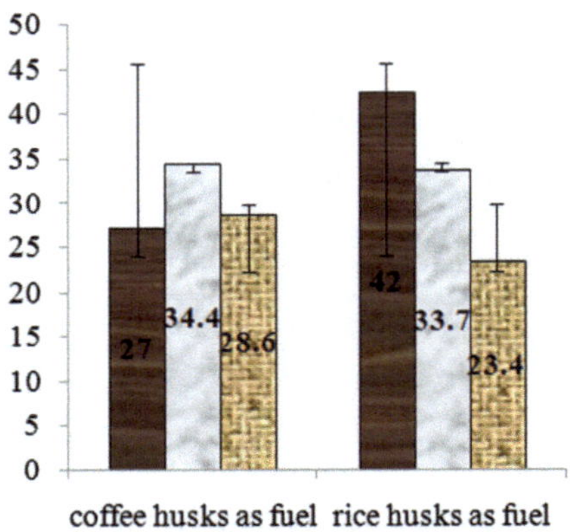

Fig. 4 Fuel conversion efficiency, net electrical power output and biomass consumption of HPS (mean ± SD, $n = 5$)

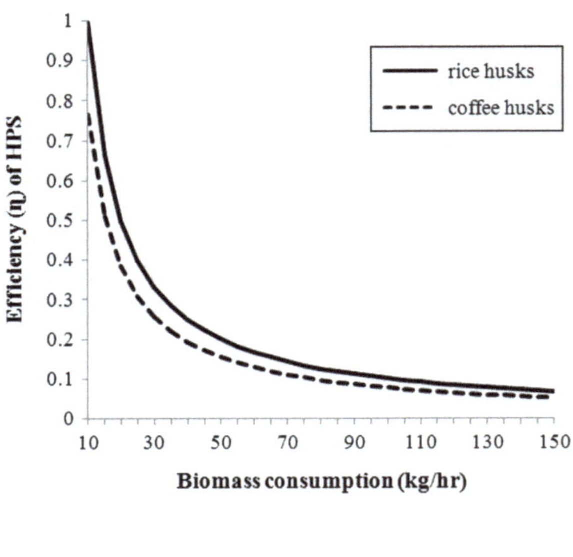

Fig. 5 Sensitivity analysis—fuel conversion efficiency versus fuel consumption of HPS

It is evident therefore that a 34 kW_e HPS is a viable option for generation of electricity at an efficiency of 28.6% using coffee husks as a feedstock.

5.2 Sustainability of Husks for HPSs in Coffee and Rice Processing Areas

Sustainability of husks for HPS refers to the amount of husks that can run an HPS at its normal operating conditions to produce electricity throughout the year. The machine milling rates were important to consider in estimating the amounts of husks generated from coffee and rice producing areas. The mean milling rates of processing machines were estimated at 1.292 and 0.765 tons of coffee and rice per hour, generating over 880 and 220 tons of coffee husks and rice husks per year per machine, respectively. An important finding was that 9 out of 17 rice producing villages in the study were not connected to the national electric grid. Yet, 22 out of 56 rice milling machines were operating on diesel at a mean milling rate lower (0.725 tons/h) than electrically driven machines (0.802 tons/h). In addition, coffee milling machines were all found operating on electricity.

Table 1 shows the amount of husks generated and number of HPSs that can be sustained in coffee and rice producing areas per year. In this study, the amount of husks that can sustain a single HPS generating 34 kW_e was estimated at 59 tons of coffee husks per year and 92 tons of rice husks per year basing on fuel consumption rate of 27 and 42 kg/h, respectively. This was calculated basing on 6 h HPS operation per day, equivalent to 2190 h per year. It was found that 36,355 tons of coffee husks and 12,591 tons of rice husks were generated each year and could sustain 615 and 136 HPSs in coffee and rice producing areas, respectively (Table 1).

Furthermore, it was found that if annual plant operational hours were increased (Fig. 6), it would result in a decrease in number of HPSs sustained but increase in power output (kilowatt-hours, kWh) per plant at a constant annual fuel (husks) supply. Operating HPS on coffee husks is highly sensitive to a change in annual operational hours than rice husks (Fig. 6), given the fact that a small change in number of annual operational hours results in a bigger change in number of HPSs sustained by coffee husks than rice husks.

5.3 Power Potential of Husk from Coffee and Rice Processing Areas

The geographical distribution of the estimated electrical power potentials of rice husks and coffee husks is depicted in Fig. 7. The study revealed that coffee

Table 1 Rice and coffee husk generated and number of HPSs sustained

Crop	Study areas (Fig. 3)	*N	*Ṙ (kg/h)	*H (h/year/mill)	*W (tons/year)	*W_i (tons/year)	*n
Coffee (CR = 0.5)	Masaka	7	1182	1414	836	5852	99
	Bukomansimbi	5	1125	1229	691	3455	58
	Ssembabule	7	1167	1592	929	6505	110
	Lwengo	12	1160	1253	727	8719	148
	Mubende	4	1998	1602	1600	6400	108
	Mityana	10	1119	970	542	5424	92
	Total	45				36,355	615
Rice (CR = 0.2)	Pallisa	26	742	988	147	3813	41
	Butaleja	30	788	1855	293	8778	95
	Total	56				12,591	136

*N = Number of processing machines; *Ṙ = Average machine milling rate of the crop; *H = Mill operational hours; *W = Average amount of husks generated per mill; *W_i = Total amount of husks generated per area; *n = Total number of HPSs sustained by husks in an area throughout the year

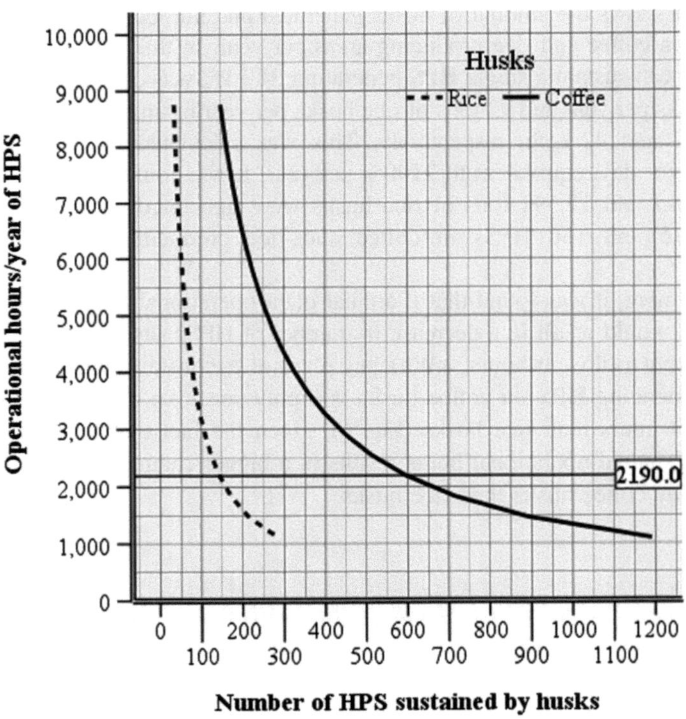

Fig. 6 A plot of annual operational hours against number of HPSs sustained by husks

processing centres have a higher electric power potential compared to rice processing centres.

In coffee processing centres, the power potentials in all sites were above 82 kW_e (Fig. 7); higher than the pilot plant size of 34 kW_e. In the rice growing areas, only 40% of the processing centres in Pallisa District depicted power potential higher than the pilot plant size of 34 kW_e compared to 100% in Butaleja District (Fig. 7).

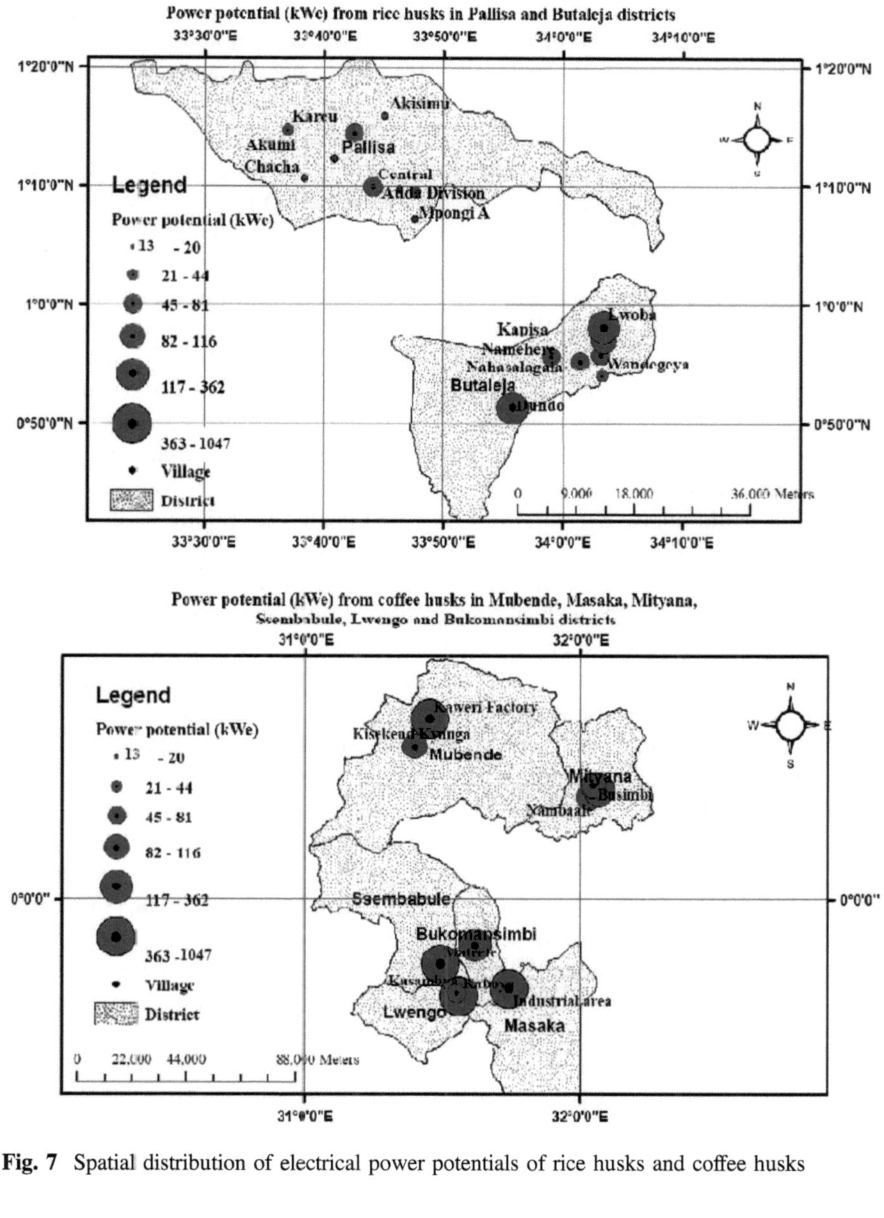

Fig. 7 Spatial distribution of electrical power potentials of rice husks and coffee husks

Generally, higher power potential was evident in coffee producing areas than rice producing areas. Though, it would be difficult to operate an HPS at coffee processing centres since such areas were already connected to the national electricity grid.

6 Discussion

Performance characterization of the pilot HPS in terms of biomass fuel consumption, the net electrical power output and overall efficiency was done to enable us ascertain the sustainability of HPS and power potential of husks generated from selected coffee and rice producing areas. The HPS was found consuming more rice husks than coffee husks, as fuel feedstock, while generating the same electrical power output. The discrepancy in biomass fuel consumption was attributed to higher ash content and lower calorific value associated with rice husks compared to coffee husks (Rahardjo 2013). Therefore, to generate the same electrical power output, a gasifier consumes more rice husks than coffee husks. Since efficiency is a function of LHV (Gómez 2010), the high conversion efficiency obtained with coffee husks was due to associated high LHV of 15.97 MJ/kg (Collins 2010) compared to 12.3 MJ/kg for rice husks (Olupot et al. 2015). The estimated overall efficiency (23–28%) of HPS operating on rice and coffee husks was comparable with similar plants of equal electrical power output elsewhere (Fischer and Pigneri 2011; Boukis et al. 2009; Bridgwater 1995; Caputo et al. 2005; Wu et al. 2001); though this could improve further with increased plant size (Boukis et al. 2009). In Uganda however, the traditional household biomass energy technologies are predominantly being utilized at efficiency <10% (Okello et al. 2013). Basing on the present work, use of HPS with greater conversion efficiency (over 20%) would, therefore, be one of the best energy options.

The sustainability of husks for HPS and power potential of husks generated from selected coffee and rice producing areas (Fig. 3) were also estimated (Table 1 and Fig. 7). Important parameters such as number of HPSs that could be sustained (optimally operated on available husks) throughout the year, amounts of husks generated and their LHVs were highly considered. The high number of HPSs sustained by coffee husks compared to rice husks was mainly due to low coffee husk consumption rate by HPS and large quantities of coffee husks available in coffee producing areas. Large quantity of coffee husks confirms the fact that coffee production is a major economic activity in Uganda, contributing the largest percentage (17.7%) of the exports (UBOS 2014). Additionally, high amount of coffee husk generated was due to electrically driven coffee milling machines operating at a higher production rate than rice processing machines. In rice producing areas, however, the sustainability potential was found lower than coffee producing areas, mainly due to high rice husk consumption rate by HPS and low levels of rice husk generated. Besides, more rice husks were generated in areas whose milling machines were electrically driven than diesel driven. Therefore, low levels of rice husks found in rice producing villages could have been attributed to limited mechanical power. In the present work, it was found that the amount of husks generated in an area affects their electric power potential. The electrical power potential of coffee husks was found higher than that of rice husks generated from their respective crop producing villages. This was mainly due to high amount of coffee husks generated annually as well as their high LHV compared to rice husks. High LHV of coffee husks compared to rice husk is also a major contributing factor to its high power potential portrayed in coffee producing areas (Collins 2010; Olupot et al. 2015).

7 Conclusions

This study has found the potential of generating and sustaining electricity supply to rural communities using a Husk Power System (HPS) run by coffee husks and rice husks. A single HPS operating for 6 h a day consumes 59 tons/year of coffee husks and 92 tons/year of rice husks, while supplying 74.46 MWh of electricity to rural communities annually. About 36,355 tons of coffee husks and 12,591 tons of rice husks were annually generated by milling machines in coffee and rice producing areas. This could sustain a total of 615 and 136 HPSs (each generating 34 kW_e) annually at a power potential of 21.1 and 4.6 MW_e for coffee and rice husk, respectively. Though coffee husks would sustain a bigger number of HPSs than rice husks, it was found more difficult to operate HPSs at coffee processing centres since such areas were already connected to the national electricity grid, which supplies sufficient power at a cheaper rate. This necessitates transportation of coffee husks to remote areas with no electricity access but higher power demand.

Acknowledgements The authors acknowledge the financial support from Renewable Energy Incubator of Makerere University and National Agricultural Research Organisation (NARO), and Pamoja Cleantech in accessing the Husk Power System installation. Special thanks go to Dr. Musinguzi Wilson for the support and guidance. The authors also acknowledge the support from government and local communities from coffee and rice growing areas of Uganda, and the HPS Company in India for providing the necessary information.

References

Boukis, I., Vassilakos, N., Karellas, S., & Kakaras, E. (2009). Techno-economic analysis of the energy exploitation of biomass residues in Heraklion Prefecture—Crete. *Renew Sustainable Energy Review, 13*(2), 362–377.

Bridgwater, A. V. (1995). The technical and economic feasibility of biomass gasification for power generation. *Fuel, 74*(5), 631–653.

Bridgwater, A. V., Toft, A. J., & Brammer, J. G. (2002). A techno-economic comparison of power production by biomass fast pyrolysis with gasification and combustion. *Renew Sustainable Energy Review, 6*(3), 181–246.

Buchholz, T., & Da Silva, I. (2010). Potential of distributed wood-based biopower systems serving basic electricity needs in rural Uganda. *Energy for Sustainable Development, 14*(1), 56–61.

Buragohain, B., Mahanta, P., & Moholkar, V. S. (2010). Biomass gasification for decentralized power generation: The Indian perspective. *Renewable and Sustainable Energy Reviews, 14*(1), 73–92.

Caputo, A. C., Palumbo, M., Pelagagge, P. M., & Scacchia, F. (2005). Economics of biomass energy utilization in combustion and gasification plants: Effects of logistic variables. *Biomass and Bioenergy, 28*(1), 35–51.

Collins, O. (2010). *Evaluating thermochemical properties and optimising densification conditions of coffee husks under low pressure* (M.Sc. dissertation, Makerere University).

Energypedia. (2015). *Uganda energy situation*. https://energypedia.info/wiki/Uganda_Energy_Situation#Biomass.5B6.5D. Accessed February 17, 2015.

EU-AITF. (2014). *Uganda rural electrification project-IG*. http://www.eu-africa-infrastructure-tf.net/activities/grants/uganda-rural-electrification-project—ig.htm. Accessed February 17, 2015.

Farley, S. E., Hawkins, R., & Wagner, C. S. (2011). *Science, technology, and innovation in Uganda: Recommendations for policy and action*. Washington, DC: World Bank Publications.

Fischer, B., & Pigneri, A. (2011). Potential for electrification from biomass gasification in Vanuatu. *Energy, 36*(3), 1640–1651.

Fracaro, G., Souza, S., Medeiros, M., Formentini, D., & Marques, A. (2011). *Economic feasibility of biomass gasification for small-scale electricity generation in Brazil*. Paper presented at the Linköping. Sweden: WREC.

Gómez, A., Zubizarreta, J., Rodrigues, M., Dopazo, C., & Fueyo, N. (2010). An estimation of the energy potential of agro-industrial residues in Spain. *Resources, Conservation and Recycling, 54*(11), 972–984.

GOU. (2010). *National development plan (2010/11–2014/15)*. Kampala: The Government of Uganda (GOU).

Hasler, P. H., & Nussbaumer, T. (1999). Gas cleaning for IC engine applications from fixed bed biomass gasification. *Biomass and Bioenergy, 16*(6), 385–395.

Hyuha, T. S., Sabiiti, E. N., & Hisal, E. (2005). *Impact of rice production on food security and women in Uganda*. Paper presented at the African crop Science Conference Proceedings.

MEMD. (2007). *The renewable energy policy for Uganda*. Kampala: Government of Uganda (GOU).

MEMD. (2012). *Renewable energy investment guide*. Kampala: Ministry of Energy and Mineral Development (MEMD).

MFPED. (2014). *Poverty status report 2014: Structural change and poverty reduction in Uganda*. Kampala: GOU.

Okello, C., Pindozzi, S., Faugno, S., & Boccia, L. (2013). Development of bioenergy technologies in Uganda: A review of progress. *Renewable and Sustainable Energy Reviews, 18*, 55–63.

Olupot, P. W., Candia, A., Menya, E., & Walozi, R. (2015). Characterization of rice husk varieties in Uganda for biofuels and their techno-economic feasibility in gasification. *Chemical Engineering Research and Design, 107*, 63–72.

Rahardjo, B. S. (2013). Influence air factor against operating parameters of biomass fluidized-bed gasifier. *International Journal of Engineering Science, 4*(2), 42–48.

Scalia, G. L., Aiello, G., Micale, R., & Enea, M. (2012). Distribution of biomass and site location of combustion and gasification power plants in western Sicily. *African Journal of Agricultural Research, 7*(2), 334–343.

Singh, S. (2010). *A comparative analysis of decentralised energy systems (DESI) power and husk power systems: small-scale biomass power generators in india rural market insight brief: empowering villages* (p. 19). Taramani, Chennai: Institute for Financial Management & Research.

Tripathi, A. K., Iyer, P., Kandpal, T. C., & Singh, K. K. (1998). Assessment of availability and costs of some agricultural residues used as feedstocks for biomass gasification and briquetting in India. *Energy Conversion and Management, 39*(15), 1611–1618.

UBOS. (2014). *Statistical abstract*. Kampala: The Republic of Uganda.

Wu, C. Z., Huang, H. T., Zheng, S. P., Luo, Z. F., Yin, X. L., & Chen, Y. (2001). *Techno economic evaluation of biomass gasification and power generation in China*. Paper presented at the proceeding of third Asia Pacific conference on sustainable energy and environmental technologies.

Author Biography

Yonah K. Turinayo holds a M.Sc. in Renewable Energy, M.Sc. in Environment & Natural Resources, and a B.Sc. in Industrial Chemistry, from Makerere University, Kampala Uganda. He works with National Agricultural Research Organization (NARO) at National Forestry Resources Research Institute (NaFORRI) as a Research Scientist (Bioenergy). He has a profound knowledge and experience in research and industrial production processes—focusing on Renewable Energy and Biomass, Sugar Technology, Forest Products, Air and Water Pollution Analysis. His areas of interest include: designing and simulation of renewable energy systems (thermochemical and biochemical energy conversion technologies); environmental implications of using renewable resources versus conventional resources for energy technology generation; life cycle assessment for biofuels; and industrial waste management.

Techno-Economic Viability of Husk Powered Systems for Rural Electrification in Uganda: Part II: Economic and Policy Aspects

Mackay A.E. Okure, Yonah K. Turinayo and Samuel B. Kucel

Abstract Large proportions of households in developing countries, like Uganda, lack modern energy services especially electricity. Yet renewable biomass resources are often available. Economic and policy barriers are blamed for this scenario. In emerging economies such as India however, cost-effective, locally suitable and environmentally sustainable solutions have been developed and deployed. One such option is the Husk Power System (HPS) which utilizes agricultural residues as feedstock for electricity generation. To evaluate the economic viability of such a system in predominantly agricultural villages of Uganda, the economic figures such as levelized unit cost of electricity, Net Present Value, Benefit-Cost ratio, Payback Period and Internal Rate of Return were determined for a 34 kW plant utilizing either coffee or rice husks as fuel. The study showed a positive Net Present Value of 20,073US$ and 27,054US$, an attractive Payback Period of 4 and 3.75 years, and a 15 year Internal Rate of Return of 14 and 17% for HPS operating on either rice or coffee husks, respectively. The system was more economical than options like diesel generators and photovoltaics, though more expensive compared to grid power. The system was also more difficult to operate and sustain in villages connected to grid electricity. Policy implications of the findings support financial subsidies and rural-based support measures for adoption of the technology.

Keywords Agricultural residues · Husk powered system · Electrification · Rural Policy · Economic viability

M.A.E. Okure
College of Engineering, Design, Art and Technology (CEDAT), Makerere University, P.O. Box 7062, Kampala, Uganda

Y.K. Turinayo (✉)
National Forestry Resources Research Institute (NaFORRI), National Agricultural Research Organisation (NARO), P.O. Box 1752, Kampala, Uganda
e-mail: t2rinayonah@gmail.com

S.B. Kucel
Faculty of Engineering, Busitema University, P.O. Box 236, Tororo, Uganda

W. Leal Filho and D. Surroop (eds.), *The Nexus: Energy, Environment and Climate Change*, Green Energy and Technology, https://doi.org/10.1007/978-3-319-63612-2_4

1 Introduction

As a country whose economy is dependent on agriculture, Uganda is richly gifted with biomass resources which contribute over 90% of total energy consumption (Zanchi et al. 2013). Biomass, a renewable resource, is expected to remain the main household energy source for a number of years due to limited access to alternative energy sources. Although agricultural residues including rice and coffee husks have been identified as potential alternative energy sources, mainly in form of briquettes and electricity, much of them are left unutilized (Okello et al. 2014; Olupot et al. 2015; Ferguson 2012). They are either openly burnt or dumped and left to rot leading to greenhouse gas (GHG) emissions and environmental problems (Olupot et al. 2015). Generating electricity from such resources for off grid remote areas could lead to their economic and social transformation, as well as environmental sustainability (Olupot et al. 2015; Zanchi et al. 2013; Okure et al. 2006). Most crop processing machines in rural areas, including rice hullers, use diesel as engine fuel to run the mills while generating significant quantities of waste husks. The cost of diesel is quite high and operating such machines on diesel may lead to higher operation costs and thus contributing to higher milling charges. Utilizing waste husks to generate electricity could therefore be a cost effective and an environmentally sustainable alternative power source.

A number of efforts have been made to promote the use of renewable energy technologies including gasification of agricultural residues for rural electrification in Uganda (Olupot et al. 2015; Buchholz and Da Silva 2010; Okure et al. 2006). Nevertheless, this has been constrained by a number of challenges including insufficient incentives for the technology development and high upfront costs of investment (MEMD 2007). Notably, the Government of Uganda put in place a Renewable Energy Policy, whose aim was to increase the use of modern renewable energy and to promote rural electrification (MEMD 2007). The conversion of agro-waste to electricity through gasification has been one of the policy strategies to enhance rural electrification. Though, little contribution has been done where the majority of the rural population (93%) has no access to electricity (IEA 2015). This may constrain poverty reduction in rural areas (Scott 2015).

Quite a few agencies are now promoting gasification technologies in Uganda, including the Husk Power System. The technology of a husk powered system (HPS) is based on the conversion of biomass such as rice or coffee husks into combustible gases that can run generators to produce electricity. Adoption of such systems focuses its attention primarily on villages that are off-grid, but have available biomass feedstock and demand for electricity. Such systems are widely adopted in a range between 5 and 250 kW in advanced developing countries like India, connecting over 250 households per system (Bhattacharyya 2014). This has improved people's livelihoods by providing light, and support to small businesses including video halls, rice mills, and small farms for irrigation at a cost of 20–30%

less than the total fuel cost for diesel-generators and pumps (Albert 2014). It has been estimated that as the number of operating HPSs increase over the years, individual rural households in India will reduce their monthly expenditure on lighting by 56–89% (Freire-Gormaly et al. 2013). However, such systems have not been extensively adopted in Uganda due to diverse economic limitations.

Limited access to grid electricity, high electricity, diesel and kerosene prices, and frequent load shading in Uganda (Zanchi et al. 2013) has been the main drivers of this study. This document therefore reports on part of a study to evaluate the economic viability of deployment of HPSs in predominantly agricultural villages of Uganda and the policy implications to support its adoption as an option for rural electrification. This was selected basing on utilizing a 34 kW pilot HPS running on rice and coffee husks. The pilot HPS was purchased by NorgesVel (The Royal Norwegian Society for Development) and implemented by the Renewable Energy Business Incubator at Makerere University. It was a pilot plant purposed to supply off-grid electricity to the community.

2 Methods

The economic viability assessment of adoption of HPSs in remote areas of coffee and rice production was done using a pilot HPS located in Mpigi District as a model. The HPS was compared with an equivalent Diesel-generator Power System (DPS) and its economic viability determined basing on DPS. Coffee and rice husks were considered as fuel feedstock for HPS and diesel for DPS. Data on the cost of the gasifier engine-generator system, diesel generator, taxes and duties, shipping, civil works, distribution network, labour, maintenance (machine, civil works, and distribution network) and system installation was sourced from the Renewable Energy Business Incubator, Makerere University. Information on the lifetime of the machine was got from the manufacture in India. The fuel cost was determined through a field survey using questionnaires administered to the coffee and rice processing farmers in Mpigi, Masaka, Lwengo, Ssembabule, Bukomansimbi, Mubende, Pallisa and Butaleja, Districts. Whereas the amount of power generated by HPS was recorded using a power analyser “Top View HT PQA 824.”

The economic viability of HPS was determined basing on levelised unit cost of electricity (LUCE), Internal Rate of Return (IRR), Net Present Value (NPV), Payback Period (PBP) and Benefit-Cost (B-C) ratio using business models developed in Microsoft Excel (Winston 2009) and realistic assumptions (Table 1). A sensitivity analysis on how LUCE, IRR and NPV respond to changes in number of annual hours a plant operates and fuel cost was also done by running iterations using models developed in Microsoft Excel (Winston 2009).

Table 1 Summary of HPS operating parameters, assumptions and project investment

No.	Parameter	Units	Value	
			HPS	DPS
1	Capital cost of HPS (gasifier, gas cleaner and engine-generator) including shipping	US[a]	60,000	8000
2	Capital cost of civil works	US$	6000	6000
3	Number of poles @ at 65US$	Numbers	100	100
4	Specific capital cost of local distribution network	US$.km^{-1}	5000	5000
5	Size of local distribution network	km	3	3
6	Generated power consumed by the auxiliaries	kW_e	1.56	1.56
7	Generated power consumed by the plant	%	4.6	4.6
8	Electrical losses in local distribution network	%	10	10
9	Useful lifetime of the plant	y	15	15
10	Useful lifetime of civil works	y	20	20
11	Useful lifetime of local distribution network	y	20	20
12	Manpower required by the plant	Numbers	2	1
13	manpower wage	US $ man^{-1} h^{-1}	1	1
14	Maintenance cost of HPS (function of its capital cost)	%	3	3
15	Maintenance cost of distribution network (function of its capital cost)	%	2	2
16	Maintenance cost of civil works (function of its capital cost)	%	1	1
17	Cost of rice husks	US$ kg^{-1}	0.03	NA
18	Cost of coffee husks	US$ kg^{-1}	0.05	NA
19	Cost of pilot fuel (petrol)	US$ l^{-1}	0.75	NA
20	Cost of diesel fuel	US$ l^{-1}	NA	1
21	Biomass fuel Consumption (rice husks and coffee husks, respectively)	kg h^{-1}	42.4 and 27.1	NA
22	Diesel fuel consumption	l h^{-1}	NA	8.9
23	Consumption of petrol (for gas engine start-up)	l h^{-1}	0.083	NA
24	Plant operational hours per day	h day^{-1}	6	6
25	Plant operational hours per year	h year^{-1}	2190	2190
26	Specific fuel consumption of diesel (referred to the electric output)	l kW_e h^{-1}	0.00245	NA
27	Specific fuel consumption of rice husks and coffee husks, respectively[b]	kg kW_e h^{-1}	1.25 and 0.80	NA
28	Specific fuel consumption of diesel (referred to the electric output)	l kW_e h^{-1}	NA	0.262

[a]1United States dollar (US$) = 3333 Ugandan currency (UGX), in April 2016 (XE 2016)
[b]Referred to the electric power output

3 Limitations

Although this study was conducted fairly well, there are a few limitations which are worth to mention. Short time period and inadequate funds were limiting factors of this study as this work was part of a Master's thesis. Much of the work including emissions analysis, biomass characterization and system optimization were not done. Instead results were based on assumptions. We could not wait to determine the demand, willingness to accept and to pay for electricity generated by the HPS. Yet these are important factors to consider for new technology adoption. This work could have probably been more detailed if time and financial resources had been adequately allocated for aforementioned activities.

4 Results

From an economic perspective, the biomass power project can only be viable when its unit of power is much cheaper than power derived from other potential sources in a given area (Leung et al. 2004). In this study, the economic analysis of HPS running on rice or coffee husk as fuel was evaluated basing on a Diesel-generator Power System (DPS) with an equivalent electric power output. Cost of power production and economic viability for both power projects was determined and compared using LUCE, IRR, PBP, NPV and B-C ratio as demonstrated in the following sub-sections. Table 1 represents summaries of operating parameters, assumptions and projects' initial investment costs used in the analysis. There was no subsidy offered by the government of Uganda to the HPS.

4.1 Cost of Power Production

The cost of power production of HPS and DPS was determined by the levelized unit cost of electricity (LUCE) following Eq. 1.

$$\text{LUCE (US\$/kWh)} = \frac{\text{Total annualized cost (US\$)}}{\text{Annual electricity generated (E, kWh)}}. \tag{1}$$

The capital cost (i.e. cost of the gasifier engine-generator system, diesel generator system, civil works, distribution network, taxes and duties, shipping and system installation), operation and maintenance cost (including wages, cost of maintenance of machine, civil works and distribution network), fuel cost, lifetime of the machine and the amount of power generated by HPS and DPS (Table 1), were used in the derivation of LUCE following expressions in the subsequent sub-sections.

4.1.1 Total Annualized Cost of HPS and DPS with Equivalent Power Output

The total annualized cost (C) of HPS and Diesel-generator Power System (DPS) was estimated from annualized capital cost, annualized operation and maintenance cost and annualized fuel cost (Eq. 2).

$$\mathrm{C} = \text{Annualized}\,(\text{Capital Cost} + \text{Operation and Maintenance Cost} + \text{Fuel Cost}). \tag{2}$$

Annualized Capital Cost (AC_c)

The capital recovery factor $R_f = \left(\frac{r(1+r)t}{((1+r)^t - 1)}\right)$ for discount rate r and life period of t years was used to annualize the capital cost of HPS and diesel-generator power systems (DPS). Assuming a constant discount rate (r) of 12%, 15 year useful lifetime (t) of HPS (gasifier and generator) and DGS, 20 year useful lifetime of civil works and distribution network, the capital recovery factor was estimated at 0.147 for HPS and diesel generator, and 0.134 for both civil works and distribution network. Hence annualized capital cost of HPS and DPS were estimated as 11,634US$ and 3990US$ following Eqs. (3) and (4), respectively.

$$\begin{aligned} \mathrm{AC_c} &= 0.147(\text{cost of HPS \& DGS}) + 0.134\,(\text{cost of civil works}) \\ &\quad + 0.134\,(\text{cost of distribution network}) \end{aligned}$$

$$\begin{aligned} \mathrm{AC_{c-HPS}} &= (0.147 \times 60,000) + (0.134 \times 6000) + (0.134 \times 15,000) \\ &= 11,634\,\text{US\$} \end{aligned} \tag{3}$$

$$\mathrm{AC_{c-DPS}} = 0.147 \times 8000) + (0.134 \times 6000) + (0.134 \times 15,000) = 3990\,\text{US\$} \tag{4}$$

Annualized Operation and Maintenance (O&M) Cost

The plant was operating for 6 h per day—equivalent to 2190 h per year (i.e. 6 h 365 days). Assuming 3% of the total cost of HPS and DPS (≈1800US$ and 240US $, respectively), 1% (≈60US$) of total cost of civil works and 2% (≈300US$) of total cost of distribution network was on maintenance per year; and 2 operators for HPS, 1 operator for DPS (each paid 1US$ h^{-1}) were employed, the annualized operation and maintenance costs ($AC_{O\&M}$) of HPS and DPS would be 6,540US$ and 2730 US$ basing on Eqs. (5) and (6), respectively.

$$AC_{O\&M-HPS} = 1800 + 60 + 300 + (2190 \times 2 \times 1) = 6540\,\text{US\$} \quad (5)$$

$$AC_{O\&M-DPS} = 240 + 60 + 300 + (2190 \times 1 \times 1) = 2790\,\text{US\$} \quad (6)$$

Annualized Cost of Fuel

The annualized cost of fuel (AC_f) depends on the cost of biomass fuel, diesel fuel and machine consumption rate. The cost of rice husk was estimated at 0.03US $ kg^{-1}, coffee husk at 0.05US$ kg^{-1}, pilot fuel (petrol) for engine startup and diesel at 1US$ l^{-1}. 42.4 kg of rice husk fuel, 27.1 kg of coffee husk fuel and about 0.083 l of petro or 8.9 l of diesel could run either HPS or DPS for 1 h. Therefore, operating the HPS on rice or coffee husk fuels, or running the DPS on diesel would incur different annual costs of 2967US$, 3149US$ and 19,491US$, respectively, as shown by Eqs. 8, 9 and 10.

$$\begin{aligned} AC_f = {} & \text{Operational hours/year} \\ & \times (\text{biomass fuel cost(US\$/h)} + \text{pilot fuel cost (US\$/h)}) \end{aligned} \quad (7)$$

$$\begin{aligned} AC_{f-HPS-rice} & = 2190 \times ((0.03\ \text{US\$/kg} \times 42.4\ \text{kg/h}) + (1\text{US\$/l} \times 0.083\ \text{l/h})) \\ & = 2967\text{US\$} \end{aligned} \quad (8)$$

$$\begin{aligned} AC_{f-HPS-coffee} & = 2190 \times ((0.05\ \text{US\$/kg} \times 27.1\ \text{kg/h}) + (1\text{US\$/l} \times 0.083\ \text{l/h})) \\ & = 3149\text{US\$} \end{aligned} \quad (9)$$

$$AC_{f-DPS-diesel} = 2190 \times (1\text{US\$/kg} \times 8.9\ \text{l/h}) = 19{,}491\text{US\$} \quad (10)$$

Total Annualized Cost of Investment

The total annualized cost of investing in HPS operating on either rice husk or coffee husk and DPS was calculated as 21,141US$, 21,323US$ and 26,211US$ using Eqs. (11), (12) and (13), respectively. This was done by substituting Eqs. (3), (5) and (8), Eqs. (3), (5) and (9) and Eqs. (4), (6) and (10) into Eq. (2).

$$C_{total\ HPS-ricehusk} = 11{,}634 + 6540 + 2967 = 21{,}141\text{US\$} \quad (11)$$

$$C_{total\ HPS-coffeehusk} = 11{,}634 + 6540 + 3149 = 21{,}323\text{US\$} \quad (12)$$

$$C_{total\ DPS-diesel} = 3990 + 2790 + 19{,}491 = 26{,}211\text{US\$} \quad (13)$$

4.1.2 Annual Electric Power Generated

The power output of the electricity generator was estimated at 34 kW. Basing on the assumptions of equivalent power output of HPS and DPS, 10% electrical power losses in the distribution network, 4.6% of generated power being consumed by auxiliaries and 2190 h of plant operation in a year, the annual electric power output (E) was projected as 63,931 kWh (Eq. 14).

$$E = 34\ \text{kW} \times (2190\ \text{h})(1 - 0.1)(1 - 0.046) = 63,931\,\text{kWh} \tag{14}$$

4.1.3 The Levelized Unit Cost of Electricity (LUCE)

The levelized unit cost of electricity (Eq. 1) for HPS running on either rice or coffee husks and DPS operating on diesel was estimated as 0.331US$/kWh, 0.334US$/kWh and 0.410US$/kWh using Eq. (15), (16) and (17), respectively.

$$\text{LUCE}_{\text{HPS-ricehusk}} = \frac{21,141\text{US\$}}{63,931\ \text{kWh}} = 0.331\ \text{US\$/kWh} \tag{15}$$

$$\text{LUCE}_{\text{HPS-coffeehusk}} = \frac{21,323\text{US\$}}{63,931\,\text{kWh}} = 0.334\ \text{US\$/kWh} \tag{16}$$

$$\text{LUCE}_{\text{DPS-diesel}} = \frac{26,211\ \text{US\$}}{63,931\ \text{kWh}} = 0.410\ \text{US\$/kWh} \tag{17}$$

The difference between the LUCE obtained for HPS operating on either coffee husks or rice husks was insignificant ($p = 0.904$), though highly significant for DPS operating on diesel ($p = 0.00$).

4.2 Economic Viability of HPS and DPS

The economic viability of power systems is determined by a number of parameters including investment cost, levelized unit cost of electricity, selling price of electricity, income generated by the power systems, IRR, PBP, NPV and B-C ratio (Table 2), as clearly discussed in the following sub-sections.

4.2.1 Investment Cost

In this study, the total investment cost was estimated at 29,000US$ for DPS (including shipping, local distribution network and tax dues) and 81,000US$ for the HPS (gasifier, gas cleaner and engine-generator) including shipping,

Table 2 Summary of economic parameters and financial evaluation of a 34 kW HPS

Parameter	Units	Values		
		HPS with rice husk	HPS with coffee husk	Diesel generator
Initial capital investment of HPS (In)	US$	81,000	81,000	29,000
Specific investment cost	US $ kW^{-1}	2382	2382	853
Total annualised cost of investment	US$	21,141	21,323	26,211
Annual Electricity generated	kWh	63,931	63,931	63,931
Levelized Unit Cost of Electricity (LUCE)[a]	US $ kWh^{-1}	0.331	0.334	0.410
Selling price of electricity	US $ kWh^{-1}	0.410	0.410	0.410
Annual Income (including briquettes for HPS)	US$	31,783	29,773	26,212
Discount rate	%	12	12	12
Payback period (PBP)	years	3.75	4	2.2
Internal rate of return (IRR)	%	14.2	17.0	24
Net Present Value (NPV)	US$	27,054	20,073	43,138
Benefit – Cost (B-C) ratio		1.05	1.18	
Uganda's reference tariffs as of the year 2012			Source:	
Household	US $ kWh^{-1}	0.19532	(UMEME 2016)	
Low voltage supply for Commercial	US $ kWh^{-1}	0.19504		
Low voltage supply for medium scale Industrial	US $ kWh^{-1}	0.18356		

[a]1United States dollar (US$) = 3333 Ugandan currency (UGX), in April 2016

local distribution network and tax dues (Table 1). The specific investment costs of HPS and DPS were then determined as the ratio of the total investment cost to the rated power output of the systems. Given equal rated power output of 34 kW for both HPS and DPS, the specific investment cost was calculated as 2382US$/kW and 853US$/kW using Eqs. 18 and 19, respectively.

$$\text{Specific Investment cost of HPS} = \frac{81{,}000\ \text{US\$}}{34\text{kW}} = 2382\ \text{US\$/kW} \quad (18)$$

$$\text{Specific Investment cost of DPS} = \frac{29{,}000\ \text{US\$}}{34\ \text{kW}} = 853\text{US\$/kW} \quad (19)$$

Figure 1 represents the relationship between specific investment cost of HPS and DPS with the unit cost of power production. It is evident that reduction in

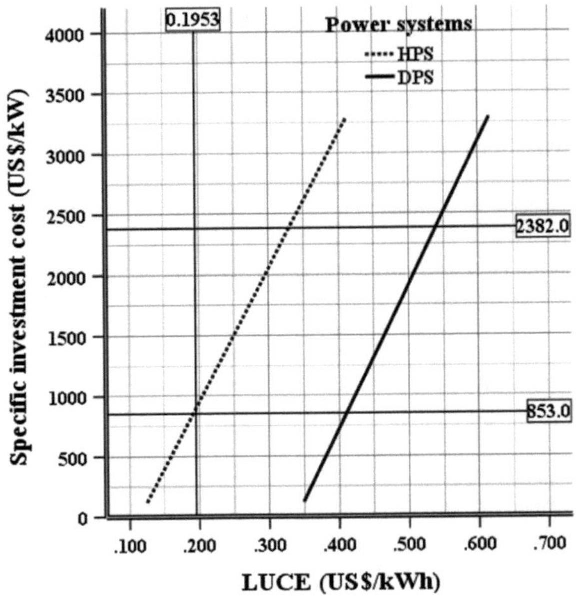

Fig. 1 Relationship between specific investment cost of HPS and DPS with LECE

investment cost of HPS to less than 880US$/kW would generate electricity cheaper than national grid tariff (<0.1953US$/kWh). Never the less, DPS would still be far more expensive than grid tariff even if investment cost is reduced to zero (Fig. 1).

4.2.2 Annual Income Generated

Annual Income from Charcoal (Bi-Product of Gasification)

Alongside power generation from biomass through gasification, charcoal is generated. This charcoal, mostly known as biochar, has been used as the main raw material for briquette making to supplement on household energy supply. More so, it is used by farmers in enhancing moisture and nutrient retention in the soils. Under this project, 20% (w/w) of the biomass feedstock gasified was generated as charcoal. Given the HPS fuel consumption rate of 27.1 kg coffee husk per hour and 42.4 kg rice husk per hour, about 5.42 kg/h or 8.48 kg/h of charcoal was generated if the plant was either operated on coffee or rice husks, respectively. Assuming all charcoal produced was converted to briquettes using a simple press machine purchased at 120US$, operated by 1 HPS operator with no additional operational cost, and replacement made every after 3 years at 12% discount rate, its total annualized cost would be 50US$ (Eq. 20).

$$C_{\text{total}-\text{BrquatteMchn}} = \left(\frac{0.12(1+0.12)^3}{\left((1+0.12)^3-1\right)}\right) \times 120\text{US\$} = 50\text{US\$/year} \tag{20}$$

Given the market price of 0.3US$ kg^{-1} (as per April, 2016), the annual income generated from briquettes using coffee and rice husk charcoal (the main briquette component) was estimated as 3,561US$/year and 5,571US$/year following Eqs. (21) and (23), respectively.

$$\begin{aligned}\text{Annual Income}_{\text{coffeehusk}} &= (5.42\ \text{kg/h} \times 2,190\ \text{h/year} \times 0.3\text{US\$/kg}) \\ &= 3561\text{US\$/year}\end{aligned} \tag{21}$$

$$\text{Net Income}_{\text{coffeehusk}} = 3561\text{US\$/year} - 50\text{US\$/year} = 3511\text{US\$/year} \tag{22}$$

$$\text{Annual Income}_{\text{ricehusk}} = (8.48\ \text{kg/h} \times 2190\ \text{h/year} \times 0.3\text{US\$/kg}) = 5571\text{US\$/year} \tag{23}$$

$$\text{Net Income}_{\text{ricehusk}} = 5571\text{US\$/year} - 50\text{US\$/year} = 5521\text{US\$/year} \tag{24}$$

4.2.3 Annual Income Generated by HPS and DPS

In this study, the price of electricity sold to households in remote villages with no access to grid electricity was determined by the cost of power from a diesel generator. This is based on the assumption that electricity generated by the DPS could be relatively cheaper and affordable compared to other possible sources nearby e.g. solar. The cost of electricity from the DPS was estimated as 0.410US$/kWh (Eq. 17), equivalent to the selling price set for HPS derived electricity. Using Eq. 25, the annual income generated by DPS was calculated as 26,212US$.

$$\text{DPS}_{\text{annual income}} = 0.410\text{US\$/kWh} \times 63{,}931\ \text{kWh/year} = 26{,}212\text{US\$/\$ year} \tag{25}$$

Furthermore, the HPS run on either coffee or rice husk has been simultaneously generating electricity and biochar for briquettes production. Once electricity and briquettes were sold, income equivalent to 29,773US$/year and 31,783US$/year was generated from HPS operated by coffee and rice husks as shown in Eq. 26 and 27, respectively.

$$\begin{aligned}\text{HPS}_{\text{annual income coffeehusk}} &= (0.410\text{US\$/kWh} \times 63{,}931\ \text{kWh}) + 3561\text{U\$/year} \\ &= 29{,}773\text{U\$/year}\end{aligned} \tag{26}$$

$$\begin{aligned}\text{HPS}_{\text{annual income ricehusk}} &= (0.410\text{US\$/kWh} \times 63{,}931\ \text{kWh}) + 5571\text{U\$/year} \\ &= 31{,}783\text{U\$/year}\end{aligned} \tag{27}$$

4.2.4 Internal Rate of Return (IRR) and Net Present Value (NPV)

The IRR of the project is the interest rate that makes the NPV equal to zero (Shmueli et al. 2011). In this study, we are comparing three choices of investments (HPS running on both rice or coffee husks and DPS) using IRR and NPV. If the NPV curves cross, then the choice of investment will depend on the discount rate. Given the annual income (A) for a specific year (t), discount rate (r) over 15 year period and zero salvage value; the net present value (NPV) of the total income was estimated as the sum of the present values of the individual amounts in the income cash flow (Eq. 28).

$$\mathrm{NPV} = \sum_{t=0}^{15} \frac{A_t}{(1+r)^t} \tag{28}$$

A plot of cumulative cash flows (NPV) of income against a range of discount rates (Eq. 28) gave a NPV for DPS as 43,138US$ - higher than HPS running on rice husk (27,054US$) and coffee husk (20,073US$), basing on 12% discount rate (Fig. 2).

The power systems were also rated basing on their IRR in the order that follows: HPS running on coffee husk (14.2%) < HPS running on rice husk (17%) < DPS (24%), at a zero NPV (Fig. 2). The discount rates for DPS and HPS operating on rice husk cross at 6%. This implies that HPS operating on rice husk becomes a better investment option at any point less than 6%. For discount rates above 6%, but below 24% (the present value of the DPS investment), DPS investment becomes a

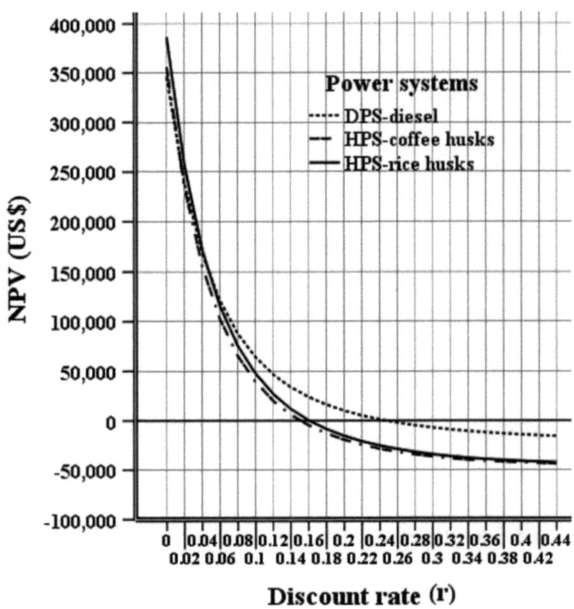

Fig. 2 Cumulative cash flows (NPV) of income against discount rates

viable option. Never the less, no business would be viable at discount rates higher than 24%, because the net present values are negative. Yet, both DPS and HPS projects are equally viable at a discount rate of less than 2% and a positive NPV of about 250,000US$.

4.2.5 Payback Period (PBP)

The PBP was obtained by a plot of cumulative cash flow (NPV) against period (t) of 15 years of project's life (Fig. 3) using Eq. (28). DPS gave a short PBP of about 2.2 years followed by HPS running on rice husk (3.75 years) and HPS running on coffee husk (4 years).

4.2.6 Benefit-Cost (B-C) Ratio

The benefit-cost (B-C) ratio was estimated as the ratio of the annual income to the total annualized cost (Eq. 29).

$$\text{B-C ratio} = \frac{\text{Annual Income (US\$)}}{\text{Total annualized cost (US\$)}} \tag{29}$$

The B-C ratios for the DPS and HPS running on either coffee or rice husk were found to be 1.00, 1.41 and 1.49, basing on Eqs. (30), (31) and (32), respectively.

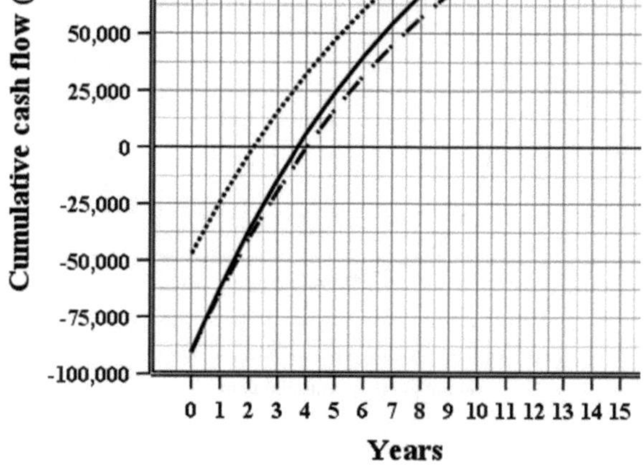

Fig. 3 Cumulative cash flows (NPV) against period (years) of cash flows

$$\text{B-C}_{\text{DPS-diesel}} = \frac{26,212\ \text{US\$/year}}{26,211\text{US\$/year}} = 1.00 \qquad (30)$$

$$\text{B-C}_{\text{HPS-coffeehusk}} = \frac{29,773\text{US\$/year}}{21,141\text{US\$}} = 1.41 \qquad (31)$$

$$\text{B-C}_{\text{HPS-ricehusk}} = \frac{31,783\text{US\$/year}}{21,323\text{US\$}} = 1.49 \qquad (32)$$

4.3 Sensitivity Analysis

Sensitivity analysis shows how the variation in the output of any mathematical model can be apportioned, qualitatively or quantitatively, to different input variation sources (Saltelli et al. 2008). Specifically, an assessment of the influence of number of hours the plant operates annually on the LUCE and annual income was done. It was found that LUCE gradually reduces with an increase in number of annual hours a plant operates (Fig. 4a).

The lower the number of hours, the higher the LUCE would be and vice versa. It is evident that at 3750 h (about 10 h/day of plant operation), LUCE for both HPS and DPS is constant at 0.250 US$/kWh. At about 6000 h per year (equivalent to 16 h/day of plant operation), LUCE for HPS competes favorably with the national grid tariff (0.195US$/kWh); whereas for DPS, it only requires 5000 annual operational hours to generate power at a cost equivalent to 0.195US$/kWh (Fig. 4a). Nevertheless, there is a linear relationship (R^2 = 0.995) between annual income generated and number of annual hours of plant operation. The annual income for

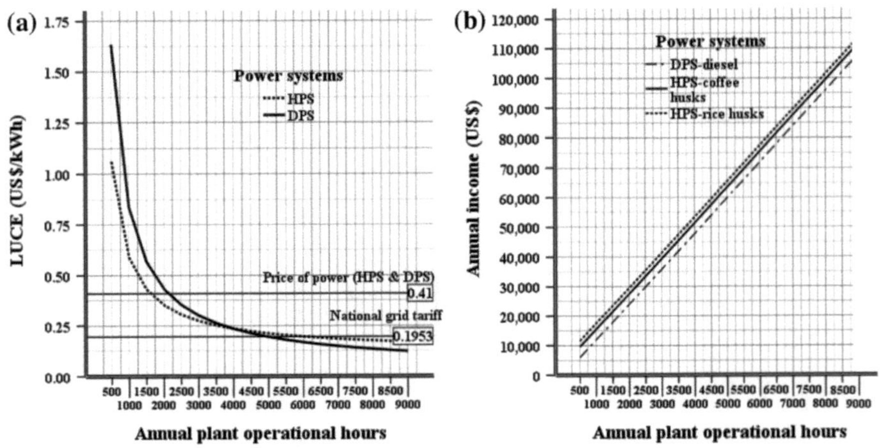

Fig. 4 Relationship between LUCE, annual income and annual hours of plant operation

both power systems is generated following the order: HPS running on rice husk > HPS running on coffee husk > DPS (Fig. 3b).

Similarly, the response of the unit cost of power to IRR and NPV was determined (Fig. 5). For DPS operating on diesel, HPS operating on rice husk and HPS operating on coffee husk to compete favorably with national grid tariff (0.195US$/kWh), their IRR would be 10, 5 and 2.5%, respectively (Fig. 5a). However, the business would not be viable if the unit cost of power is less than 0.225US$/kWh, 0.275US$/kWh and 0.325US$/kWh for DPS and HPS operating on either rice husk or coffee husk, respectively, basing on the NPV (Fig. 5b).

Figure 6 shows the variation between LUCE, cost of biomass fuel and plant size. When costs of rice husk and coffee husk fuels increase, the LUCE also increase but

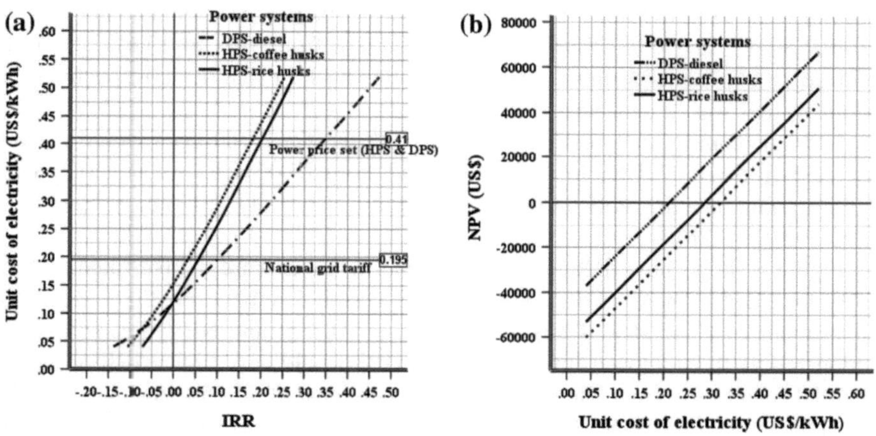

Fig. 5 Response of the unit cost of electricity to IRR (**a**) and NPV (**b**)

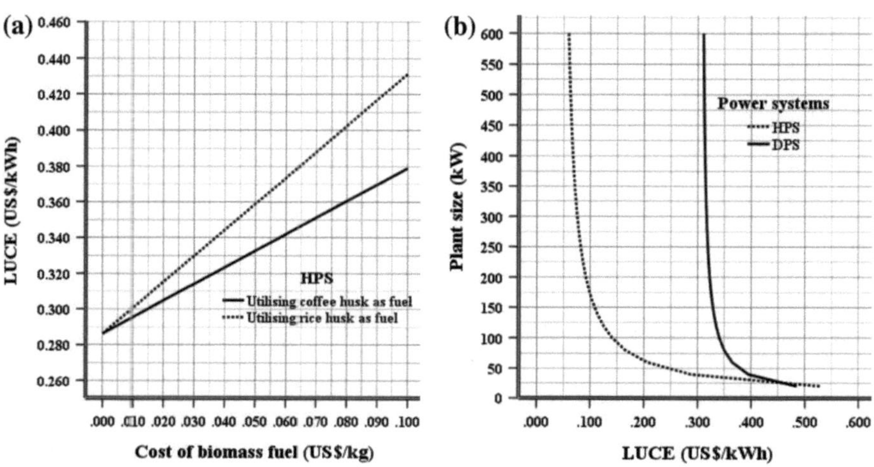

Fig. 6 Variation between LUCE, cost of biomass fuel (**a**) and plant size (**b**)

at different rates (1.45 and 0.93 kg/kWh, respectively), as depicted by the gradients. For instance (Fig. 6a), the increase in cost of biomass fuel from zero to 1US$ kg^{-1} significantly increases the LUCE from 0.286US$/kWh by 0.145US$/kWh and 0.093US$/kWh for rice husk and coffee husk, respectively. The difference is significant ($p = 0.03$).

Assuming the HPS was utilizing disposed of biomass fuel at zero cost (Olupot et al. 2015), the LUCE would be 0.286US$/kWh (Fig. 6a), which is still higher than the current national grid tariff for households (0.195US$/kWh), but much cheaper than diesel-derived electricity (0.410US$/kWh). However, increase in plant size given other factor constant, would further reduce the LUCE for both HPS and DPS (Fig. 6b). Specifically, if the size of HPS is doubled from 34 kW to 68 kW, its power would be cheaper than the grid tariff (0.1953US$/kWh) although DPS would remain expensive (above 0.33US$/kWh). More so, IRR at biomass fuel cost of 0.05US$ kg^{-1} and 0.03USD kg^{-1} would increase from 14.2 to 15.6% and from 17.0 to 18.3% for HPS running on either coffee husk or rice husk, respectively, if biomass fuel was at zero cost (Fig. 7a).

NPV at biomass fuel cost of 0.05US$ kg^{-1} and 0.03USD kg^{-1} would as well increase from 20,073US$ to 22,750US$ and from 27,054US$ to 29,550US$ for HPS running on either coffee husk or rice husk, respectively, if biomass fuel was at zero cost (Fig. 7b).

Figure 8 shows how IRR and NPV respond to the increase in investment cost of HPS and DPS. There is a strong negative correlation between the increase in investment cost and IRR (Pearson correlation: −0.755) and NPV (Pearson correlation: −0.910). At a lower investment cost, HPS would have higher IRR and NPV than DPS and vice versa. Beyond the investment cost of 77,250US$, 102,500US$ and 111,500US$ for DPS, HPS operating on coffee husk and HPS operating on rice husk, respectively, the business would not be viable basing on the IRR and NPV that would be less than zero. Otherwise, both power system investments would be viable basing on a positive NPV (Fig. 8).

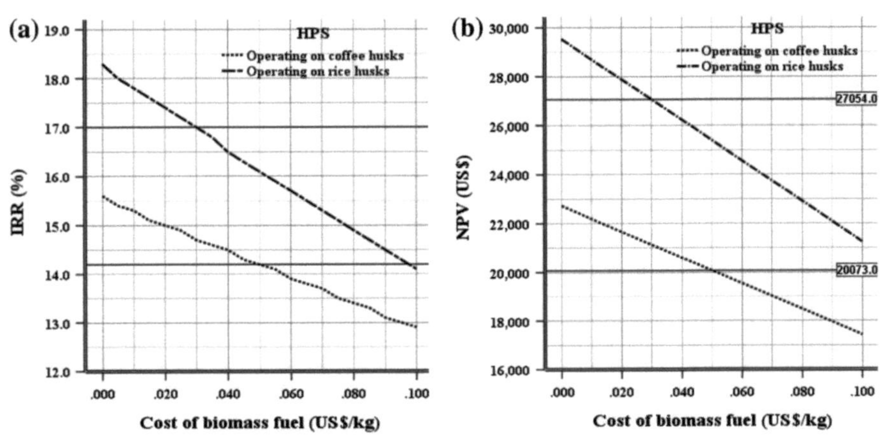

Fig. 7 Response of IRR and NPV to cost of biomass fuel

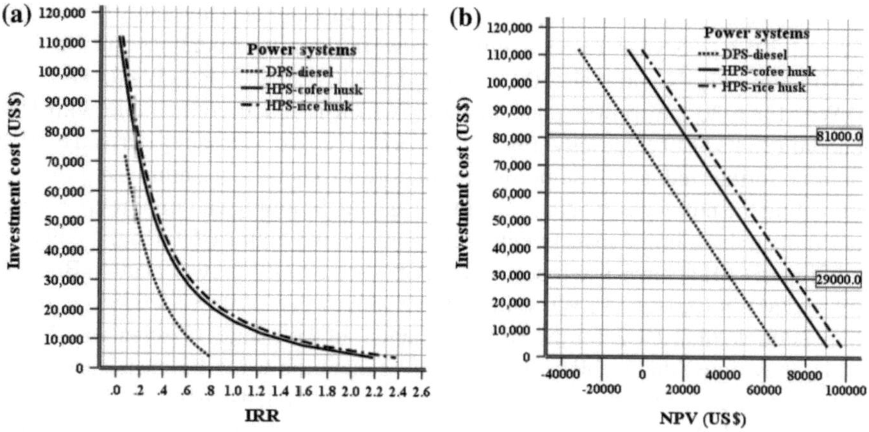

Fig. 8 Response of investment cost to IRR and NPV

5 Discussion

The economic viability of adoption of HPS operating on either coffee or rice husks in Uganda was evaluated basing on the cost of power production (LUCE), IRR, NPV, B-C ratio and PBP as the main financial indicators. The study found that the cost of electricity production by HPS running on either rice husk (0.331 US$/kWh) or coffee husk (0.334 US$/kWh) is much cheaper than for DPS (0.410 US$/kWh) but more expensive than national grid tariff (0.1953US$/kWh). Given the fact that specific investment cost of HPS (2382US$/kW) is much higher than DPS (853US$/kW), high LUCE for DPS could be mainly attributed to its high fuel cost of about 19,491US $/year compared to HPS running on either rice husk or coffee husk (2967US$/year and 3149US$/year, respectively). Notably, similar studies on economic viability of biomass gasification power plants were done in several areas including Uganda, India and Brazil. Olupot et al. (2015) recently found that running a 250 kW power plant on rice husk in Uganda, would generate electricity at a cost of 0.22US$/kWh compared to 0.36US$/kWh for equivalent power output using diesel-engine generator. Basing on the present work (Fig. 6b), increasing the size of HPS and DPS to 250 kW would generate electricity much cheaper (0.10US$/kWh for HPS and 0.34US$/kWh for DPS) than power plant reported by Olupot et al. (2015). Buchholz and Da Silva (2010) also estimated electricity production costs of an installed capacity of 25.1 kW wood-based bio-power system in Uganda as 0.11 US$/kWh, which is over 3 times lower than the cost of electricity generated by HPS in the present work. The high value of LUCE by HPS could be mainly attributed to its low annual hours of plant operation (25%), as illustrated in Fig. 4a, compared to wood-based bio-power system (50%) reported by Buchholz and Da Silva (2010). Apart from operation and maintenance cost (6,540US$/year) which is much higher than 3,500US$/year reported by Buchholz and Da Silva (2010), other factors that may influence LUCE of HPS, specifically investment cost (2382US$/kW) and fuel cost (2967US$/year and

3149US$/year for rice and coffee husk, respectively), compares favorably (2300 US$/kW and 3053 US$/year, respectively) with work done by Buchholz and Da Silva (2010). Similarly, Nouni et al. (2007) reported a LUCE at 13.72 Rs/kWh (0.31083 US $/kWh) for a 30 kW power plant in India, which is much comparable with HPS generating 34 kW of electricity (Fig. 6b). However, cost of electricity derived by HPS could be further reduced if shipping and import tax charges are minimized through fabricating the bulky gasification system in Uganda. Shipping and import tax charges contribute to high initial investment costs resulting into high LUCE (Fig. 1). Fischer and Pigneri (2011) have also reported a LUCE for a 100 kW power plant in Brazil of 459.83€/MWh (0.64061$/kWh using an exchange value of 0.7178€/US$) which is almost double the LUCE for the HPS in Uganda. In their study (Fischer and Pigneri 2011), they noted that small-scale electricity generation using biomass gasification technology is economically unfeasible in Brazil. In general, it can be surmised that the LUCE of the HPS in this study is comparable with the LUCE generated by biomass gasification power plants in the previous studies by other researchers.

Other financial figures used in the evaluation of viability of HPS in the present work were determined. Income generated by electricity from both HPS operating on rice and coffee husks was constant at 26,212US$/year, equivalent to income generated by DPS (a model that was used to assess economic viability of HPS). But due to extra income resulting from charcoal briquettes (a bi-product of gasification), the total income generated by HPS operating on rice or coffee husks increased to 31,783US$/year and 29,773US$/year, respectively, making it higher than DPS. The income difference is attributed to high feeding rate of rice husk fuel (42.4 kg/h) which generated more charcoal (8.42 kg/year) for briquette production, compared to coffee husks (27.1 kg/h feedstock and 5.42 kg/h charcoal). On this note, the more charcoal produced and sold in form of briquettes at constant price, the more income would be generated—hence supplementing on annual income from electricity sales. Generally, the annual income generated by HPS is higher than income from DPS. However, HPS depicts a lower IRR (14.2% for coffee husk and 17% for rice husk) and NPV (20,054US$ for coffee husk and 27,054US$) than DPS (24% IRR and 43,138US$ NPV). This is mainly due to high investment cost of HPS (81,000US$) compared to DPS (29,000US$) that gives a strong negative correlation of −0.755 and -0.910 basing on the IRR and the NPV, respectively (Fig. 8). However, a positive NPV and a B-C ratio greater than 1 (1.49 and 1.41 for HPS operating on rice husk or coffee husk, respectively) at a PBP of 3.75 years and 4 years for rice and coffee husk, respectively, clearly indicates that HPS is an economically viable investment.

6 Policy Implications of Adoption of HPS in Uganda

Renewable energy constitutes over 92% of the energy consumption in Uganda, including biomass (Okello et al. 2013). In particular, biomass is massively used in its traditional form as fire wood and charcoal leading to high deforastation rate (Obua et al. 2010; MEMD 2007). Utilization of agricultural residues for energy could be an alternative option, though may be constrained by unavailability of appropriate conversion technologies (Olupot et al. 2015). Large quantities of agricultural residues such as rice husks are burnt without energy recovery. They are openly dumped and left to decompose, leading to GHG emissions and environmental health related problems (Olupot et al. 2015). In Uganda, use of agricultural residues to meet modern energy services, particularly electricity, is limited mainly due to a number of challenges. Some of which include: making electricity reachable to the rural population; expanding access to affordable, reliable and sufficient energy supplies to address the poverty issues; insufficient specific incentives for the technology development; the disposal of biomass waste by burning without extracting the energy content; and high upfront costs of investment, which have held up biomass energy technology expansion.

This study provides an entry point for the government, the entrepreneurs, and other promoters of renewable energy technologies (RETs) to support investments in modern biomass energy technologies. This is in accordance with the renewable energy policy of Uganda; whose one of the objectives is "to establish an appropriate financing and fiscal policy framework for RET investments (MEMD 2007)." According to Leung et al. (2004), biomass power projects can only be economically feasible when they are much cheaper than other kinds of energy technologies. A report by Ministry of Energy and Mineral Development (MEMD 2007) outlines the investment costs of various renewable energy technologies. The investment costs of solar PV, solar water heating and small hydro were estimated in a range of 12,000–15,000 US$/kW, 810–1500 US$/kW and 2500–5000 US/$kW, respectively. In this study however, the investment cost of HPS (2382 US$/kW) competes favorably with small hydropower, though its unit cost of power (0.331 – 0.334 US $/kWh) is more expensive than the national grid power tariff (0.1953 US$/kWh). The high unit cost of power is due to high investment cost, operation and maintenance costs and unreliable biomass fuel costs associated with HPS. A study by Szabo et al. (2011) revealed that a unit cost of electricity was about 0.420 and 0.455 US$/kWh for solar and diesel, respectively, in Uganda. Relating this to the current work, HPS-derived electricity using rice husks and coffee husks as fuel source is therefore cheaper than solar and diesel-derived electricity.

The HPS in the present work has proven a favorable competitor among other energy technologies in terms of reliability, economics, and the potential to conserve the environment. The conversion of biomass into electricity by the HPS can increase greatly the potential usefulness of biomass, a renewable resource, which could make it a considerable part of the national energy mix. This can be achieved by efficient utilization of biomass resources for energy, which is one of the policy objectives contributing to sustainable management of biomass resources. Due to the fact that coffee husks and rice husks are available on a renewable basis, it may be surmised that a 34 kW HPS is sustainable in most of the crop processing centers in rural areas. The electricity generated can be used for lighting and small businesses, leading to rural transformation. This is particularly the case in Uganda were the availability of electricity is seen as a key factor in assisting development, and transformation of livelihoods in Uganda's rural areas.

7 Conclusions

This work presents findings on economic viability assessment of Husk Power System (HPS) adoption in rice and coffee production areas of Uganda. Using coffee and rice husks as fuel feedstock, the unit cost of power (0.331US$/kWh and 0.334US$/kWh, respectively) for a 34 kW HPS was found cheaper than Diesel-generator power system (DPS) of equivalent size (0.410US$/kWh), though much expensive than national grid tariff (0.1953US$/kWh). This makes it unpopular and uneconomical to operate and sustain a 34 kW HPS in areas having access to grid electricity. However, increase in either plant size (above 70 kW) or operational hours (above 16 h/day), or reduction in initial investment cost to less than 880US$/kW, would make HPS electricity much cheaper than grid tariff. Nevertheless, it is possible to improve its hours of operation if additional uses of the power are developed in the villages. Utilizing either rice husk or coffee husk to run HPS gave a positive NPV of 27,054US$ and 20,073US$, a pay-back period of 3.75 and 4 years, and a 15 year IRR of 17 and 14%, respectively – indicating a successful investment. Yet, its economic viability would be enhanced further if it is manufactured locally. Indeed, if the gasifier is fabricated within Uganda, the investment cost would reduce since there would be no importation costs resulting from the system. This necessitates policy options in form of financial subsidies and other support measures in the area of business and technical development for the HPS adoption.

Acknowledgements The authors acknowledge the support from Renewable Energy Incubator of Makerere University and Pamoja Cleantech in accessing the HPS installation and essential Data. Special thanks go to Dr. Musinguzi Wilson for the support and guidance. The authors also acknowledge the support from the government and local communities from coffee and rice growing areas of Uganda, and the HPS Company in India for the necessary information.

References

Albert, P. J. (2014). In P. H. Werhane & T. Rolph (Eds.), *Global poverty alleviation: A case book*. Berlin: Springer.

Bhattacharyya, S. C. (2014). Viability of off-grid electricity supply using rice husk: A case study from South Asia. *Biomass and Bioenergy, 68*, 44–54.

Buchholz, T., & Da Silva, I. (2010). Potential of distributed wood-based biopower systems serving basic electricity needs in rural Uganda. *Energy for Sustainable Development, 14*(1), 56–61.

Ferguson, H. (2012). *Briquette businesses in Uganda. The potential for briquette enterprises to address the sustainability of the Ugandan biomass fuel market*. London: GVEP International.

Fischer, B., & Pigneri, A. (2011). Potential for electrification from biomass gasification in Vanuatu. *Energy, 36*(3), 1640–1651.

Freire-Gormaly, M., Metcalfe, M., & Cheng, Y. L. (2013). Modeling Rural Electrification in Bihar, India Towards 2030 CCTC 2013 Paper Number 1569695607.

IEA. (2015). *World energy outlook: Energy access database*. Retrieved 8th May 2016, from http://www.worldenergyoutlook.org/resources/energydevelopment/energyaccessdatabase/.

Leung, D. Y. C., Yin, X. L., & Wu, C. Z. (2004). A review on the development and commercialization of biomass gasification technologies in China. *Renewable and Sustainable Energy Reviews, 8,* 565–580.

MEMD. (2007). *The renewable energy policy for Uganda*. Uganda: Government of Uganda (GOU).

Nouni, M. R., Mullick, S. C., & Kandpal, T. C. (2007). Biomass gasifier projects for decentralized power supply in India: A financial evaluation. *Energy Policy, 35*(2), 1373–1385.

Obua, J., Agea, J. G., & Ogwal, J. J. (2010). Status of forests in Uganda. *African Journal of Ecology, 48*(4), 853–859.

Okello, C., Pindozzi, S., Faugno, S., & Boccia, L. (2013). Development of bioenergy technologies in Uganda: A review of progress. *Renewable and Sustainable Energy Reviews, 18,* 55–63.

Okello, C., Pindozzi, S., Faugno, S., & Boccia, L. (2014). Appraising bioenergy alternatives in Uganda using strengths, weaknesses, opportunities and threats (SWOT)-analytical hierarchy process (AHP) and a desirability functions approach. *Energies, 7*(3), 1171–1192.

Okure, M., Ndemere, J. A., Kucel, S. B., & Kjellstrom, B. O. (2006). Suitability of agricultural residues as feedstock for fixed bed gasifiers. In *Proceeding of the first international conference on advances in engineering and technology* (pp. 756–763).

Olupot, P. W., Candia, A., Menya, E., & Walozi, R. (2015). Characterization of rice husk varieties in Uganda for biofuels and their techno-economic feasibility in gasification. *Chemical Engineering Research and Design, 107,* 63–72.

Saltelli, A., Ratto, M., Andres, T., Campolongo, F., Cariboni, J., Gatelli, D., et al. (2008). *Global sensitivity analysis: The primer*. New York: Wiley.

Scott, A. (2015). Building electricity supplies in Africa for growth and universal access. Background paper for *Power, People, Planet: Seizing Africa's energy and climate opportunities*. New Climate Economy, London and Washington, D.C. Available at http://newclimateeconomy.report/misc/working-papers.

Shmueli, G., Patel, N. R., & Bruce, P. C. (2011). *Data mining for business intelligence: Concepts, techniques, and applications in microsoft office excel with xlminer*. New York: Wiley.

Szabo, S., Bódis, K., Huld, T., & Moner-Girona, M. (2011). Energy solutions in rural Africa: Mapping electrification costs of distributed solar and diesel generation versus grid extension. *Environmental Research Letters, 6*(3), 034002.

UMEME. (2015). Help center, tariff. Retrieved April 19, 2016, from http://www.umeme.co.ug/help-centre/downloads/tariff.html.

Winston, W. L. (2009). *Microsoft® excel data analysis and business modeling*. Newton: O'Reilly Media Inc.

XE. (2016). Tools, XE currency converter. Retrieved April 19, 2016, from http://www.xe.com/currencyconverter/convert/?Amount=1&From=USD&To=UGX.

Zanchi, G., Frieden, D., Pucker, J., Bird, D. N., Buchholz, T., & Windhorst, K. (2013). Climate benefits from alternative energy uses of biomass plantations in Uganda. *Biomass and Bioenergy, 59,* 128–136.

Author Biography

Yonah K. Turinayo is a Scientist who holds a M.Sc. in Renewable Energy, M.Sc. in Environment & Natural Resources, and a B.Sc. in Industrial Chemistry, from Makerere University, Kampala Uganda. He works with National Agricultural Research Organization (NARO) at National Forestry Resources Research Institute (NaFORRI) as a Research Scientist (Bioenergy). He has a profound knowledge and experience in research and industrial production processes—focusing on Renewable Energy and Biomass, Sugar Technology, Forest Products, Air and Water Pollution Analysis. His areas of interest include: designing and simulation of renewable energy systems (thermochemical and biochemical energy conversion technologies); environmental implications of using renewable resources versus conventional resources for energy technology generation; life cycle assessment for biofuels; and industrial waste management.

Stand-Alone Wind or Photovoltaic Power System: A Holistic Approach for System Design and Selection

Yatindra Kumar Ramgolam and Roobesh Thumiah

Abstract Solar energy and wind energy are the most reliable energies that remain yet to be exploited in our rapidly developing country. The main objective of this study is to develop a holistic approach for the design and selection of stand-alone wind, solar renewable energy (RE) systems, or a combination of both, that is, a hybrid system for an industrial site in Mauritius. Sizing of the RE system is performed such that energy and load requirement of the industrial site are met for the whole operation time. A load and electrical energy audit was performed and weather data for the corresponding location was collected from the meteorological station, Meteonorm and NASA SSE. The AS/NZS 4509.2010 standard is used to size each system and for every possibility of RE solutions. The solutions are simulated on the system advisor software from NREL and optimized so that a higher benefit to cost ratio is obtained. The most appropriate solutions are short-listed and a multi-criteria analysis as well as a cost–benefit analysis is carried out so as to assess each system thoroughly while taking into consideration all the possible factors that may affect the yield of the system, the neighborhood, and even the environment. A simple financial analysis is performed to evaluate the return on investment of the preferred solution.

Keywords Solar energy · Wind energy · Multi-criteria analysis · Cost–benefit analysis

1 Introduction

According to Central Electricity Board (2014), with rapid development in the country, electrical energy consumption has been increasing by 2.9% yearly. In 2012, the island has had a peak load demand of 430.1 MW and a total energy

Y.K. Ramgolam (✉) · R. Thumiah
Department of Electrical and Electronic Engineering, University of Mauritius, Reduit 80837, Mauritius
e-mail: y.ramgolam@uom.ac.mu

W. Leal Filho and D. Surroop (eds.), *The Nexus: Energy, Environment and Climate Change*, Green Energy and Technology, https://doi.org/10.1007/978-3-319-63612-2_5

production of 2796 GWh compared to 1844.1 GWh in 2003 (Central Electricity Board 2014). 79% of the electrical energy was generated from fossil fuels and the rest from renewable energy sources (RES). RES included bagasse, waste-to-energy, hydro, PV, and wind sources. Having ratified the Kyoto protocol, Mauritius has committed to participate in the global effort of mitigating climate change. In addition, with major concerns about the increasing cost of unit energy from fossil fuels and their impacts (Brizmohun et al. 2013) on the environment, the Government of Mauritius is emphasizing more on sustainable development through higher integration of renewable energy (RE) technologies in the local grid. The question of which technology to choose has been a difficult to answer with proponents of each technology strongly supporting their technology. But market penetration and grid integration of a renewable energy technology requires a good balance of the four major factors, namely (1) available resource, (2) maturity of technology, (3) grid readiness and eventually, and (4) promotion mechanisms set up by the government.

During the last decade, advances in RE technologies have demonstrated that potential of RES can be untapped (REN21 2014). In 2014, a total of 51 GW of Wind, 40 GW of solar PV, and 37 GW of hydro power plants were installed globally (REN21 2015). Locally, the RE sector development has been slightly different. Total energy production from local renewable sources, hydro, wind, landfill gas, photovoltaic, bagasse, and fuelwood, went down by 3.2% in 2014 (Statistics Mauritius 2014). It was largely due to a decline of 4.1% in the production of bagasse in 2014. The reason behind the decrease is that Mauritius has seen constant infrastructural development during the last decade and has hence witnessed a decrease in the effective area under sugar cane cultivation by nearly 5000 hectares from 2007 to 2009 only (Statistics Mauritius 2011). This is having a direct impact on bagasse production. Consequently, the government's target to reach 35% of renewable energy is at risk. But other renewable energy sources such as solar and wind energy have great potential to alleviate the country's electrical need. According to energy and water digest 2013 (Statistics Mauritius 2013), integration of photovoltaic power systems (PVPS) has been increasing exponentially since 2009 and has reached the 3 MW limit in 2013. Increases were noted in the energy production of landfill gas as well (+5.9%) and photovoltaic (+808%) with respect to 2013 (Statistics Mauritius 2014). With introduction of renewable energy promotion mechanisms such as feed in tarrif, tendering, and net metering, the installed PV capacity grew to almost 18 MW in 2014 and energy production from 2.7 to 24.7 GWh (Statistics Mauritius 2014). Solar PV increase is due to the geographical location of Mauritius in the tropical zone which contributes to high insolation levels throughout the year. Numerous research works have been carried out to quantify the solar resource potential in Mauritius. A peak irradiance exceeding the 1 kW/m^2 bar for most of the months and an average insolation of nearly 4 kWh/m^2 per day have been reported (Ramgolam and Soyjaudah 2015).

On the other hand, wind power system (WPS) has not seen any development locally in the last decade. Palanichamy et al. (2014) accounted the latter mainly due to occurrence of cyclones, insufficient, and reliable data as well as low public interest. Only recently, a 9.35 MW WPS farm has been installed in the eastern part of Mauritius. The latter consists of 11 turbines (Hamuth 2016). Wind power has raised concerns about negative environmental and social impacts and claimed impacts pertain to visual and aesthetic obstruction, noise generation, as well as wildlife mortality (REN21 2014). In Mare Chicose, Sotravic Ltd is engaged in the conversion of energy into gas from waste (Hamuth 2016). Sustainability and livelihood concerns associated with the use of biomass continue to be debated, especially where linked with deforestation, and where land and water used for energy crop production such as elephant grass competes with food and fiber crops such as sugar cane. In addition, there is uncertainty about the use of biomass being truly “carbon neutral” (REN21 2014).

Mauritius has had concerns with the hydrological years lately. Unavailability of water at higher altitudes, topography of the country makes the island limited in terms of inland water resources for energy conversion. Therefore, new development in hydro-energy sector is not foreseen. In 2014, solar power and wind were the leading technologies by far in terms of dollars committed, with solar power (mostly solar PV) accounting for more than 55% of new investment in renewable power and wind power taking 36.8% (REN21 2015). WPS and PVPS have also proved to be very cost-competitive and WPS the least cost option. With decreasing cost of PV panels and solar energy already quantified, local contractors have been promoting market penetration of PVPS while preferring wafer-based module technologies such as mono- and multi-silicon.

RE systems are designed to meet a given load requirement and system sizing involves determining the rating and size of all components in the system. Two approaches are generally adopted depending on the level of detail (Solanki 2012): Approximate design and precise design. In the approximate design, several simplifying assumptions are made with respect to the component performance, solar data, seasonal variation in load, and performance variation of panel. While in precise design all the latter factors are attended with detail. It also requires proper system simulation for optimization. In general, local contractors design PVPS to meet energy requirements without emphasizing on demand, solar energy available, and do not perform system optimization through simulation.

Numerous standard practices have been developed for design of RE systems. AS/NZS 4509 details a methodology that can be used for sizing of stand-alone power systems such as wind and PV. IEC 62548 (2013) sets out design requirements for photovoltaic (PV) arrays including DC array wiring, electrical protection devices, switching, and earthing provisions. The scope includes all parts of the PV array up to but not including energy storage devices, power conversion equipment, or loads. Numerous software packages exist for modeling of RE systems which include PV SYST, NREL SAM, and Retscreen. For the purpose, meteodata arexxx used which can be obtained from sources such as Meteonorm and NASA SSE (NASA Surface meteorology and Solar Energy 2005). Ramgolam et al. (2013)

developed a methodology for system sizing for Mauritius adapted from AS/NZS 4509 (Australian/New Zealand Standard, AS/NZS 4509 2010) and proposed system configuration that can increase yield by 6%.

1.1 Need for Holistic System Selection and Design

In order to implement clean development mechanisms in Mauritius, numerous renewable energy promotion mechanisms have been implemented since 2010. Initially, Small-Scale Distributed Generations (SSDGs) with feed in tariff were introduced for connection of RE systems to the grid. The scheme ended in 2013 and currently SSDGs, as RE systems with rating up to 5 kW, are being considered under net metering scheme (CEB 2016). Medium-Scale Distributed Generations (MSDGs), representing RE systems between 50 kW and 2 MW, are invited through tenders (CEB 2016). In the case of net metering, the RE system must be designed in such a way that total energy produced by the system is equal to total annual energy consumed by the load. In case annual production is greater than annual consumption, it implies that the system is over engineered and not cost-effective. If annual production is less than consumption, the user will be purchasing required energy units at cost which is higher than conventional unit energy cost. Therefore, annual energy bill may end up being as it was before the RE system was installed. On the other hand, MSDGs require that the levelised cost of electricity produced is least such that the project is competitive in the bid such that it can be considered for connection. As for stand-alone RE systems, they need to meet the power and energy demand of the load throughout the autonomy time. In such cases information on the available resource, seasonal variations in load and user preferences are of utmost importance. Potential RE system users require reliable information on the performance of various RE systems and their returns on investment prior to any investment in RE systems. Therefore, a holistic approach for the selection and choice of RE systems is required. The approach should consider the technical aspects of system design and choice of RE system should be based on cost–benefit analysis (CBA) and multi-criteria analysis (MCA). Moreover, it has been proved that multi-criteria analysis (MCA) enables the proper assessment of all factors whether the criteria are environmental, economic, or social (Alsayed et al. 2014). While many works have been carried out in the design and optimization field of hybrid RES (Lin et al. 2014; Diaf et al. 2007; Hongxing et al. 2009; Chen 2013; Bhattacharjee and Acharya 2015; Bernal-Augustìn and Dufo-Lòpez 2009; Ekren and Ekren 2009), the aim of this research is to provide a holistic step–by-step methodology for RE system design (PVPS, WPS and Hybrid), their optimization through NREL SAM and finally selection of the most cost-effective system using MCA and CBA. As WPS and PVPS have a better scope and potential for exploitation in the local context, the two technologies are investigated and compared in terms of effectiveness and value for money.

2 Methodology

The research work was conducted in four main phases. During phase I, a site analysis and an energy audit were performed to warehouse information related to electrical load requirements, profiles, and site details. RE resource data were also gathered. During Phase II, Wind and PV RE systems were designed according to AS/NZS 4509 and optimized using NREL SAM in Phase III. Finally, in Phase IV, a MCA and CBA were performed for selection of the most cost-effective RE system.

2.1 Phase 1: Energy Audit and Data Warehousing

A load analysis was carried out at the industrial site so as to define the load profiles, surge demand, and monthly energy consumption. Table 1 shows the monthly energy consumption at the site as well as the surge and maximum demand. Solar radiation data were obtained from Meteonorm in the form of a weather file and from related previous research works (Ramgolam and Soyjaudah 2015; Ramgolam et al. 2013). Wind data were obtained from the Mauritius Meteorological Station (MMS) and the NASA Langley Research Center Atmospheric Science Data Center Surface meteorological and Solar Energy (SSE). Both are monthly average values but the former are measured values for the ground level only, while the latter is a satellite-based dataset containing wind profiles for 0, 50, 100, 150, and 300 m. Wind data gathered from MMS and NASA were expressed in the form of an epw weather file for simulation in NREL SAM.

2.2 Phase 2: System Design and Selection

Table 2 shows the type of RE systems as well as the equipments that were taken into consideration in the design process. Lead-acid batteries were used in the design due to their profitability in the long run (Bruch and Müller 2013), while the inverters' size was chosen according to the surge and the 50% oversize recommended by Ramgolam and Soyjaudah (2015). All the RE systems were designed according to the AS/NZS 4509.2010 standard for yearly averaged performance.

Since the wind values are monthly averages, as shown in Fig. 1, the WPS output is consequently averages and hence, are constant throughout the month. Therefore, in the case of hybrid systems, the WPS was first simulated and then the PV systems were used to fill in the difference in the monthly load. Figure 2 shows the system design and optimization methodology adopted during the research work.

For PVPS, soiling effects were considered to be 10% (Roberts and Guariento 2009) and a 93% derate factor was assumed for both the wind and PV systems. The derate factor included energy conversion, transmission, and storage losses.

Table 1 Monthly load variation

Month	January	February	March	April	May	June	July	August	September	October	November	December
Load (kWh)	291	349	397	559	440	382	347	398	385	400	526	490
Surge Demand: 32208 VA							Maximum Demand: 8576 VA					

Table 2 Type of systems and equipment considered

Type of system	Equipment considered
PV-only with batteries	PV panels, charge controllers, batteries and inverters
Wind-only with batteries	Wind turbine, tower, rectifiers, charge controllers, batteries, inverters, dump loads
Hybrid system with both using batteries	PV panels, wind turbine, tower, rectifiers, charge controllers, batteries, inverters and dump loads

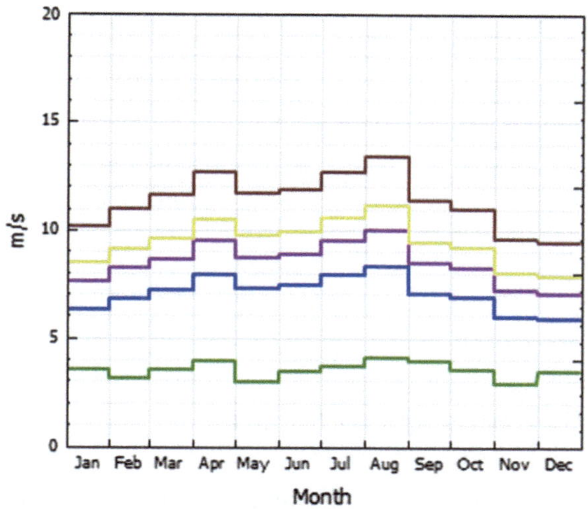

Fig. 1 Wind speed variation (m/s) for altitude 0, 50, 100, 150, and 300 m

For WPS tower height ranging from 10 to 45 m, complementary PV systems were designed to meet total energy requirements. Wind turbines used during this study were 1 kW Bergey XL.1, 3 kW Kestrel e400i, 5 kW Evance R9000, 10 kW Bergey Excel S.60, and 11 kW Gaia 133.

2.3 *Phase 3: Optimization*

The RE systems were optimized separately by using different parameters offered by the NREL SAM software such as tilt angle, azimuth angle, number of PV modules, size of array and module power ratings for the PV system, WPS tower heights, and turbine selection for the wind system. The swept area and the wind profile are limited to the choice of the wind turbine itself. Since any of the wind turbine parameters could not be changed, only the PV systems were optimized. The optimizations were done in the following order: decrease the array size, lower the panels rating, and finally optimize the tilt angle so as to boost the annual production while decreasing the self-shading effect. For tower heights between the obtained altitudes, the NREL SAM interpolates the data through the power law by using the

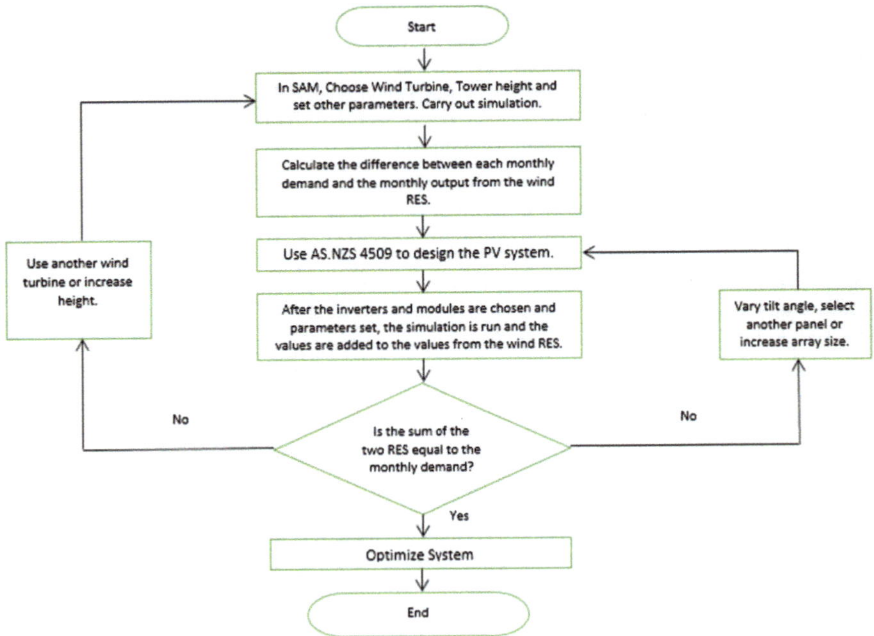

Fig. 2 Stand-alone hybrid RES design

upper and lower datasets. Since the values are monthly averaged, it is important to note that the simulations done did not take into consideration a no-wind scenario, wind gusts, and wind turbulences as well. For practical reasons, the maximum tower height used in the simulations was limited to 45 m.

2.4 Phase 4: Multi-criteria Analysis and Cost–Benefit Analysis

From among 40 solutions, 20 RE systems were selected according to performance. The selected systems were then evaluated using MCA and CBA and assessed with the following criteria: tilt angle, azimuth angle, self-shading, number of panels required, PV area required, panel efficiency, availability, turbine rating, tower height, cut-in speed, swept area, turbine availability, noise, wind and PV maintenance, obstruction and view, weight and structure, overtime, data reliability, lifespan, season variations, and finally payback time and cost. Each criterion was assigned weight and the evaluation matrix was filled. For the MCA, the systems with the highest score are considered best for the site, whereas in the CBA, the systems found at the bottom right corner of the CBA graph are the ideal ones. Both analyses were carried out so as to get a two-directional approach for the selection process.

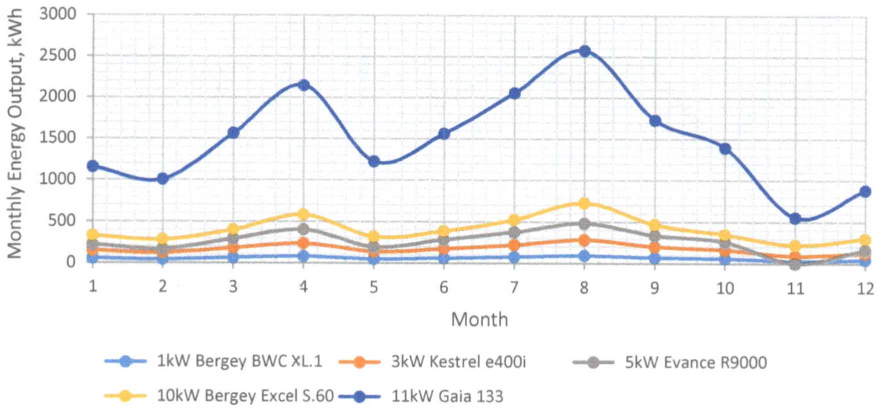

Fig. 3 Wind turbine output

3 Results and Discussions

3.1 Technical Evaluation of PVPS and WPS

Since wind speed varies with height, some of the wind turbines did not produce any energy because of their low cut-in speed. Simulations also show that the wind turbines have an energy output curve as shown in Fig. 3 which indicates that during the winter season, the energy output is greater than in the summer season. Thus, for hybrid systems, the PV systems have to complement the energy output from the wind turbines. Hence, a downward-facing parabola shape energy output curve is preferred for PV systems. So, the tilt and azimuth angles were varied and every change in energy output noted.

It was found through simulations that a tilt angle of 22° and an azimuth angle of 8° facing north were the optimal positions for the maximum annual yield. Hence, for all hybrid and PV systems, a 22° tilt angle was initially selected. But simulation of a 5.31 kW PVPS for different tilt angles (Fig. 4) shows that a tilt angle of less than 22° is more appropriate for summer season which is from October to December and tilt angles greater than that is appropriate in winter season which is from April to September. Hence, for the hybrid systems, tilt angles less or equal than the 22° threshold were hence used depending on the level of contribution of the wind turbines. Wind systems having a smaller power rating required more support from the PV modules than the larger ones; hence, PV panels of the latter generally used tilt angles less than the former. Referring to the Bergey BWX XL.1 tower height curve given in Fig. 5, it can be deduced that the tilt angle required for PVPS gradually decreases as the wind to PV ratio increases. For the case of the PV-only system, the 22° tilt angle was found to cater for the monthly production and maximum annual production, which is the ideal case. The performance of the latter was further evaluated with respect to self-shading effects as shown in Figs. 5 and 6.

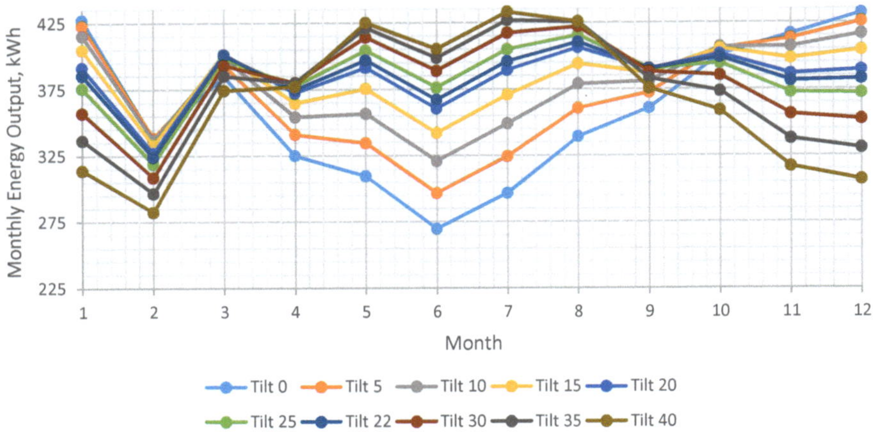

Fig. 4 Effect of tilt angle on energy output

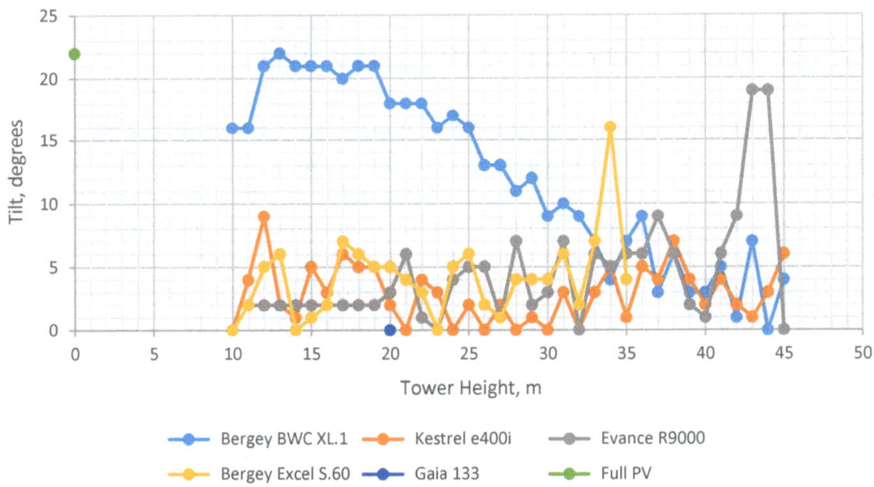

Fig. 5 Optimum PVPS tilt angle for WPS tower height

As tower height increases, WPS energy output increases, implying that number of panels required to meet the overall energy requirements decreases. In order to reduce power loss due to self-shading, PV panels having lower tilt angles could be mounted nearer to each other. Figures 7 and 8 show the variation of the total system rating and number of panels used as the tower height was varied. It must also be noted that if a greater number of panels are used or the swept area and tower height are large, the investment, maintenance, and repairs will be higher and the payback time greater.

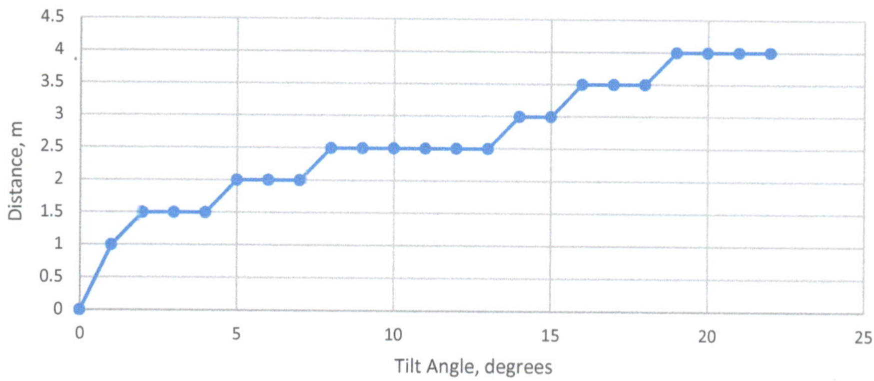

Fig. 6 Self-shading recommended distance

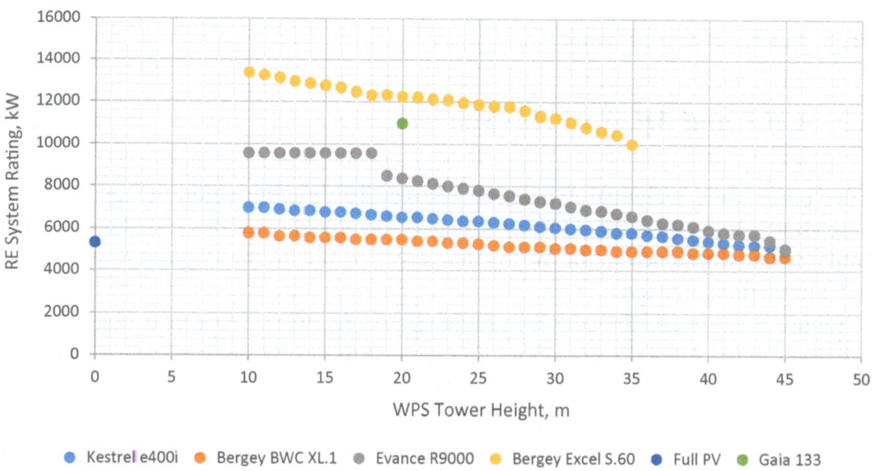

Fig. 7 System ratings

The negative gradient of the graphs shows that as the tower height was increased, the RE system rating required to meet energy requirements decreased. This is due to the fact that at higher altitudes, the wind turbines operate nearer to its rated speed, hence improving its efficiency and therefore, resulting in more energy output. Consequently, the PV system rating required to complement the energy requirements decreased. It is also noticed that at a certain height, the hybrid systems proved to be more power-efficient than using the PV system solely. This proves that regions with a good wind speed regime can make better use of their hybrid systems than regions bound with low wind speed. The system from the Gaia 133 at 20 m and the Bergey S.60 at a tower height of 35 m are wind-only systems and required no complementation from the PV panels.

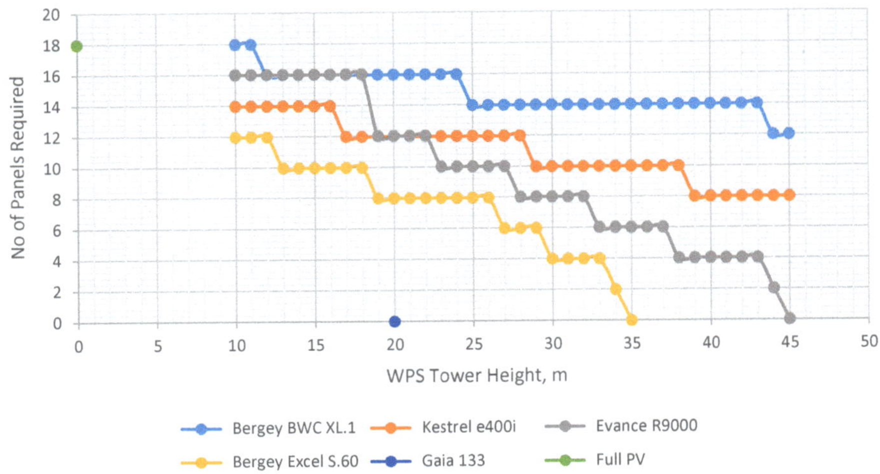

Fig. 8 Number of panels required

3.2 *CBA and MCA of RE Systems*

After performing technical evaluation of over 40 different solutions, 20 solutions were shortlisted with respect to performance. Details of solutions tested are given in Table 3. The difference in the WP system costs' is due to the larger turbine, blades, and foundations needed.

Figure 9 expresses the cost for all the simulated RE systems. It is observed that only the Evance R9000's cost trend line seemed to have a steeper negative gradient than the others. This is because the PVPS array size in the hybrid RE system could be further decreased compared to other hybrid systems. Besides, in some cases, the system could not be further optimized and hence, although the tower height was increased, the same equipment were being used. The payback time variation for the different systems will follow the same trend since a flat-rate electricity tariff was used. Currently, installers tend to set up wind turbines at 10 m above the highest obstacle within 500 m. Thus, a height of 17 m or greater for the wind turbine was preferred due to 7-m building's height. Wind turbines are systems that permanently emit noise from the moment they start running. Hence, it is necessary that the noise levels comply with authorized level and not be a nuisance for the surroundings. The same applies for the visual obstruction. The noise level and visual discomfort increase as the swept area increases and decrease with a lower tower height. The noise level for each system was calculated using the inverse square law. Also, as the swept area and tower height increases, the wind turbine tends to capture more energy by stopping more wind; therefore, more stress is exerted on the system. Since Mauritius is a tropical island, it is prone to cyclones. So, the structure must be able to support the turbines and the PV panels (weight considered in cases where panels are installed on the roofs).

Table 3 Selected systems (WB—with battery; H—hybrid; PVPS—PV only)

Solution	PV	Wind	Total rating (kW)	Comments	Cost, US$
1	13 X CS6X-295 M	–	5.31	PVPS WB—Tilt 22	15,780.8
2	16 X CS6X-280 M	1 kW at 17 m	5.48	H WB—Tilt 20	19,772.5
3	14 X CS6X-305P	1 kW at 25 m	5.27	H WB—Tilt 16	19,307.2
4	14 X CS6X-285P	1 kW at 40 m	4.99	H WB—Tilt 5	19,219
5	12 X CS6X-310P	3 kW at 17 m	6.72	H WB—Tilt 5	23,431.3
6	12 X CS6X-295P	3 kW at 21 m	6.54	H WB—Tilt 4	23,443.4
7	12 x CS6X-280P	3 kW at 23 m	6.36	H WB—Tilt 0	23,395.5
8	10 X CS6X-310 M	3 kW at 29 m	6.10	H WB—Tilt 0	23,019.8
9	10 X CS6X-295P	3 kW at 33 m	5.95	H WB—Tilt 5	22,997.9
10	15 X CS6X-285P	5 kW at 17 m	9.56	H WB—Tilt 2	29,075.1
11	12 X CS6X-280 M	5 kW at 20 m	7.80	H WB—Tilt 6	28,189.9
12	10 X CS6X-290P	5 kW at 24 m	7.90	H WB—Tilt 5	28,153.5
13	10 X CS6X-255P	5 kW at 27 m	7.55	H WB—Tilt 7	27,594.2
14	8 X CS6X-300 M	5 kW at 28 m	7.40	H WB—Tilt 2	26,958.6
15	8 X CS6X-275 M	5 kW at 30 m	7.20	H WB—Tilt 0	26,861.6
16	10 X CS6X-265P	10 kW at 16 m	12.65	H WB—Tilt 7	38,967.8
17	8 X CS6X-290P	10 kW at 19 m	12.32	H WB—Tilt 5	38,780.1
18	8 X CS6P-260P	10 kW at 23 m	12.1	H WB—Tilt 5	38,717.2
19	–	10 kW at 35 m	10	Wind-only—WB	36,118.8
20	–	11 kW at 20 m	11	Wind-only—WB	38,636.5

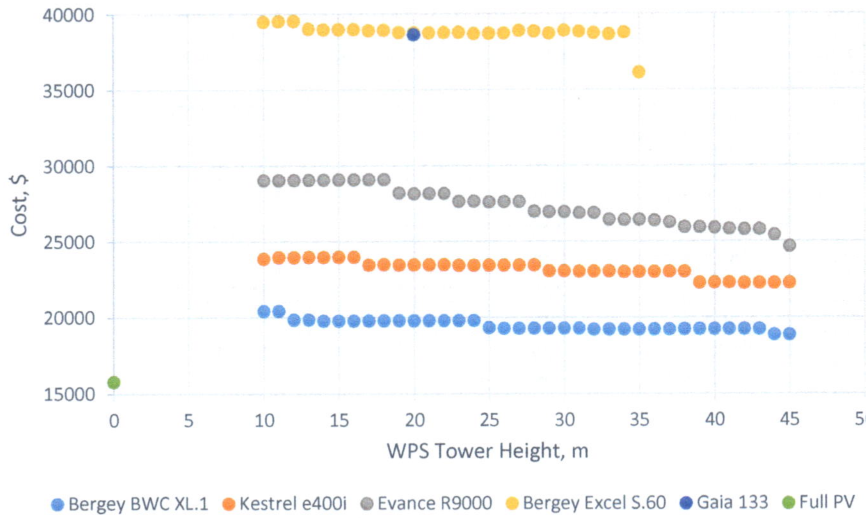

Fig. 9 System costs

With the criteria established, the most appropriate solutions are selected from the simulation results as described in Table 2. The 20 systems were then subjected to MCA and CBA for system selection. Based on facilitators' advice, an evaluation matrix was then filled with weight for each criterion and score for each solution, in order to compute the MCA and CBA scores that each system obtained with respect to each criterion. The scores are given in Table 4.

The CBA was then used to ensure that there was no disparity among the ranks of the systems obtained and to know which system was more cost-effective. The benefits were summed and plotted against the RE system cost as shown in Fig. 10.

Table 5 summarizes the results of the cost–benefit analysis. From the latter, it is confirmed that system 1 (PVPS with battery backup) is best-suited for the site and provides most of benefits at a lower cost. Since the system is also cheaper, the payback time will also be less than the others. Therefore, the return on investment will be much greater than the others. Also, the owner will not have to worry about the noise and visual pollution as the system is pollution-free. However, the system will require a larger surface area to cater for the self-shading effect and for the relative larger number of panels.

It is also important to note that the CBA and MCA did not end up with the same choices for the fifth and fifth solutions. This is because although some systems had good benefit scores, the cost or payback score were poor. Hence, the resulting scores were found to be average (Example: solution 19 with MCA). The same situation aroused with systems with poor benefits and good cost score. With CBA, the system with the highest benefit is chosen with respect to the budget. The cost–benefit analysis provides a better see-through approach for the system selection for

Table 4 MCA and CBA scores

Solution	MCA	CBA	Cost (US$)
1	61.50	49.35	15,780.8
2	46.56	36.40	19,772.5
3	49.75	39.35	19,307.2
4	53.32	42.88	19,219
5	52.94	44.59	23,431.3
6	54.00	45.67	23,447.5
7	54.45	46.09	23,395.5
8	55.16	46.61	23,019.8
9	52.89	44.33	23,006
10	47.29	41.76	29,075.1
11	47.93	41.93	28,138.2
12	50.30	44.04	27,628.9
13	49.20	42.93	27,601.9
14	51.60	45.01	26,958.6
15	50.96	44.35	26,934
16	44.40	43.80	38,991.8
17	44.85	44.14	38,780.1
18	42.37	41.66	38,779.9
19	51.64	49.61	36,118.8
20	50.06	49.28	38,636.5

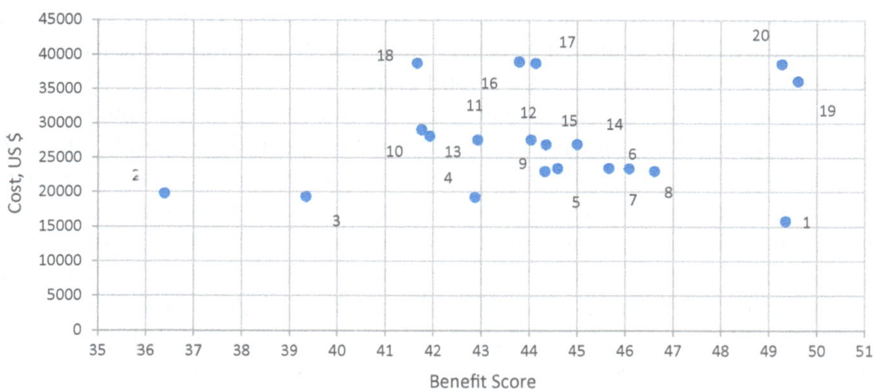

Fig. 10 Cost–benefit analysis

this reason. Finally, a simple financial analysis showed that the return on investment for solution 1 was 17 years if an industrial flat-rate tariff of 0.18$ per kWh was used and 9 years only if a commercial flat-rate tariff of 0.34$ per kWh was used assuming an annual tariff increase of 3.75 and 4.77%, respectively.

Table 5 CBA system ranks

Rank	Solution in terms of cost	Solution in terms of benefits	Cost–benefit analysis
1	1	19	1
2	4	1	8
3	3	20	7
4	2	8	14
5	9	7	5

4 Conclusion

A holistic approach for stand-alone RE system design and selection has been proposed and implemented. System design was based on AS/NZS 4509:10 and system optimization was performed on NREL SAM. Over 40 solutions were evaluated and 20 solutions were shortlisted. Shortlisted solutions were assessed using CBA and MCA in order to identify the solution which meets the requirements in the best way.

It was found that a 22° tilt and 8° azimuth angles of PVPS turned out to produce the maximum annual yield and that a decrease in the tilt angle would favor the summer season, while increasing it would favor the winter season. WPS produce more energy in the winter season than in the summer season. Therefore, wind systems have to be oversized to cater for the summer load demand which leads to an increase in investment since the energy cannot be exported to the grid. Being more productive in summer, PV panels allow PVPS to complement the energy output of the wind turbines for the hybrid systems.

It was also found that some wind turbines could not operate at low height or performed poorly due to the low wind speed. Consequently, the hybrid systems as well as the wind-only systems performed poorly and resulted in being more expensive than the PV-only systems. At higher heights, as the wind speed increased, the hybrid systems proved to be more efficient than the PV system.

But, the PV-only systems were much cheaper than the other solutions and since the latter provided no noise or visual pollution, from MCA analysis, it was concluded that PV-only systems can be recommended for areas with low wind speed profiles. It is also noticed that both wind-only systems obtained high scores in the benefit analysis which indicates that wind turbines have great potential to alleviate the country's energy needs but require a larger investment compared to the PV systems. PVPS is more cost-effective and will have lower payback years if the system is grid-tied since the system can export the excess energy to the grid. MCA and CBA nevertheless concluded with almost the same output which proves their effectiveness in the selection process.

References

Alsayed, M., Cacciato, M., Scarcell, G., & Scelba, G. (2014). Design of hybrid power generation systems based on multi criteria decision analysis. *Solar Energy, 105,* 548–560.

Australian/New Zealand Standard, AS/NZS 4509. (2010). *Stand alone power system.* s.l.:s.n.

Bernal-Augustin, J. L., & Dufo-Lòpez, R. (2009). Simulation and optimization of stand-alone hybrid renewable energy systems. *Renewable and Sustainable Energy Reviews, 13,* 2111–2118.

Bhattacharjee, S., & Acharya, S. (2015). PV–wind hybrid power option for a low wind topography. *Energy Conversion and Management, 89,* 942–954.

Brizmohun, R., Ramjeawon, T., & Azapagic, A. (2013). Life cycle assessment of electricity generation in Mauritius. *Journal of Cleaner Production,* 1–11.

Bruch, M., & Müller, M. (2013). Calculation of the cost-effectiveness of a PV battery system. *Energy Procedia, 46,* 262–270.

CEB. (2016). *Central Electricity Board* (Online). Available at http://ceb.intnet.mu/. Accessed April 5, 2016.

Central Electricity Board. (2014). *Integrated electricity plan 2013–2022.* (online). Available at http://ceb.intnet.mu/CorporateInfo/IEP2013/Chapter5_Power%20Generation%20Plan.pdf. Accessed January 2015.

Chen, H.-C. (2013). Optimum capacity determination of stand-alone hybrid generation system. *Applied Energy, 103,* 155–164.

Diaf, S., et al. (2007). A methodology for optimal sizing of autonomous hybrid. *Energy Policy, 35,* 5708–5718.

Ekren, B. Y., & Ekren, O. (2009). Simulation based size optimization of a PV/wind hybrid energy conversion system with battery storage under various load and auxiliary energy conditions. *Applied Energy, 86,* 1387–1394.

Hamuth, S. (2016). [Online] Available at http://defimedia.info/renewable-energy-pillar-of-the-next-budget-23551/. Accessed April 9, 2016.

Hongxing, Y., Wei, Z., & Chengzhi, L. (2009). Optimal design and techno-economic analysis of a hybrid solar–wind power. *Applied Energy, 86*(2), 163–169.

IEC TS 62548. (2013). *Photovoltaic (PV) arrays—Design requirements.* s.l.:s.n.

Lin, L., Yang, H., & Ma, T. (2014). A feasibility study of a stand-alone hybrid solar–wind–battery system. *Applied Energy, 121,* 149–158.

NASA Surface meteorology and Solar Energy. (2005). *Atmospheric Science Data Center* (Online). Available at https://eosweb.larc.nasa.gov/. Accessed August 2012.

Palanichamy, C., Sundar Babu, N., & Nadarajan, C. (2014). Renewable energy investment opportunities—An investor's perspective. *Renewable Energy, 29,* 709–710.

Ramgolam, Y. K., & Soyjaudah, K. M. S. (2015). Unveiling the solar resource potential for photovoltaic applications in Mauritius. *Renewable Energy, 77,* 94–100.

Ramgolam, Y. K., Sunnoo, D., & Chummun, U. (2013). Evaluating and optimizing the potential of photovoltaics in Mauritius. In *IEEE AFRICON 2013*, Mauritius.

REN21. (2014). *Renewables 2014 global status report*. Paris: REN21 Secretariat.

REN21. (2015). *Renewables 2015 global status report*. Paris: REN 21 Secretariat.

Roberts, S., & Guariento, N. (2009). PV maintenance. In: *Building integrated photovoltaics: A handbook.* s.l.: Springer, p. 60.

Solanki, C. S. (2012). *Solar photovoltaics: Fundamentals, technologies and applications* (2nd ed.). New Delhi: PHI Learning Private Ltd.

Statistics Mauritius. (2011). *Housing and population census* (Online). Available at http://www.gov.mu/portal/goc/cso/ei915/esi2011.pdf. Accessed 2013.

Statistics Mauritius. (2013). *Digest of energy and water statistics.* (Online). Available at http://statsmauritius.govmu.org/English/Publications/Documents/Regular%20Reports/energy%20and%20water/Energy2013.pdf. Accessed January 2015.

Statistics Mauritius. (2014). *Digest of energy and water statistics.* (Online). Available at http://statsmauritius.govmu.org/English/Publications/Documents/Regular%20Reports/energy%20and%20water/Energy2014.pdf. Accessed January 15, 2016.

Author Biographies

Mr. Yatindra Kumar Ramgolam holds a Masters degree in Electronics and Microlectronics Engineering and Technology and is currently enrolled in his Ph.D. in the field of Electrical and Electronic Engineering with a focus on photovoltaic system modeling. He was the Chair of the Energy Management Standards Committee at Mauritius Standards Bureau (MSB), Technical Member of the Energy Standards and Labelling Committee at the Energy Efficiency Management office, and is Chair of numerous Electrical Engineering Standards Sub-Committees on Safety of electrical appliances and equipments at MSB. He is a member of Institute of Electrical and Electronic Engineers and the International Solar Energy Society.

Mr. Roobesh Thumiah holds a Bachelor degree in Electrical and Electronic Engineering. He carried out research work in photovoltaic system and wind system design and modeling and intends to pursue further studies in the field of renewable energy.

Energie Hill Neu Wulmstorf—Reuse of a Former Landfill Site for Energy Production from Renewable Energy

Stefan Lübben and Tilmann Wolfsteller

Abstract To minimize the use of fossil fuels, the Stadtreinigung Hamburg operates, in addition to optimizing energy production from trash incineration, the production of energy and heat from renewable energy sources. At the site of a former municipal waste landfill, all possibilities are put to action to produce electricity from renewable energy sources for the city's public electricity grid. Combining landfill gas collection and its combustion at a combined heat and power plant, with the operation of three wind turbines and a 7900 m^2 module sizes photovoltaic systems, an annual production of between five and six million kilowatt hours of electricity is achieved. This prevents the release of more than 3000 Mg of CO_2, because this produced energy does originate from fossil fuels processed in other power plants.

Keywords Landfill · Wind turbine · Photovoltaic · Landfill gas · Combined heat and power plant · Methane · Renewable energy · Electricity · Solar energy Wind energy

1 Introduction

Until 1999, the Stadtreinigung Hamburg (SRH) disposed Hamburg's municipal waste in landfills. The former landfill Neu Wulmstorf was also used for this purpose until 1986. This site is located southwest of Hamburg, between the townships of Wulmstorf, Ketzendorf and Neu Wulmstorf in Lower Saxony. This borrow pit landfill was established at the site of a former gravel pit and was filled with 3.4

S. Lübben (✉)
Climate Protection Management, Stadtreinigung Hamburg, Bullerdeich 19, 20537 Hamburg, Germany
e-mail: s.luebben@srhh.de

T. Wolfsteller
Landfill Technical Operations, Stadtreinigung Hamburg, Bullerdeich 19, 20537 Hamburg, Germany

W. Leal Filho and D. Surroop (eds.), *The Nexus: Energy, Environment and Climate Change*, Green Energy and Technology, https://doi.org/10.1007/978-3-319-63612-2_6

million m^3 municipal waste from Hamburg. Thus, a hill with a height of 25 m developed and today stands out against the surrounding landscape. Because this landfill is owned by the SRH and drawn-out voting procedures with other owners and co-owners of the site would not occur, the SRH decided to transform the former landfill, after its recultivation, into an "Energy Hill" to implement a useful reuse of the after all 32 ha area. As a public company, the SRH is closely watched by the public eye and seeks to lead by good example, especially by implementing notably sustainable projects. The transformation of a former municipal waste landfill into an innovative Energy Hill had started already in the late 90s as a model project and is run as such to this day. To make sure the site would not end up as an urban barren land, but could be designed in an ecological meaningful and climate protective way, the first deliberations of the future use of the site already began during the landfill recultivation. Besides planned cultivation of meadows and wood areas, renewable energy systems were to be installed. With this, the SRH wanted to make their own effective contribution to the reduction of CO_2 emissions from fossil fuels. Due to the height of the hill and the orientation of the hillsides in a southern direction, ideal conditions for the construction and operation of wind turbines and photovoltaics systems existed.

2 History

The area of the later landfill was used as a gravel quarry since the mid-1950s. Early on uncontrolled trash was disposed of there. After the gravel mining stopped in 1970, the SRH started disposing municipal waste in 1972. This was done in the not yet sealed area at the southern part of the landfill, which was in line with the then valid permit. In 1982, a 10 ha enlargement, built at the north side of the landfill with a layer of a compacted mineral barrier, was used by the SRH until 1986 to dispose most of Hamburg's municipal waste. After the disposal of municipal waste ended, the area was covered with soil.

Starting in fall of 1989, foliage and green waste was collected at the closed down landfill site, and starting in early 1990 this material was composted at a makeshift composting plot (LBHSR 1991). Composting was thought to be an intermediate use of the landfill site to process foliage and green waste, collected through the SRH, into fresh compost for the upcoming recultivation of the former landfill (LBHSR 1989). Between 1990 and 1994, approximately 30,000 Mg foliage and shredded green waste was composted at the composting plot (LBHSR 1991, 1992, 1993; SRH 1994, 1995). Composing at the site had to be abandoned after the start of the landfill recultivation.

In July 1995, the recultivation process of the landfill started and was concluded in 1998. More than 130,000 m^3 of trash had to be transferred and more than 1 million m^3 of soil moved, to profile the former landfill site. More than 300,000 m^2 surface sealing was placed and an additional 23 km of pipes were installed for the collection of surface water, leachate, and landfill gas. Since then, a compressor station draws out the landfill gas from 122 gas wells and conveys it to the main collection station. In the beginning, the gas was burned at 1200 °C in a high-temperature flare system. Altogether, 33 million euros were invested in the recultivation, which includes the service building and leachate pump station (see Figs. 1 and 2).

Because of the surface seal, the amount of leachate originating from the landfill was reduced within a few years from 24,000 m^3 annually to considerably less than 1000 m^3 (see Fig. 3).

After the successful recultivation, a service building was added to the landfill site, which contains a 90-m^2 apartment and a 90 m^2 level with an office, an engineering room and the sanitary facilities. The tenants of the apartment supervise the technical systems of the landfill. In 1998, the service building was built as a low-energy building from native wood. To underline the environmental focus of the SRH, the building was furnished with a wooden facade and green roof. The green

Fig. 1 Aerial photograph of the landfill Neu Wulmstorf during installation of the surface seal in 1996

Fig. 2 Construction of the landfill cover from drainage mats, plastic liner sealing, and bentonite mats

roof should retain rainwater, bind CO_2, and isolate the interior. Rainwater collected from the green roof is used as raw water inside the building and the building ventilation has a built in heat recovery system.

3 Use of Landfill Gas

Since 1997, the generated landfill gas was burned in two 307 kW-combined heat and power (CHP) plants, powering generators (see Fig. 4). During peak periods up to 3.95 million kilowatt hours (kWh) of electricity were produced annually, which were fed into the city's public electricity grid.

Already 4 years after begin of operation of the two CHP plants, the amount of landfill gas generated fell drastically and starting in 2001 only one of the two CHP plants remained in operation to produce electricity from landfill gas. Based on the advanced decomposition of organic material inside the landfill and the continued drying-up of the landfill (due to the surface seal), the landfill gas yield is continuously declining.

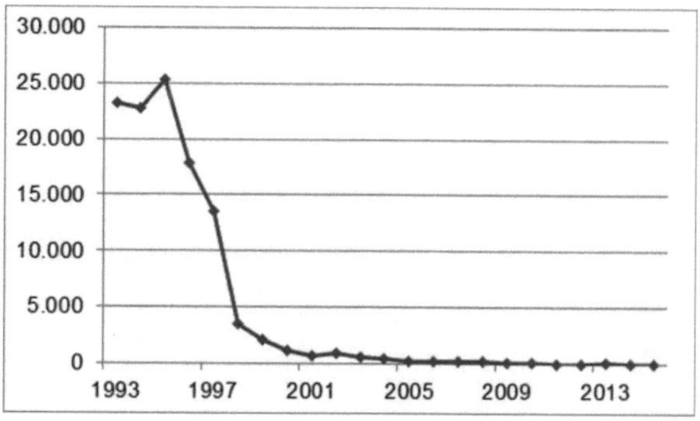

Fig. 3 Reduction of annual landfill leachate due to installation of the surface seal (data in m^3/a)

In 2006, the operative CHP plant was temporarily shut down, because the quality of the landfill gas was insufficient and an optimization of the gas extraction system was required. Although contrary to the goals of sustainable waste management, it was inevitable in 2006 to destroy a portion of gas using the landfill gas high-temperature flare system. However, burning landfill gas is still considered an environmentally friendly action, because methane (which is the largest constituent of landfill gas) is converted to CO_2. With CO_2 being overall 28 times less harmful to the climate than methane.

Due to optimization adjustments done in 2006, the operative CHP plant was successfully run until 2011. However, this was done with increasingly longer periods of downtime, because the CHP plant was constructed for much larger gas yields (percent by volume methane). Therefore, the 307 kW CHP plant was replaced by a 124 kW CHP plant in 2011. With this, a continuous operation of the plant was ensured during 2012–2014. In 2015, the problems with the gas volume and the gas quality started again, the electricity yield dropped below 600.000 kWh in this year (see Fig. 5).

The use of CHP plants can lead to an especially high efficiency, if the heat generated through combustion is also used. Ideally, the heat is fed into a district heating system. However, due to the distance of the landfill Neu Wulmstorf to the nearest residential development, using a district heating system is economically not feasible. Nevertheless, the 180 m^2 living area at the service building and the hot water supply are heated using the heat generated at the CHP plant. In addition, heat from the CHP plan is used to keep the compressor station and leachate pump station frost-protected. Unfortunately, the majority of heat generated has no recipient and is lost to the environment.

The length of time in which the landfill gas can be used economically is currently unknown.

Fig. 4 Buildings for the combined heat and power plant, compressor station, landfill gas high-temperature flare system, and leachate transfer station

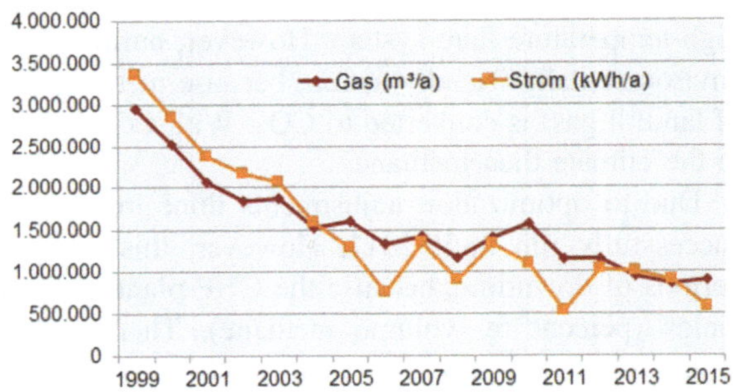

Fig. 5 Trend of electricity produced from landfill gas, 1999–2015

4 Use of Wind Energy

Already during the time of recultivation, the plan was developed to use the former landfill site for the generation of renewable energy. Due to the exposed location of the landfill (40–65 m above mean sea level), a survey done in 1995 predicted a wind speed of approximately 6 m per second (m/s) at a height of 65 m above the site. Based on this survey, output projections for wind turbines were calculated and it was concluded to build three turbines at the landfill site. The distance to the nearest residential area was relatively far; therefore, problems for a permit regarding noise emission were not anticipated.

Because of the unstable substrate of the landfill site, extensive foundation work had to be performed. Due to groundwater protection, pile foundations, which penetrate the surface and base of the landfill, are problematic. Therefore, only raft

Fig. 6 Dynamic compaction of the landfill site in preparation for the installation of wind turbines

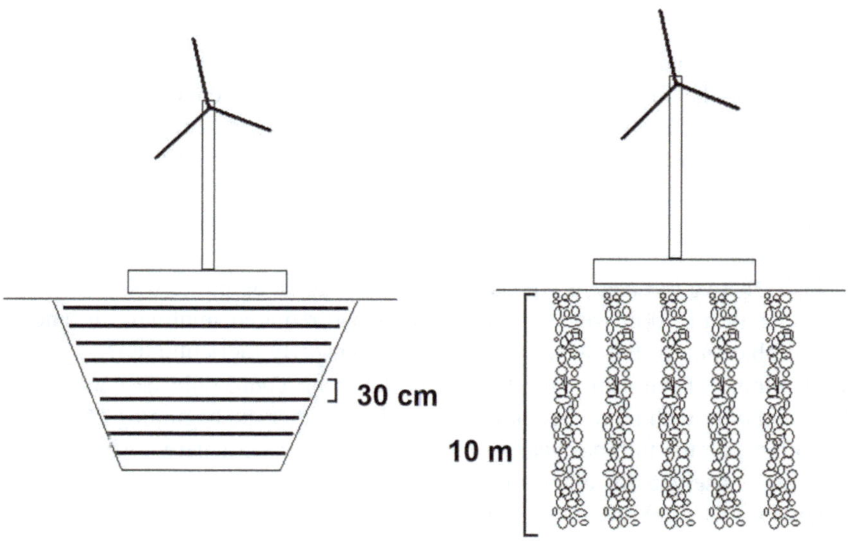

Fig. 7 Different methods for substrate compaction at depth. *Left side* compacting each layer, *right side* compaction by vibrating

foundations were considered a possibility. Already before the surface sealing was put in place, the landfill site was compacted at the predetermined locations of the planned wind turbines using dynamic compaction (see Fig. 6). This was done to simulate the anticipated naturally occurring settlement of the landfill site of the following years.

Only shortly before construction of the wind turbines, additional compacting work was done at the site, to ensure the required stability of the landfill substrate. For the turbine to be built in the south, a pit with a depth of approximately 7 m deep and a diameter of 20 m was filled with sand, which was installed and compacted in 30–cm-thick layers (see Fig. 7, left graphic). For the turbines to be built in the center and to the north, the soil was compacted using a vibration technique. Here a 10-m-deep hole is drilled using a vibrating drill, the hole is simultaneously filled with gravel, and the gravel in turn is compacted through the vibrations of the drill. This leads to compaction of the substrate to a depth of approximately 10 m (see Fig. 7, right graphic and Fig. 8). To increase safety, each wind turbine was fitted with a 14-m-diameter concrete foundation plate, 3 m larger than generally used for this type of turbine. The thickness of these plates was 1.55 m (see Fig. 10).

In the first 10 years after closing the landfill, the surface of the landfill site settles by approximately 3 cm per year, which is due to decomposition of organic material and compaction. During the last years this settlement decreased to 1–2 cm per year or less. Accordingly, the wind turbines also settle. The vertically inconsistent decomposition of organic material at depth can result in the tilting of the concrete foundation plates and in turn the wind turbine. To measure the potential tilting of the concrete foundation plate, sensors were installed at four locations of the plate. Until 2008, the maximum tilting of the turbines measured only 0.07°, with a maximum of 0.11° measured in 2015 (see Fig. 9). Only at tilting rates above 0.4° does the manufacturer see a risk for structural problems and fear of damage to the bearings of the gearing mechanism. The settlement of the landfill substrate and slight tilting are currently not seen as a problem, but are diligently monitored for safety reasons.

In 2001, two turbines of type DEWIND D4 with an output of 600 kW each and a hub height of 70 m were installed. Due to the required minimum distance and formal objection of the owner of the adjacent property (agricultural use), the third wind turbine of the same type located in the south could only be constructed in 2002 and only with a hub height of 55 m. Shortly after the beginning of operation, problems with the turbines gearing mechanisms were encountered, which however were also observed for other wind turbines in recent years. Based on the manufacturer's guarantee all three gearing mechanisms were replaced. Although the gearing mechanisms were replaced, problems arose again and the gearing mechanisms had to be replaced a second time. The costs of 100,000 € per replacement were covered by the insurance. Therefore, the maintenance work resulted in recurring turbine downtimes. Aside from the early damages to the gearing

Fig. 8 Substrate compacted, using a vibration technique, as preparation for the construction of wind turbines

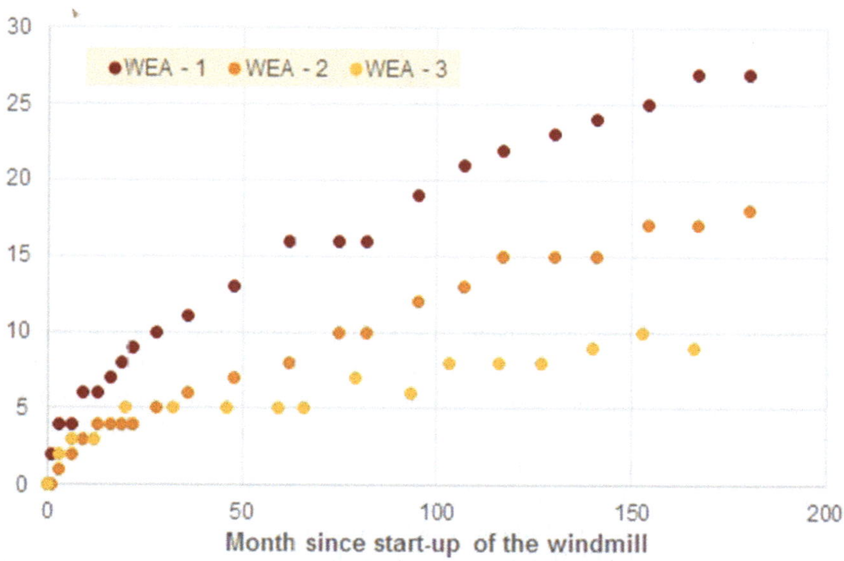

Fig. 9 Difference in settlements (mm) of the respective concrete foundation of the three wind turbines (Wolfsteller 2013)

Fig. 10 Foundation of one of the wind turbines

Fig. 11 Development of energy production from 2001 to 2015, sum of all three wind turbines. Production drop in 2009/2010 was a result of the gearing mechanisms technical difficulties

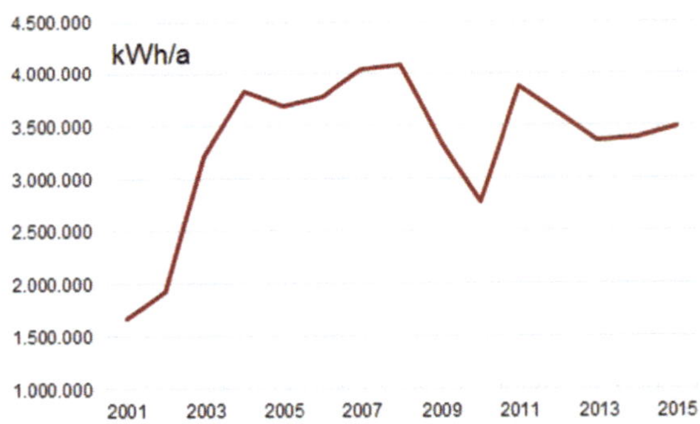

mechanisms, the turbines have been functioning reliably and produce approximately 3.5 million kWh of electricity annually. This electricity is fed into the city's public electricity grid and based on the Renewable Energy Sources Act (EEG) with a buyback price of 9.1 cent/kWh. Figure 11 illustrates the development of electricity production since installation of the turbines.

Looking back, the three wind turbines with a 600 kW capacity are comparatively small. Since 2012, the turbines amortized and are only now profitable. The first request made by the SRH to the licensing authority in regards to repowering of the turbines, replacing the old turbines with newer, larger turbines, was rejected based on the regional development plan 2025 (RROP 2014). The township Neu Wulmstorf is planning to move the development of their residential area toward the former landfill site. Therefore, they are looking to protect this new development from any possible noise or shadows cast by wind turbines at low sun position. With this, the old turbines have the right to continuance, but cannot be replaced by new turbines. The SRH filed an objection in regards to the regional development plan 2025, with the goal to replace the 3-old wind turbines at the end of their 20-year service life with new higher capacity turbines (repowering), using EEG funding.

Currently—summer of 2016—the technical problems with the approximately 15-year-old turbines are considerably increasing. Periods of downtime and extensive repairs are accumulating. At this time, it is questionable, if the three turbines will reach the planned 20-year run-time for which the EEG compensation is legally guaranteed.

5 Use of Solar Power

Figure 12 shows an aerial photograph of the two photovoltaics systems at the landfill site. The southern system is located on an elongated area sloped toward the east. In 2005, this area was covered with a large open-land photovoltaics system, constructed from 3132-polycrystalline modules with 160-W output each from the manufacturer SHARP. The size of the active module is 4103 m^2 with an up to 501-kW peak production (maximum output at full solar radiation). Already several modules have developed hotspots, due to heat accumulation inside the modules. These affected modules were replaced by the manufacturer under their warranty and goodwill. The produced electricity is fed into the city's public electricity grid via a double inverter from manufacturer SMA. The annual electricity production of this system lies at approximately 475,000 kWh. Refer to Fig. 15 for the illustration of annual data.

At the northern side of the landfill site, a photovoltaics system with a top performance of 551 kW peak production was built in 2010 (see Fig. 13). The 2323 polycrystalline modules with 237 W output each are from the manufacturer YINGLI. This system has an overall module size of 3786 m^2 and produces approximately 535,000 kWh of electricity annually. A central SMA inverter transfers the produced electricity into the city's public electricity grid. The annual values are shown in Fig. 15. The electricity produced of both systems is compensated according to the EEG.

The construction of these systems was subject to a specialty structural analysis due to the particular substrate. On one hand, damage to the surface sealing and gas collection system had to be prevented and on the other hand, the different levels of substrate compaction at the site had to be taken into account for the installation of

Fig. 12 Aerial photograph of the landfill Neu Wulmstorf (Google 2014). The picture shows an exact north–south alignment. The wind turbines and both photovoltaics systems are easily recognizable

Fig. 13 Photovoltaics system at the northern part of the landfill site, completion date end of 2010

Fig. 14 Installation of the module sub-construction on strip foundations. Each frame is set on two foundation blocks and carries two rows of eight modules

the systems. It was decided to use 30-cm flat strip foundations, which were only buried to about half their heights to keep a safety distance to the surface sealing located at approximately 1 m below surface. The foundation strips (surface 300 × 75 cm) protruded from the turf by approximately 10 cm, but were covered with soil to prevent draining rainwater from flowing past the concrete foundations and producing soil erosion gullies. To prevent this from happening, the lower edge of the modules was placed at a height of 50 cm above the turf. Thus, enough light

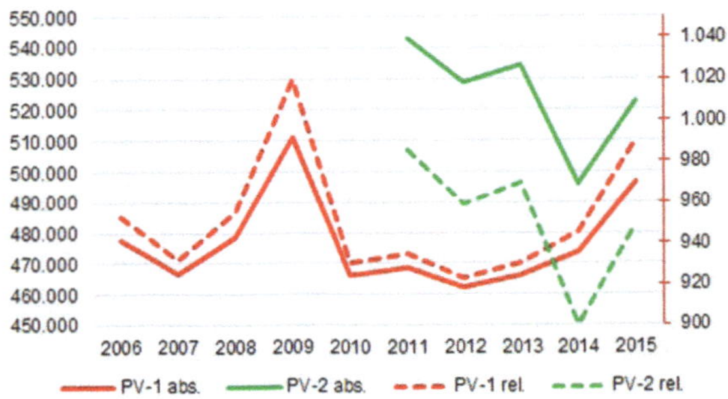

Fig. 15 Development of the electricity production of both photovoltaics systems. The *solid lines* show the absolute annual yield in kWh (*left axis*), the two *dashed lines* illustrate the annual yield in kWh per kilowatt at peak capacity of the systems (*right axis*)

can reach the turf below the module to ensure its growth and prevent possible erosion caused by heavy rains (see Fig. 14).

To prevent the long rows of modules to be damaged by tension and twisting of the assembly stand, caused by different compaction levels of the substrate beneath, the sub-construction was designed as independent 10-m-long frames. Each individual frame holding two rows of eight modules was placed onto two foundation blocks. If settlement of an area occurs, the frames can move into different directions without causing excessive tension within the assembly stands. The cable connections running between adjacent frames were designed with respective tolerance to endure movement of the frames without the cables being damaged.

To ensure the rows of modules do not obscure each other in regards to solar radiation when the sun is lower on the horizon, a distance between rows is maintained (distance rear edge of front row to front edge of next row), which is equal to the threefold heights of the front row.

Though the former landfill site is still secured by a fence, a large number of modules were stolen due to their comparatively high value. In the meantime, the modules have been secured by an alarm system.

Both systems average electricity production per kW installed performance and, based on several years, lies above the average yield in northern Germany (951 kWh/kWp older system, 953 kWh/kWp new system).

6 Summary and Outlook

Already, before the start of recultivation, the clear goal was set to find an ecologically valuable reuse for the landfill site. The landfill gas collection system was state of the art in the mid-1990 and was therefore considered in the recultivation plans. The subsequent construction of the initially two CHP plants for conversion of

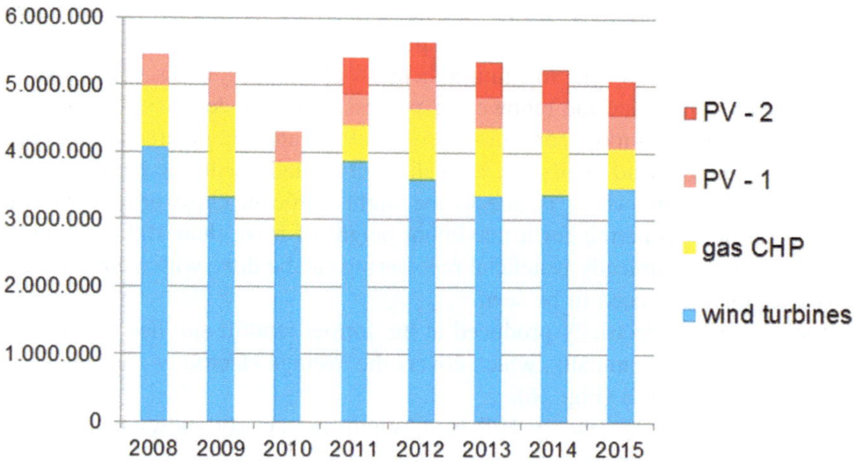

Fig. 16 Annual electricity yield in kWh from the different renewable energy systems at the Energy Hill Neu Wulmstorf. A drop in electricity production in 2010 was a result of the gearing mechanisms technical difficulties of two wind turbines

landfill gas to electricity, right after completion of recultivation, was the first step in the transformation of the landfill site to an "Energy Hill".

Because of the buyback price for electricity from wind power (EEG 2000), installation of wind turbines became economically interesting. The area of the landfill was then built-up with the number of wind turbines permitted at the time. Following the amendment to the EEG in 2004 (EEG 2004), the production of electricity through solar power also became economical more and more interesting. The photovoltaics system is first built at the site and was considered the largest open-land system in northern Germany and the Hamburg metropolitan area. Due to the permanent buyback prices based on EEG, a second photovoltaics system was installed in 2010. However, an advance request for construction of a third photovoltaics system in the western part of the landfill site was rejected by the township Buxtehude.

With this, all possibilities to produce renewable energy at the landfill site and to feed "green" electricity into the city's public electricity grid were maximized and exhausted in the previous years. Every produced kWh, regardless if produced from landfill gas, wind power, or solar power, receives a 20-year term fixed buyback price according to the EEG. The legally guaranteed buyback price has assured the SRH aspiration to expand the production possibilities for renewable energy.

From today's perspective, it can be determined that very few formal obstacles were given for the construction of the early wind turbines. The three wind turbines were erected without noteworthy resistance or legal problems. Today's legal requirements are considerably stricter, especially in regards to minimum distances to residential areas or the maximum heights of these systems. These stricter requirements complicate the repowering process, since modern wind turbines are around 200 m tall. As more constant wind conditions are encountered at this height, accordingly larger wind yields can be realized. First talks with the permit granting agency have shown, which is due to the flight traffic in close proximity to the existing turbines; a permit for a maximum height of more than 100 m cannot be granted. If an economically beneficial repowering can be done with a turbine height of 100 m, then it is need to be seen.

The amount of electricity produced at the former landfill site lies between five and six million kWh annually, which covers the average electricity consumption of up to 2000 households (Fig. 16).

The example of the Energy Hill Neu Wulmstorf clearly shows that the SRH produces electricity from environmental-friendly renewable energy sources, in addition to the energy and heat production at their waste incinerator plants, which greatly reduces the release of climate-damaging CO_2. Annually, the release of more than 3000 Mg can be prevented, because of the amount of electricity produced at the landfill site and the produced electricity does not come from fossil fuels processed in other power plants (BSU 2013). Since the development of the former landfill site in the middle of the 90s, 88,958,000 kWh of electricity has been produced from renewable energy sources at the Energy Hill, leading to a saving of an overall approximately 50,350 Mg CO_2.

The regulatory framework now prevents the further development of renewable energy at the Energy Hill Neu Wulmstorf. The amendments to the EEG (2014) in the last years have first lead to a massive decline and then in 2014 to the cancelation of governmental guaranteed, fixed buyback prices for electricity produced by open-land photovoltaics systems with more than 500 kW performance. With this regulation in place, the economic values of these kinds of systems could not be safely calculated and construction of these systems was determined to be rather unprofitable for the SRH. The strategy of the SRH to possibly repower the three existing wind turbines at the landfill site seems to be limited against the backdrop of the 2025 regional development plan.

In the past, the SRH has tried to install wind turbines at other former landfill sites. However, in all of these cases, these plans have failed due to the lack of permitting capabilities. Considerations to build larger photovoltaics systems at other former landfill sites were rejected by the SRH itself, because those former landfill sites did not have permanent personnel and risks of theft were especially high. In addition, embankments at those sites were either too steep or the reflecting surface of photovoltaics systems would potentially have interfered with the surrounding landscape. Two construction requests for wind turbines to be constructed at this

Fig. 17 Overview of Energy Hill Neu Wulmstorf 2006, with the township Neu Wulmstorf in the background

landfill site were rejected, because the turbines were deemed to interfere with the existing radio relay system and the natural scenery of the area. A request for one or more potential wind turbines, to be erected directly adjacent to the autobahn A24 (between Hamburg and Berlin) at the former landfill Stemwarde-1 near Reinbek, was instantly rejected by the rural district.

Based on the many positive experiences with the Energy Hill Neu Wulmstorf, the SRH is proud to have already decided in 1996 on the right goal of developing the former landfill site consequently as an Energy Hill. In addition, within the framework of the International Building Exhibit Hamburg (IBA 2013), the former landfill site Georgswerder in Hamburg was extended in its function as an Energy Hill, using wind power, solar power, and bioenergy (Esser 2015). Against the backdrop of the vociferously demanded energy transition from 2 to 3 years ago, it is sobering to see that renewable energy systems located near residential areas almost instantly lead to citizen movements against these same systems and their ultimate rejection. Furthermore, the new EEG has become very complex and has made it extremely difficult for corporations, whose main business is waste disposal management, to work through the regulatory thicket and estimate if these system can be operated with economic benefit. Unfortunately, reservation and fear of this topic seem to be increasing, rather than the support for an intense pursuit of a sensible reuse of former landfill sites.

The experience of the last 20 years with two concrete examples has shown, that is, it technically unproblematic and economically beneficial, to build different renewable energy systems at sites of closed down landfills. In most cases, such systems are perceived sympathetically by the public and are supported. Most legal issues arise from the very strict agency interpretation of the legislation, which makes the construction of additional systems at other closed down landfills impossible. In addition, compacting of the current system locations at the existing sites is another hindrance. However, the SRH will continue to take every future possibility that is economically viable, to build new renewable energy systems (Fig. 17).

References

BSU. (2013). *Emissionsfaktoren für CO2*. Leitstelle Klimaschutz, Hamburg: Behörde für Stadtentwicklung und Umwelt der Freien und Hansestadt Hamburg.

EEG. (2000). Gesetz für den Vorrang erneuerbarer Energien – Erneuerbare-Energien-Gesetz vom 29. März 2000,- Bundesgesetzblatt I, p. 305.

EEG. (2004). Gesetz für den Vorrang erneuerbarer Energien – Erneuerbare-Energien-Gesetz- vom 21. Juli 2004,- Bundesgesetzblatt I, p. 1918 (EEG-Novelle).

EEG. (2014). Gesetz für den Ausbau erneuerbarer Energien – Erneuerbare Energien Gesetz vom 1.08.2014, Link vom 9.09.2014. http://www.bmwi.de/BMWi/Redaktion/PDF/G/gesetz-fuer-den-ausbau-erneuerbarer-energien,property=pdf,bereich=bmwi2012,sprache=de,rwb=true.pdf.

ESSER, D. (2015). Der Energieberg Georgswerder – Von der positiven Transformation einer Altlast.- Müll und Abfall 6, p. 313 – 315, Erich Schmidt Verlag Berlin.

Google. (2014). Luftbild der Deponie Neu Wulmstorf, Link vom 9.09.2014. https://www.google.de/maps/place/Neu+Wulmstorf/@53.4492149,9.7821091,1499m/data=!3m1!1e3!4m2!3m1!1s0x47b19b78c0e21e15:0x4554e71dddc5e43b.

LBHSR. (1989). Landesbetrieb Hamburger Stadtreinigung, Maßnahmenplan für die Jahre 1990–1994, Freie und Hansestadt Hamburg, Baubehörde, Landesbetrieb Hamburger Stadtreinigung, p. 125.

LBHSR. (1991). Landesbetrieb Hamburger Stadtreinigung, Geschäftsbericht 1990, Freie und Hansestadt Hamburg, Umweltbehörde, Landesbetrieb Hamburger Stadtreinigung, p. 56.

LBHSR. (1992). Landesbetrieb Hamburger Stadtreinigung, Geschäftsbericht 1991, Freie und Hansestadt Hamburg, Umweltbehörde, Landesbetrieb Hamburger Stadtreinigung, p. 58.

LBHSR. (1993). Landesbetrieb Hamburger Stadtreinigung, Geschäftsbericht 1992, Freie und Hansestadt Hamburg, Umweltbehörde, Landesbetrieb Hamburger Stadtreinigung, p. 58.

RROP. (2014). Entwurf des regionalen Raumordnungsprogramms 2025 für den Landkreise Harburg – Stand Mai 2014. Link vom 9.09.2014. http://www.landkreis-harburg.de/portal/seiten/beteiligungsverfahren-regionales-raumordnungsprogramm-rrop-2025-901000668-20100.html?s_sprache=de&rubrik=901000076.

SRH. (1994). Geschäftsbericht 1993, Freie und Hansestadt Hamburg, Stadtreinigung Hamburg AöR, p. 62.

SRH. (1995). Geschäftsbericht 1994, Freie und Hansestadt Hamburg, Stadtreinigung Hamburg AöR, p. 66.

Wolfsteller, T. (2013). Construction and operation of windmills and solar power plants on landfills. ISWA-Tagung Hamburg. www.iswa.org/uploads/tx_iswaknowledgebase/Wolfsteller.pdf.

Assessing Energy Potential from Waste Tyres in Mauritius by Direct Combustion, Pyrolysis and Gasification

Akshaye Rekhaye and Pratima Jeetah

Abstract Stock piling of used tyres in the landfill is a problematic issue all around the world. The bridging of ecology and energy can be done through the waste tyres. Direct combustion, pyrolysis and gasification are technologies available to reduce the rubber waste disposed at Mare Chicose and at the same time, relieve the energy crisis. The government wants to implement a gasification plant at La Chaumiere and a foreign investor wants to invest in a pyrolysis plant near Mare Chicose. This will benefit the country by a large extent. Our importation of fossil fuel will decrease and the country will save in terms of currency. To tackle this problem in Mauritius, three methods are chosen: Direct combustion, pyrolysis and gasification. It was observed that direct combustion will increase the CO_2 emission by 26% if shredded waste tyres as a replacement for coal. Pyrolysis gave 35% of char 33% of heavy pyrolytic oil, 17% of light pyrolytic oil and 15% of gas by mass percent. The GCV of the char was 30.64 MJ/kg and that of heavy pyrolytic oil was 42.69 MJ/kg while light pyrolytic oil was 41.92 MJ/kg. The char can be compared with a good quality coal which has a GCV of 30–33 MJ/kg and the pyrolytic oil was compared with commercial diesel having a GCV of 46.0 MJ/kg. Gasification, however, was not chosen as the best Waste-to-Energy (WTE) conversion technology due to its high temperature which increases the cost of operation.

Keywords Waste tyres · Pyrolysis · Gasification · Energy · Light pyrolytic oil

1 Introduction

Tyre is that spare part of a vehicle that is subjected to the most rapid wear and tear. In 2011, the European Tyre and Rubber Manufacturer Association (ETRMA 2015), estimated the annual global production capacity of tyres to be 1.4 billion units.

A. Rekhaye (✉) · P. Jeetah
Department of Chemical and Environmental Engineering, University of Mauritius, Reduit, Mauritius
e-mail: akshaye.rekhaye1@umail.uom.ac.mu

W. Leal Filho and D. Surroop (eds.), *The Nexus: Energy, Environment and Climate Change*, Green Energy and Technology, https://doi.org/10.1007/978-3-319-63612-2_7

China, USA, Japan, India and the European countries are among the largest producers of tyre waste in the world and its production has increased considerably in the last 20 years. The rise in the standard of living and the economic development has contributed to the growth of motor vehicle ownership, thus increasing the tyre production across the world. Consequently, the accumulation of waste tyres is also escalating in the world (Quek and Balasubramanian 2013). According to Williams (2013), 3.4 million tonnes of waste tyres are generated every year in Europe, 4.6 million tonnes in the USA, more than 1 million tonnes in Japan and approximately 5 million tonnes in China.

For instance, in Mauritius, 454,426 vehicles were registered at the end of June 2014 according to the Central Statistical Office (CSO) and in 2012, 11.3 million tonnes of tyres were imported to Mauritius. After its usage in the automobile world, the tyres are of no use and are hence disposed of. Out of 1200 tonnes of waste received daily at Mare Chicose landfill, 95% were domestic waste, the rest were wastes like tyre and rubber-based wastes, medical wastes, poultry wastes and sludge wastes. In 2012, the Ministry of Local Government and Outer Island recorded around 371.56 tonnes of waste tyres in the landfill. However, this value does not match with the amount of tyres that are imported since most of the tyres are burnt illegally in crematorium grounds after their prime usage.

Disposal of tyres is a main environmental issue. Stock piling of waste tyre in an open space has a high environmental impact as it becomes a breeding ground for mosquitoes thus affecting human health. Moreover, stockpiles of tyres are potential fire hazard that can cause uncontrollable blaze, and hence contribute towards the air pollution where toxic gases including carcinogens are released in the atmosphere (Birkholz et al. 2003). Due to their void space, tyres takes lot of space in landfilling and this has urged many countries to banned whole tyre landfilling or they may be charged higher tipping fees as compared to other waste (Roy et al. 1994). Subsequently, this paper explores alternative routes for the wastes tyres by converting them into usable products namely oil and gases via pyrolysis and gasification.

1.1 Life Cycle Assessment of Tyres

Assessing sustainability of tyres is a pursuit leading to complexity and subjectivity. Life Cycle Assessment (LCA) is a method for assessing the overall environmental impact of a product from cradle to grave (ETRMA 2014). From studies made by the ETRMA organization in 2013, it is noted that the average end of life scenario has a relatively low contribution to the overall load of the life cycle. The contribution of the average scenario is less important than the production and use phase and finally landfilling is the least attractive option for waste tyres. Some of the recovery routes for end of life tyres are as follows (Lecouls and Klöpffer 2010):

- Retreading.
- Recovery in Civil Engineering applications such as mixing with concrete, use as retention and infiltration basin.
- Material recycling such as foundries, steel works, equestrian floors and synthetic floors.
- Energy recovery option such as direct combustion, pyrolysis and gasification.

1.2 Waste Tyres Legislation in Mauritius

In view to protect the environment, the Ministry of Environment was created in 1992 following the endorsement of the Environment Protection Act (EPA) in 1991 which has been replaced by a new EPA in 2002. The EPA (2002) is a comprehensive piece of legislation which focuses on the coordination of the environmental matters through administrative framework. Subject to Part X of the Environmental Protection Act 2002, an environment protection fee of 50 MUR per unit shall be levied on pneumatic tyres meant for all vehicles except motorcycle, bicycles and wheel chairs at the point of entry in the country.

Waste tyres are used to be regarded as a nuisance for the environment. With development in technology, scrap tyres have now become an opportunity. Manufacturing of tyres is costly, and only one-third is subjected to wear and tear on the road surface. The rest of the material used remains intact and are subjected as waste. Pyrolysis of waste tyres gives good yield of oil and gaseous products. Waste tyres contain different constituents with different thermal properties. Some components are cellulose, hemicellulose and lignin. Upon heating, the carbohydrates (cellulose and hemicellulose) break down to provide low molecular weight volatile products. The lignin is charred to a carbonaceous residue (Sadaka 2015). When these components are heated during pyrolysis, adequate energy is available for pyrolysis reactions at elevated temperatures. Increasing the temperature reduces the char yield. When temperature exceeds 500 °C and extended residence times prevails, lighter products such as condensable aromatic vapours and gases.

2 Products from Thermochemical Conversion of Waste Tyres

2.1 Pyrolysis Products

The main pyrolysis products of waste tyres are pyrolytic oil, char and volatiles. The yield depends on parameters such as temperature, residence time, pressure and heating rate. Low-temperature and high residence times favour the production of char and tars (heavy oil). Medium temperatures (200–600 °C) and high partial pressures favour the combinations of volatiles and light tars. Under such circumstances, the molecules are in close proximity for significant periods of time in the condensed phase and frequent collisions also occur in the vapour phase (Sadaka 2015). The products of pyrolysis can be used as fuels or as feedstock in chemical industries. More refined products of pyrolysis can be used with greater efficiency.

2.2 Gasification

Gasification is the partial oxidation process whereby a carbon source such as coal, biomass or waste is broken down into carbon monoxide (CO), hydrogen (H_2), carbon dioxide (CO_2) and possibly hydrocarbon molecules such as methane (CH_4). The mixture of the gas is called syngas or producer gas. Syngas can be used in internal combustion engines as a replacement for fossil fuels. The gasification technology has been used for over 200 years to produce combustible from organic feeds. Due to shortage of petrol to run vehicles in both world wars, gasification technology was been used to run vehicles. By 1945, this technology used to power trucks, buses, agricultural and industrial machines. Nearly, 9,000,000 vehicles were running on producer gas (Leung and Wang 2003). After the World War II, lack of strategic impetus and the availability of cheap fossil fuels led to lower interest in the syngas technology. However, Sweden continued to use the syngas technology and it became again at their aid in the 1956 Suez Canal crisis. The Swedish decided to include gasification process in their strategic emergency plans. Moreover, the 1973 oil crisis had urged the contemporary interest of the US in small-scale gasifier. As from that time, intense research had been carried out to finally make gasification a mature technology (Nariphaltan 2014).

When waste tyre is burnt at high temperatures (above 600 °C) in a limited amount of oxygen, gasification reaction takes place. It involves the following four sub-processes—drying, pyrolysis, combustion and reduction. To increase the efficiency of the gasification process, different types of gasifier are designed. The type of gasifier depends on the type of biomass or waste used for gasification since there is an interaction between the air and the waste to be gasified; the gasifiers are classified according to the way air is introduced in it, namely downdraft gasifier, updraft gasifier and cross-draft gasifier.

3 Materials and Methods

3.1 Sample Collection

Tyres are discarded in nature as a waste without changing its structure. It is a solid material made up with rubber, steel belts and textile overlays. The rubber is cured along with the steel belts and the textile overlays at high temperature. Thus, their separation becomes a major problem. The first step to recover energy from used tyres consisted of separate rubber from the whole tyre. The rubber powder size must be between 1 and 2 mm. For the purpose of this study, shredded tyre powders were collected from local tyres retreading plants. The collected powder was of different sizes. It was then sieved to 1–2 mm size.

3.2 Pilot-Scale Direct Combustion

Homogenized powdered rubber is burned in a controlled atmosphere. The use of powdered rubber as a potential fuel through direct combustion is investigated. The ultimate analysis of the scrap tyre is analysed thoroughly. Moisture and ash content is studied from the proximate analysis. The products are calculated by the carbon, hydrogen, nitrogen and sulphur content available in the ultimate analysis. The results are tabulated and discussed with the actual coal issues pertaining in the country.

3.3 Pilot-Scale Pyrolysis

A special reactor was used for pyrolysis. The reactor of 400-mm long and 70 mm in diameter was made up of stainless steel to resist high temperature in the furnace. The reactor was designed in such a way that no air would enter the reactor during pyrolysis. A long tube of 600 mm was connected with the reactor to remove the pyrolytic oils and gas formation during the reaction. The reactor was fed with 100 g of powdered tyre ground mechanically which was free from metal reinforcing. The bolt of the reactor was sealed tightly so that there is no leakage of air entering during the reaction. To ensure the safety of the reaction, the reactor was placed in a bucket of water and compressed air was injected at high pressure. The reaction was continued only if there was no bubble of air forming in the bucket of water. As per literature, pyrolysis occurs in the absence of oxygen (Basu 2013). To remove oxygen present in the compressed air that was injected, an excess amount of nitrogen gas was purged into the reactor. The temperature was varied from 350 to 650 °C (Dufton 2001). The yield and GCV of the products were further analysed to see the variation at different temperatures, thus obtaining an optimum temperature for pyrolysis.

3.4 Pilot-Scale Gasification

In contrast with pyrolysis, gasification by air is an exothermic process and it is a very attractive method to recover energy and materials efficiently (Gang et al.

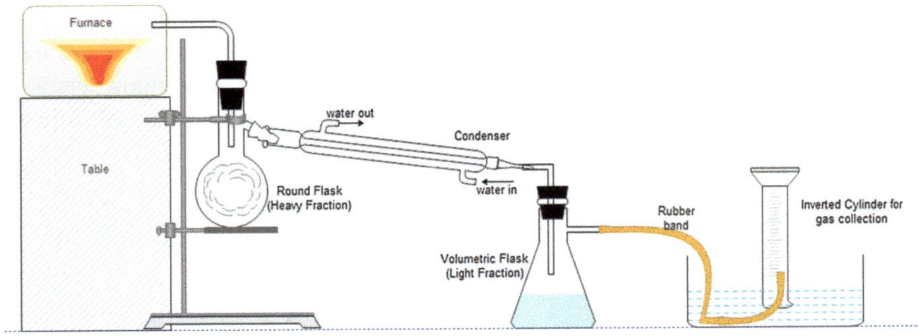

Fig. 1 Pyrolysis setup used for the experiment

2008). According to various researches, it has been observed that gasification usually takes place from 550 to 950 °C. In an economic point of view, the lower the temperature of gasification, the better it is. For the purpose of the study and taking the material of construction of the reactor into account, a low-temperature gasification was used. Leung and Wang (2003) reported that the yield of carbon black and conversion ratio of fixed carbon lied in the range of 24–37 and 14–33%, respectively. Lee et al. (1995) investigated the effect of oxygen concentration and found that it can hardly affect energy conversion ratio which is limited in the range of 0.3–0.35. To check upon the feasibility of gasification from powdered rubber, the same setup as in the Pyrolysis experiment was used (Fig. 1). Although the setup was similar, some factors had to be changed for gasification process. According to literature, a higher temperature is needed for gasification reaction to occur. In order to favour gasification, a limited amount of oxygen is needed. Thus, only 25 g of powdered rubber were placed in the reactor so that there is enough air as medium for the gasification process. Leakage was checked by inserting the reactor in a bucket of water and insert compressed air. The reactor was then placed in the furnace and the experiment was repeated at the following temperatures; 450, 500, 550, 600, and 650 °C. The yield was analysed to see the variation at different temperatures, thus obtaining an optimum temperature for the gasification process.

3.5 *Limitations*

The main limitations with respect to the analysis occurred basically in pyrolysis reaction, whereby nitrogen gas was purged once in the reactor before placing it in the furnace. To ensure complete removal of oxygen, the inert gas had to be continuously fed to the reactor during the reaction which was not the case as the nitrogen cylinder was far from the furnace. Moreover, for the gasification process, the temperature was not varied above 650 °C due to safety reasons.

4 Results and Discussions

4.1 *Direct Combustion*

The GCV of coal is 27.0 MJ/kg while that of waste tyre powder is 35.0 MJ/kg. The rise of 8 MJ/kg is due to the presence of organic compounds such as lignin, hemicellulose and cellulose which are higher in waste tyre than in coal. These organic compounds are used in the manufacture of tyre as it makes the tyre surface more durable (Bortman 2003) (Table 1).

Combustion process generates lots of gaseous pollutants and solid waste materials as products. In Mauritius, the Ministry of Environment has strict regulations on air pollution controls. For instance, some NGOs are questioning on the EIA License given to an emerging coal power station plant in Albion. According to them, the emission level is too high and this will harm the environment. They are asking the government to implement green energy technologies. The future use of shredded scrap tyre as a replacement to coal in the Mauritian power plants will remain a disputable matter.

4.2 *Pyrolysis*

At 450 °C, 40% of carbon black powder (CBp) was formed. When the temperature was increased by 50 °C, the yield decreased to 38% and by further increasing the temperature to 550 °C, the yield of CBp was approximately 34%. At 600 and 650 °C, no major difference was noted in the CBp yield. When the temperature was increased from 450 to 550 °C, 5% decreased in CBp yield was noted. In pyrolysis, the solid char is not the most desirable products although it has certain energy potentials. The lesser the amount of CBp formed, the more efficient is the system. This decrease in yield can be explained by further decomposition of waste tyre at higher temperatures (Sienkiewicz et al. 2012). 550 °C is the temperature where the lowest yield of CBp is noted. Rising the temperature after 550 °C in pyrolysis reaction will not further decrease the yield of CBp (Fig. 2).

Table 1 Gaseous products formed during combustion of 1 kg of shredded scrap tyre

Products	Amount of products (kg/kg of tested sample)	
	Coal	Waste tyre
Carbon dioxide	2.22	3.01
Water vapour	0.313	0.698
Sulphur dioxide	0.0322	0.0338
Nitrogen dioxide	0.0166	0.0269
Nitrogen	10.32	9.99

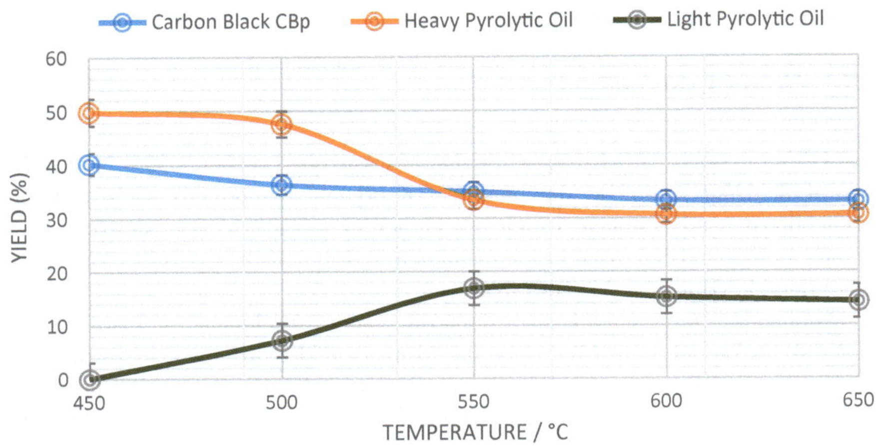

Fig. 2 Variation of amount of products formed with temperature

At 450 °C, 50% of heavy pyrolytic oil (HPO) was formed. As the temperature was increased to 500 °C, the yield was decreased to 48% and at 550 °C it was further decreased to 33%. At 600 and 650 °C, the yield was reduced to 30%. The largest reduction was between 500 and 550 °C being 31% down. As temperature rises, the heavier hydrocarbons are cracked and it gives rise to the formation of lighter oil fractions (LPO), methane and hydrogen. Thus, at higher temperature, less HPO is formed. Pyrolytic oil is the most anticipated products of pyrolysis. Recovering more oil at a lower temperature is more economic for a process. However, the chemical composition of the oil has to be determined. After 550 °C, a linear yield is observed when temperature increases; therefore, by further increasing the temperature, there is no change in the yield of HPO.

At 450 °C, no lighter oil fractions were formed because Polybutadiene Rubber (BR) and Styrene-Butadiene Rubber (SBR) bonds were not decomposed at 450 °C since the heat provided was not enough to break the bonds to form light products (Limbachiya and Roberts 2004). Normally, pyrolysis should be carried out at higher temperature to achieve higher yields of products. The first light fraction was recorded at 500 °C; at this temperature, only 8% of the light fraction was condensed. At 550 °C, around 18% of LPO was formed, showing that a 50 °C rise in temperature decreased the yield of LPO by 10% (from 16.89 to 15.19%). Finally, at 650 °C, the yield was around 14%. Between 500 and 550 °C, there was a rise of 56% because heavier hydrocarbons were being cracked to produce lighter oil fraction and consequently, the amount of HPO at 550 °C was drastically reduced. At 550 °C, good yields of HPO and LPO were formed. When the temperature was increased above 550 °C, no increase in the yield of HPO and LPO was observed. At 450 °C, only HPO was produced and the gross calorific value was calculated to be 41.264 MJ/kg. As temperature was increased by a further 50 °C, both LPO and HPO were formed with calorific values of 39.780 and 41.720 MJ/kg, respectively.

LPO was formed after the hot gas was passed through the condenser. The GCV of LPO was 4.6% lower than the GCV of HPO at 500 °C. This may be due to a lower carbon balance in the LPO. As heat is supplied to the powdered rubber, it breaks down to simpler organic compounds forming HPO and LPO. The HPO will have a higher carbon content than LPO as it is the first product which is basically formed during the pyrolysis reaction when the temperature rises.

4.2.1 Variation of Liquid Fractions with Temperature

The highest values of GCV obtained were 42.69 MJ/kg with Heavy Pyrolytic Oil (HPO) and 41.29 MJ kg with Light Pyrolytic Oil (LPO). However, the difference in GCV between LPO and HPO was only 1.8%. For maximum recovery of oil, it is more appropriate to have a close GCV of LPO and HPO. The small difference was due to improper decomposition of organic compounds such as lignin, hemicellulose and cellulose to simpler compounds at high temperatures. This affirmation again brings us to the conclusion that 550 °C is the optimum temperature for pyrolysis of waste tyres as the maximum yield of oils and the maximum GCV of oil takes place at this temperature (Fig. 3).

This GCV value of pyrolytic oil exceeds the heating value of sub-bituminous coal which is used in Mauritian power plants (Approx. 28 MJ/kg), OMNICANE (2014), wood charcoal (Approx. 31 MJ/kg), Stefano (2014), and even higher than GCV of powdered rubber (35.0 MJ/kg). The pyrolytic oil has a lower ash content, sulphur content and Conradson Carbon Residue (CCR) when compared to a marketed diesel fuel (CIMAK-B10) (Roy et al. 1994). To confirm the statement of Roy et al. (2014), some physiochemical properties of the pyrolytic oil were compared

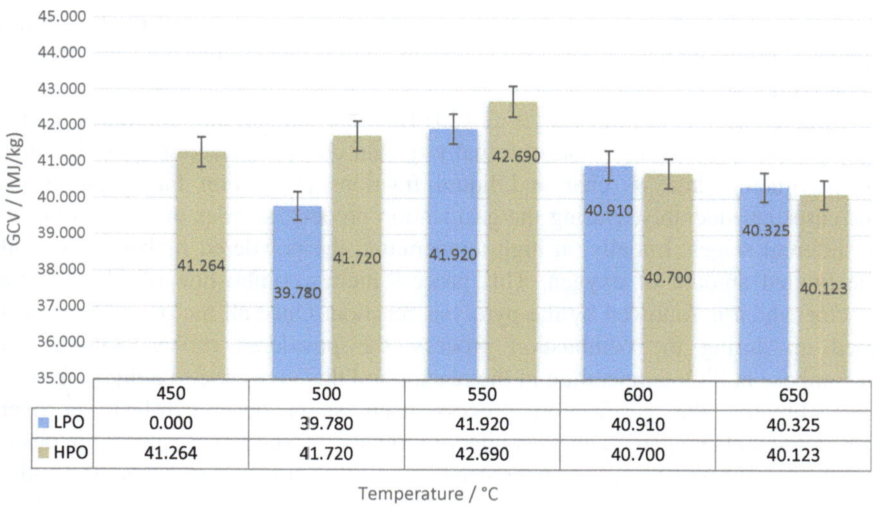

Fig. 3 The variation of liquid fractions with temperature

Table 2 Physiochemical properties of pyrolytic oil and diesel oil

Properties	Pyrolytic oil	Diesel oil
GCV (MJ/kg)	42.69	46.15
Density, kg/m^3 25 °C	870	802
Viscosity, cSt, 50 °C	6.1	9.7
Moisture content, wt%	36.1	29.8

with a normal diesel fuel available on the Mauritian market. The pyrolytic oil used in Table 2 was the one obtained at pyrolysis temperature of 550 °C. The high calorific value of the liquid fractions from pyrolysis makes it suitable to be used as a fuel. The high density of the fuel confirms that the oils are high in aromatics; this is further supported by low hydrocarbon atomic ratio and high carbon residue (Roy et al. 1994).

4.3 *Gasification*

There was no gas formation at 450 °C and gas formation was observed at 500 °C. When the temperature was increased to 550 °C, a rise of 7.5% was observed in the gas formation. Between 550 and 600 °C, a rise of 23% was obtained and between 600 and 650 °C, a rise of 14% was recorded. The temperatures above 650 °C were not investigated due to safety reasons. The R^2 value of 0.860 indicates that there is a significant correlation between the stated equation in the above figure and the values. This shows that there is a general increase in the yield of syngas fraction as the operating temperature increases. This is explained by the fact that at higher temperatures the complex SBR decomposes to simpler organic compounds. As a result, the carbon number will shift to lighter fractions (light oil and syngas) as the operating temperatures increase. At higher temperatures, a better quality of the syngas is produced having a calorific value of 46.0 MJ/kg (Limbachiya and Roberts 2004) (Fig. 4).

In agreement with Martinez et al. (2014), if the temperature continues to rise, char production decreases and eventually remain constant. According to this thermogravimetric data, as char and liquid fractions production falls, that of gas increases considerably. During the gasification process, a series of reaction occurs at different stages. Initially, at high temperature, the powdered rubber reacts with the limited amount of oxygen. This process incites combustion reactions in the reactor which is followed by the pyrolysis reaction. Once all the oxygen has been used up during the combustion process, the powdered rubber continues to decompose at high temperature in the reactor and this gives rise to a liquid fraction (oil). Finally, the gasification process itself starts once pyrolysis is over. Gasification also regarded as "secondary pyrolysis", as the process is continuous of pyrolysis, whereby the reaction proceeds to higher temperature keeping all other conditions during pyrolysis stagnant. In gasification, the remaining char is fractionated to a larger extent. According to certain authors, pyrolysis stops at 550 °C.

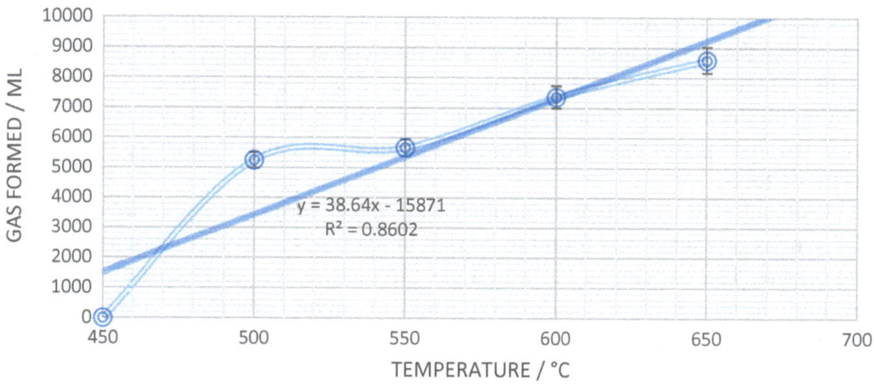

Fig. 4 Gas formation with increasing temperatures

Above this temperature, the gasification reaction takes place to yield a syngas consists of hydrogen, (H_2) and carbon monoxide (CO) as the main product. The decomposed organic products are methane (CH_4), propane (C_3H_8) and butane (C_4H_{10}).

5 Conclusion

Direct combustion of waste tyres was found to be the cheapest among the other two wastes to energy method. Shredded tyres could have been used as a replacement to coal in Mauritius, as it is locally available without involving much money. Comparing the calorific value of waste tyre which is 35.0 MJ/kg and coal which is 27.0 MJ/kg, waste tyre has been found to be far much better in terms of energy content. However, the major problem which rejects direct combustion of waste tyre as an alternative waste to energy is the high emission level of CO_2, where it has been found that the level of CO_2 is 26% higher with waste tyre as compared to coal. As regards the products of the pyrolysis process, the yield and the quality of the products depends on the final pyrolysis temperature which was found to be 550 °C. Carbon black powder is a carbonic residue during pyrolysis reaction but it has a good calorific value which is in the range of 29–32 MJ/kg. The high calorific value of carbon black powder makes it suitable to be used as fuel or even as a replacement to coal. However, the high emission level and the presence of sulphur and chlorine in the residue can jeopardize its use as a fuel if proper exhaust cleaning equipment is not installed. Moreover, the liquid fraction produced also has a high calorific value that changes with varying temperature. The physiochemical properties of the pyrolytic oil were compared with the commercial diesel fuel. According to the chemical composition, the pyrolytic oil can be blended with the commercial diesel oil since they possess almost similar characteristics with some unimportant degree of variations. Blending of these two fuels can reduce viscosity and therefore improving the atomization which will in turn ensure complete combustion in the engine. Hence, pyrolytic oil from waste tyres can be considered as valuable petroleum products. The blending of the fuel will reduce the dependency on the importation of petrol base fuel, ultimately reducing the dependency of our country on fossil fuels.

Gasification of waste tyres is another way of recovering energy from scrap tyres. The syngas produced was found to be rich in hydrogen, carbon monoxide, propane and butane. The gas can be burnt in a gas turbine to generate electricity. However, the high temperature of gasification leads to higher operating cost. At higher temperature, the quality of the gas increases and treatment of the syngas is very costly. For this reason, gasification of waste tyre can be regarded as the second best waste-to-energy technology after pyrolysis. During pyrolysis at 550 °C, a small amount of non-condensable gas is also produced. The characteristic is very close to that of a syngas. By virtue of its calorific value, this gas can be used to supply heat for the pyrolysis process, thus increasing the efficiency of the pyrolysis system and lowering its operating cost. Subsequently, it can be concluded that the most economical and efficient waste-to-energy method of waste tyres is pyrolysis, followed by gasification. Direct combustion is neglected due to its rejection for high emission level in the environment.

Acknowledgements The achievement of this paper was completed under guidance, reinforcement and kindness of many persons that have to be mentioned and received appreciation. I owe my deepest gratitude to the head of Chemical Engineering Department of the University of Mauritius, Dr. D. Surroop for allowing me to use his design concept of pyrolytic reactor. Dr. P. Jeetah for her supervision, advice and guidance from the very early stage of this project. Last but not least, I extend my warmest thanks to the laboratory staffs of the chemical engineering department.

References

Basu, P. (2013). *Biomass gasification, pyrolysis and torrefaction*. Amsterdam: Academic Press.

Birkholz, D., Belton, K., & Guidotti, T. (2003). Toxicological evaluation for the hazard assessment of tire crumb for use in public playgrounds. *Journal of the Air and Waste Management Association, 53*(7), 903–907.

Bortman, M. (2003). *Environmental encyclopedia*. Detroit: Gale.

Dufton, P. (2001). *End-of-life tyres: Exploiting their value*. Shrewsbury: Rapra Technology.

Etrma.org. (2014). *ETRMA: Statistics* [online]. http://www.etrma.org/statistics-2. Accessed March 21, 2015.

Etrma.org. (2015). *ETRMA: ETRMA-European Tyre & Rubber Manufacturers' Association* [online]. http://www.etrma.org/. Accessed March 21, 2015.

Lecouls, H., & Klöpffer, W. (2010). Note on the critical review of the study "Life Cycle Assessment for the different used tyres recycling methods" prepared for Aliapur by Ecobilan. *The International Journal of Life Cycle Assessment, 15*(9), 893–895.

Lee, J., Lee, J., Kim, J., & Kim, S. (1995). Pyrolysis of waste tires with partial oxidation in a fluidized-bed reactor. *Energy, 20*(10), 969–976.

Leung, D., & Wang, C. (2003). Fluidized-bed gasification of waste tire powders. *Fuel Processing Technology, 84*(1–3), 175–196.

Limbachiya, M., & Roberts, J. (2004). *Sustainable waste management and recycling*. London: Thomas Telford.

Martínez, J., Murillo, R., García, T., & Arauzo, I. (2014). Thermodynamic analysis for syngas production from volatiles released in waste tire pyrolysis. *Energy Conversion and Management, 81,* 338–353.

Nariphaltan. (2014). *Nimbkar Agricultural Research Institute (NARI) home page* [online]. http://www.nariphaltan.org. Accessed March 21, 2015.

Quek, A., & Balasubramanian, R. (2013). Liquefaction of waste tires by pyrolysis for oil and chemicals: A review. *Journal of Analytical and Applied Pyrolysis, 101,* 1–16.

Ralph, R. A., Haseneder, R., Gerlach, D., Repke, J., & Otto, M. (2014). Analysis of pyrolysis liquids from scrap tires using comprehensive gas chromatography–mass spectrometry and unsupervised learning. *Journal of Analytical and Applied Pyrolysis, 109,* 234–243.

Roy, C., Pakdel, H., Zhang, H., & Elliott, D. (1994). Characterization and catalytic gasification of the aqueous by-product from vacuum pyrolysis of biomass. *Canadian Journal of Chemical Engineering, 72*(1), 98–105.

Sadaka, D. (2015). *Pyrolysis* [online] Department of Agricultural and Biosystems Engineering Iowa State University. http://bioweb.sungrant.org/NR/rdonlyres/57BCB4D0-1F59-4BC3-A4DD-4B72E9A3DA30/0/Pyrolysis.pdf.

Sienkiewicz, M., Kucinska-Lipka, J., Janik, H., & Balas, A. (2012). Progress in used tyres management in the European Union: A review. *Waste Management, 32*(10), 1742–1751.

Williams, P. (2013). Pyrolysis of waste tyres: A review. *Waste Management, 33*(8), 1714–1728.

Xiao, G., Ni, M.-J., Chi, Y., & Cen, K.-F. (2008). Low-temperature gasification of waste tire in a fluidized bed. *Energy Conversion and Management, 49*(8), 2078–2082.

Author Biographies

Mr. Akshaye Rekhaye is Chemical Engineer and is working in a tyre retreading plant as Engineer.

Dr. Pratima Jeetah is Senior Lecturer at the University of Mauritius in the Department of Chemical and Environmental Engineering. She has Ph.D. in Bioprocess Engineering and M.Sc. in Sustainable Energy Engineering.

A Review of Thermochemical Technologies for the Conversion of Waste Biomass to Biofuel and Energy in Developing Countries

Y.P. Rago, R. Mohee and D. Surroop

Abstract Rising energy security concerns associated with the depleting fossil fuel reserves have triggered renewed interests in the utilisation of biomass for energy. Highly vulnerable to the energy instability, developing countries are turning to thermochemical technologies for the conversion of biomass to biofuels to displace the use of the traditional fossil fuels. A comparative analysis of the potentials of thermochemical conversion technologies of biomass to biofuels and energy in developing countries was conducted. The current status of biomass combustion, cogeneration, gasification, pyrolysis and torrefaction were assessed in different emerging economies, namely, South Africa, Ghana, Malaysia, China, India, Thailand, Tanzania, Brazil and Mauritius. Biomass combustion and cogeneration held the foothold at commercial level both in industrialised and developing countries based on technology maturity and reliability. Biomass gasification is evolving as an eminent method for the production of energy and power in various developing countries. India and China have already embarked on the commercialisation of biomass gasification projects, while other nations have enrolled on pilot-scale schemes. The effectiveness, versatility and environment-friendly aspect of biomass gasification have favoured its preferred integration in the energy system in contrast to combustion and other techniques. Biomass pyrolysis is as promising as gasification in the production of pyrolysis oil to generate transportation fuels. Presently limited at demonstration scale, pyrolysis is being extensively investigated to promote its rapid scale-up and commercial application. Pyrolysis oil represents a crucial resource for the generation of other biofuels, chemicals and power. Torrefaction is a desirable pre-treatment prior to advanced thermal treatment of biomass as it enhances the energy density of processed biomass. Still in its infancy, torrefaction is recurrently being studied to facilitate its large-scale application. The reluctance to adopt biomass thermochemical conversion technologies in developing countries is accounted by economical, technical and social issues. Thermochemical conversion of biomass to biofuel can become important Clean Development

Y.P. Rago (✉) · R. Mohee · D. Surroop
Department of Chemical and Environmental Engineering, University of Mauritius, Réduit, Mauritius
e-mail: yogeshwari.rago@umail.uom.ac.mu

W. Leal Filho and D. Surroop (eds.), *The Nexus: Energy, Environment and Climate Change*, Green Energy and Technology, https://doi.org/10.1007/978-3-319-63612-2_8

Mechanism (CDM) projects with mitigated greenhouse gas (GHG) emissions in the long run.

Keywords Biomass · Biofuel · Thermochemical conversion · Gasification Pyrolysis · Torrefaction · Developing countries

1 Introduction

In face of the dwindling fossil fuel reserves, soaring demand for petroleum products and environmental concerns about fossils, it is imperative to adopt more energy-efficient and economical processes for the sustainable production of fuels (Huber et al. 2006). Developing countries are particularly susceptible to this energy security problem as they are expected to contribute to over two-third of the increase in the world's energy use in the near future (Birol 2006). Biomass has been the primary energy source that has sustained human development prior to the fossil energy regime. The use of wood or dung for heating and cooking purposes is a testimony. Currently, 14% of the world final energy consumption comes from biomass (Dermibas 2009). Biomass sources can either be natural or derived from materials: they include dedicated energy crops, agricultural residues, wood, Municipal Solid Wastes (MSW), animal wastes and sludge (McKendry 2002). In contrast to energy crops, waste biomass is easily available, less costly and does not cause market imbalances as with agricultural products. Being an environment-friendly energy source with great potentials, biomass plays a fundamental role in the sustainable management of wastes and in the production of high-quality carbon-neutral biofuels.

In line with the Kyoto Protocol, biofuels are gaining great eminence as their level of greenhouse gas (GHG) emission is significantly lower than that of fossil fuels (Huber et al. 2006; Mohee and Mudhoo 2012). Today, a wide range of technologies and energy-efficient devices are available for the successful conversion of waste biomass to biofuels. These technologies can be broadly categorised into direct combustion, thermochemical processes, biochemical processes and mechanical extraction, respectively. Selection of the conversion route depends on the end-use application of the biofuels generated: whether they would be used for heat/power generation or employed as feedstock for the chemical industry (McKendry 2002). Thermochemical processes are being viewed as promising and viable techniques in developing countries due to their improved productivity and compatibility with existing infrastructure facilities. Nevertheless, they are still under development and are exposed to several challenges in terms of technical limitations, infrastructure, process efficiency, costs and social acceptance (Verma et al. 2012). In this review, the thermochemical conversion of waste biomass to biofuels and energy in developing countries is compared and analysed. Different emerging economies (including BRICS, IMF and other emerging markets) have been covered to provide an overview of the current situation around the globe.

2 Thermochemical Conversion Processes

Thermochemical transformation of biomass is highly dependent on the nature and composition of waste biomass (Verma et al. 2012). The presence of impurities influences the efficiency of the processes and related devices and the quality of biofuels produced. Moisture content is another important factor. Waste biomass in developing countries generally have high moisture content >50% rendering thermal treatment more challenging (Mohee and Mudhoo 2012). Owing to the heterogeneous nature of waste biomass, pre-treatment of the biomass feedstock is often necessary prior to thermochemical conversion. Typical thermochemical conversion processes that have been investigated in developing countries are combustion, cogeneration, gasification, pyrolysis and torrefaction.

3 Direct Combustion and Cogeneration

Combustion is an ancient technique for recovering and converting the chemical energy stored in biomass to heat and/or power via the steam Rankine cycle (Bridgwater 2004). This reaction between biomass and oxygen is exothermic, releasing significant amount of heat energy and producing stable gaseous, liquid and solid end-products (Arena 2012). Generally, biomass with a moisture content <50% is appropriate for effective combustion (McKendry 2002). A major drawback of the combustion process is the emanation of toxic pollutants in the atmosphere. Pollution prevention equipment is therefore annexed to combustion units for the treatment of flue gases and secondary wastes (Verma et al. 2012). However, combustion of biomass is less polluting compared to combustion of fossil fuels. McKendry (2002) reported a low net efficiency of 20–40% for electricity generation from biomass combustion. Co-combustion of waste biomass in coal-fired power plants boosts the energy efficiency of the process, resulting in reduced emissions of GHG and other pollutants compared to coal combustion alone (McKendry 2002; Senneca 2007). Cogeneration focuses on using excess heat stemming from the electricity production process to increase the efficiency of total energy use by up to 80% in contrast to conventional power generation systems. 15–35% less fuel is consumed during cogeneration, making the process less costly and more environment-friendly in lieu of producing electricity and heat separately (Dermibas 2009).

4 Biomass Gasification

Over the past 30 years, biomass gasification has been broadly investigated as a flexible, efficient and environmentally sound method for meeting future energy requirements especially in rural areas in many countries (Dermibas 2009; Naik et al. 2010). Gasification is the partial oxidation of biomass at temperatures beyond 800 K to generate a combustible gaseous mixture, often contaminated with small amounts of char and condensable products. The fuel gas generated consists primarily of a carbon monoxide (CO) and hydrogen (H_2) mixture, carbon dioxide (CO_2), methane (CH_4), water vapour (H_2O) and nitrogen (N_2). The quality and energy value of the gas are influenced by the gasifying agents used, the biomass composition and the gasification process. Air gasification generates a producer gas that is largely diluted by N_2 and that has a low calorific value of 4–7 MJ/N m^3. Oxygen gasification generates a gas of higher calorific value of 10–15 MJ/N m^3 and H_2/CO ratio (Arena 2012). Direct gasification occurs in air or oxygen-blown systems and is exothermic, whilst indirect gasification is steam-based and endothermic (Belgiorno et al. 2003). Syngas can either be combusted conventionally or used as feedstock for the production of valuable chemicals and fuels. It is utilised in more energy-efficient gas reciprocating engines and gas turbines for electricity production via the integrated gasification combined cycle (IGCC) (McKendry 2002; Arena 2012). The main routes for the conversion of syngas to transportation fuels involve the production of hydrogen via the water–gas shift (WGS) reaction, generation of hydrocarbons through the Fischer–Tropsch (F–T) synthesis and formation of hydrocarbons or oxygenated liquid fuels through methanol synthesis (Naik et al. 2010; Canabarro et al. 2013). The principal types of gasifiers used include updraft and downdraft fixed bed gasifiers, bubbling fluidised bed (BFB) and circulating fluidised bed (CFB) and entrained flow gasifiers, respectively (Arena 2012; Belgiorno et al. 2003; Canabarro et al. 2013). Pollution control devices are integrated in gasification systems to contain environmental impacts (Belgiorno et al. 2003). Biomass pre-treatment is a prerequisite for handling heterogeneous biomass and removal of unwanted species from the feed streams (Bosmans et al. 2013). Biomass gasification can either operate as stand-alone applications or be an integral part of modular systems combining combustion and pyrolysis (Arena 2012). As with combustion, co-gasification of biomass with coal improves process efficiency. Co-gasification with coal is currently being practised in Netherlands (Larson 2008). Biomass gasification is a strong competitor of conventional combustion processes for extracting clean bioenergy from wastes (Arena 2012). Main challenges of gasification include efficient reactor design, biomass pre-treatment, energy efficiency, economic viability and effective gas cleanup (Verma et al. 2012; Larson 2008).

5 Pyrolysis

Pyrolysis can successfully and efficiently convert a wide variety of biomass—MSW, agricultural and forestry residues—to commercially viable biofuels and chemical feedstock (Verma et al. 2012). It is a precursor of combustion and gasification processes and is followed by partial or complete oxidation of primary products (Bridgwater 2004). Pyrolysis is the thermal decomposition of biomass in the total absence of oxygen at moderate temperatures around 600–1000 K to generate pyrolysis bio-oil, solid char and pyrolysis gas (Canabarro et al. 2013; Jahirul et al. 2012). The relative proportion of each product is determined by the characteristics of the biomass feed, the pyrolysis method and the reactor configuration. Based on the operating conditions, pyrolysis can be undertaken via three approaches: slow/conventional, fast and flash pyrolysis. The primary difference lies in the heating rate, solid residence time and process temperature (Bridgwater 2004). Table 1 summarises the operating conditions and characteristics of the different pyrolysis processes.

Pyrolysis is highly energy intensive. Bridgwater (2012) stated that char retains about 25% of the energy of the feedstock out of which 75% is used to fuel the pyrolysis process through combustion, whilst the pyrolysis gas of low calorific value is entirely used to supply thermal energy for pyrolysis reactions (Verma et al.

Table 1 Summary of operating conditions and characteristics of different pyrolysis processes

Pyrolysis process	Temperature (K)	Heating rates (K/s)	Residence time (s)	Characteristics	References
Slow	(Low) 550–950	(Low) 0.1–1	(High) 45–550	• High char formation compared to liquid and gaseous products • Energy intensive due to low heat transfer	Canabarro et al. (2013), Jahirul et al. (2012)
Fast	(High) 850–1250	(High) 10–200	(Short) 0.5–10	• 60–75 wt% of liquid bio-oil • 15–25 wt% of solid char • 10–20 wt% of non-condensable gases • Most preferred method: low investment cost and high energy efficiencies	Naik et al. (2010), Jahirul et al. (2012), Mohan et al. (2006)
Flash	(Very high) 1050–1300	(Very high) >1000	(Very short) <0.5	• Currently at pilot stage • Bio-oil yield of 75%	Naik et al. (2010), Jahirul et al. (2012)

2012). Bio-oil is most preferred as a cleaner and more stable intermediate energy carrier (Venderbosch and Prins 2010). It has found successful applications in furnace/boiler system, diesel engines and turbines (Bridgwater 2004; Balat et al. 2009). Bio-oil upgrading to transportation fuels improves biomass handling and treatment by tackling problems associated with bio-oil such as high viscosity, corrosive nature and polymerization. Upgrading is achieved through catalytic pyrolysis (Balat et al. 2009). Calcined dolomite, nickel and alkali-based metals are common catalysts used in biomass pyrolysis (Han and Kim 2008). Different pyrolysis reactors have been developed to heighten pyrolysis reactions, namely fixed bed, fluidized bed, ablative, auger, vacuum, rotating cone and entrained flow reactors, respectively (Verma et al. 2012; Jahirul et al. 2012; Jones et al. 2009). Each reactor is highly specific, having its advantages and drawbacks. The pyrolytic conversion of biomass to biofuel still has to face innumerable trials before its commercial use. Advanced near-commercial applications in terms of rector configuration and plant set-up have been demonstrated by the Biomass Technology Group (BTG) in Malaysia, DynaMotive in Canada and Ensyn in USA and Canada, respectively (Venderbosch and Prins 2010).

6 Torrefaction

Torrefaction is an energy densification process that employs mild pyrolysis. Early use of torrefaction was restricted to different types of wood that were combusted, gasified or co-fired (Van der Stelt et al. 2011). It is essentially a pre-treatment where waste biomass is heated in an inert or nitrogen atmosphere at relatively low temperatures of 400–600 K in the absence of air, producing a solid biofuel product (Chen and Kuo 2010; Chew and Doshi 2011; Kleinschmidt 2011). The partial decomposition of biomass produces a brittle and dried solid product, free from volatile species. The calorific value of biomass is increased, while the torrefied product becomes hydrophobic with higher fixed carbon content (Kleinschmidt 2011). Storage of torrefied products is improved as the latter are less prone to decay. Despite some minor loss in mass of the torrefied biomass, their energy density, ignitability, grindability and reactivity are far better than the parent one, making them more appropriate for transportation (Van der Stelt et al. 2011; Chen et al. 2012). Van der Stelt et al. (2011) reported that 70% of the mass is retained as solid product. Ninety percentage of its initial energy content is retained, whilst the remaining 30% of the mass is converted to torrefied gas containing 10% of the energy of the biomass. The solid biofuel has superior handling, milling and co-firing capabilities in contrast to other biofuels and is used in the production of chemicals. Researchers are expressing huge interest in the potentials and large-scale application of torrefaction for improving the properties of biomass. Lately, the demand for briquetted torrefied products has risen due to their upgraded properties that allow more effective thermal treatment. Prins et al. (2006) probed into the application of torrefaction of wood as pre-treatment prior to gasification in an entrained flow gasifier. Deng et al. (2009) deduced that the properties of torrefied agricultural residues are similar to those of coal and is apt for gasification. Although presently unfledged, torrefaction has wide scope as a leading biomass pre-treatment in the near future (Van der Stelt et al. 2011). Technical limitations in terms of reactor technology and the lack of process optimization are the main hurdles (Kleinschmidt 2011). Major gaps that have to be addressed prior to technology scale-up entail investigations on torrefaction by-products and analysis of different kinetic approaches to confirm the reliability and consistency of kinetic studies (Chew and Doshi 2011). Topell Energy established its first commercial torrefaction plant in Netherlands in 2010 with a production capacity of 60,000 tonnes/year. The plant employs torrefaction coupled with pelletisation of woody and agricultural biomass to generate bio-coal pellets which are co-fired with coal up to a range of 80%. The French Engineering company, Thermya produced torrefied biomass using directly heated moving-bed reactors. The bio-coal generated retained 95% of the initial energy content and 90% of the original mass of the feedstock with a moisture <1%. Bio-coal has a net estimated calorific value of around 20–21 MJ/kg compared to 7.4–11.4 MJ/kg for untreated crushed wood (Chew and Doshi 2011).

7 Research Status in Developing Countries

Developing countries are now turning to biomass for the production of clean and sustainable biofuels in order to curb energy costs, energy insecurity and environmental problems associated with petroleum products (Larson 2008). Various factors are impeding the development of thermochemical conversion of biomass in emerging economies even though the latter are abundant in biomass resources. An analysis of the status of thermochemical conversion of biomass to biofuel and energy was carried out in developing countries to grasp their evolution.

1. South Africa

Biomass is a predominant renewable resource in South Africa. Commercial biomass comes mainly from the sugarcane industry (field residues and bagasse), forest industry, sawmills and the pulp and paper industry. Combustion of waste biomass and black liquor from the paper and pulp industry and cogeneration of sugarcane bagasse are practised commercially in South Africa for combined heat and power (CHP) (Bradley 2006). Future projects include co-firing of biomass in coal-fired power plants using high-pressure biomass boilers to improve CHP efficiency and promote cogeneration in almost all sugar mills (Bole-Rentel and Bruinsma 2013). Total sugarcane biomass is estimated to 11.47 Mt, out of which 6.1 Mt is from bagasse and 5.3 Mt is from trash. A total energy content of 22.33 TWh is thus expected (Bradley 2006). Presently, no large-scale pyrolysis plants are available in South Africa, though intense research works are being carried out at the University of Stellenbosch in collaboration with other institutes. The South African authorities are planning to come up with pilot-scale biomass pyrolysis units for the production of bio-oil that would be sold to the European market (Bole-Rentel and Bruinsma 2013). Likewise, no torrefaction plants are operational in South Africa. Two South African energy companies, namely Eskom and Sasol, have expressed their interests in performing co-firing and co-gasification tests with torrefied biomass. Eskom has proposed to come up with a pilot-scale biomass torrefaction plant of 60 000 tonnes per annum. The first plant is to be set into operation by 2015 and is estimated to substitute 10% of coal requirement (Bole-Rentel and Bruinsma 2013). Biomass gasification is not yet feasible commercially, whilst coal gasification is a common practice (Bole-Rentel and Bruinsma 2013). A micro-scale biomass gasification project of 300 Nm^3/h was initiated in the village of Melani in Eastern Cape by Eskom and the University of Fort Hare. The project consisted of a downdraft gasifier and produced a gas with an average heating value of 6.4 MJ/Nm^3. The gasification project was successfully implemented to meet the electricity requirement of the village (Mamphweli and Meyer 2009). Furthermore, generation of F–T liquid from locally available biomass and their subsequent gasification are being considered in South African countries (Bole-Rentel and Bruinsma 2013).

2. Ghana

Though cogeneration represents a promising sector for energy management in Ghana, major obstacles include cheaper power supply sources from grid electricity, financial issues and lack of regulatory frameworks to motivate investors to generate and sell electricity to the grid. The 6 MW cogeneration plant installed in Ghana, using sawmill and palm oil wastes as feedstock, has powered industries and surrounding communities without grid electricity (Ahiataku-Togobo and Ofosu-Ahenkorah 2009). No biomass pyrolysis, torrefaction and gasification plants are established in Ghana. Biomass gasification is limited to research and development only, whilst a 6-tonnes pyrolysis plant using sawdust was set up in Kumasi in 1980 as an alternative for power generation. Nevertheless, poor supply and drying of feedstock and low yield of products resulted in the closure of the pyrolysis plant (Atakora 2000). It has been proposed that a pilot-scale pyrolysis plant project would be launched for the production of bio-oil using agricultural crop residues and wood processing wastes in Ghana (Duku et al. 2011).

3. Malaysia

As one of the world's biggest producers of palm oil, Malaysia is exposed to a substantial amount of biomass that is generated from the palm oil industry. Oil palm biomass is estimated to have a total energy potential of about 15.81 Mtoe (million tonne of oil equivalent). Considering an efficiency of 50%, energy generated from oil palm biomass may reach up to 8 Mtoe (Shuit et al. 2009). If tapped correctly, it can be utilised to meet the energy requirement of the country. Combustion of the oil palm residue empty fruit bunch (EFB) is commercialised in Malaysia to power palm oil mills. Around 300 palm oil mills operate with self-generated electricity from oil palm biomass, whilst surplus electricity is supplied to the surrounding areas. 62 such projects were approved out of which 25 are operational (Shuit et al.

2009). Likewise, many biomass-based power generation and cogeneration projects have been initiated and implemented in the Malaysian palm oil industry so as to reduce GHG emissions and encourage clean development in line with the Kyoto Protocol. Co-firing of biomass with coal in boilers for the production of electricity is also common. Generally, the proportion of oil palm biomass being co-fired with coal should not exceed 10% owing to the high ash and potassium content in oil palm biomass (Ericsson 2007; Junker 2005). Biomass gasification is still being developed, whilst the only commercial gasification project in Malaysia uses coal as feedstock (Higman 2014). Palm oil biomass gasification for generating hydrogen is under review. High moisture and energy contents of biomass are suitable for supercritical water gasification. Though this project seems to be economically viable, its technical feasibility still has to be assessed (Shuit et al. 2009). Pyrolysis of oil palm residues for producing bio-oil has bright prospects for replacing petroleum products in Malaysia. The first pyrolysis plant established by the BTG is a breakthrough step to generate 1.2 tonnes/h of pyrolysis oil from palm oil residues (Bradley 2006). As per the U.S. Department of Energy (2006), pyrolysis oil can substitute up to 60% of transportation fuel. The BTG is currently in the final stages of commissioning the plant.

4. India

With around 1000 million tonnes of biomass available in India—mainly from agricultural residues and wastes—the thermal conversion of biomass to biofuels and power can favourably be applied to reduce dependence on oil and curb carbon emissions (Naik et al. 2010). Biomass combustion and cogeneration are commercially viable for generation of power. Various decentralised electricity generation projects involving biomass combustion and cogeneration have been effected especially in energy-deficit rural areas. 17000 MW power is harnessed from cogeneration with 6000 MW coming from the sugar industry (Shukla 2001). Biomass gasification is at present practised on a small scale in India. However, over 1600 small gasifier systems with a total capacity of 16 MW generate 42 million kWh in industries and rural zones. Large-scale gasifiers are still at pilot stage (Shukla 2001; Sharholy et al. 2008). Numerous works undertaken in the field of gasification have probed into operational problems so as to help in developing gasification technology in India. These challenges include tar removal, use of by-products of gasification and evaluation of the internal combustion engines on producer gas amongst others (Sharma 2011). Production of bio-oil from biomass pyrolysis so as to generate transportation fuel is still evolving. Pyrolysis has not yet found large-scale application in India, whilst extensive research is ongoing to render the technology commercially feasible (Mohanty et al. 2011). Biomass torrefaction is emerging as an efficient technique for upgrading biomass for co-firing purposes. The technology is at demonstration stage, but experimental works have shown that the grindability of torrefied biomass can be compared to coal with combustion reactivity comparable to wood (Patel et al. 2011).

5. China

Being a large agricultural country, China is abundant in biomass resources. Grate and fluidized bed combustion are the most popular combustion techniques applied in the biomass power industry. Around 125 biomass combustion power generation projects have been realised between 2007 and 2010. Most biomass power plants in China have relatively small installed capacity of 30 MW. Nonetheless, the installation of a new biomass-fired power plant including two steam turbines, each rated a capacity of 50 MW, has been reported as the world's largest capacity of biomass power generation unit (Fei 2012). Biomass fast pyrolysis for bio-oil production has received great attention in China. A fast pyrolysis demonstration plant with a capacity of 3000 tonnes of bio-oil per year was established in 2007 using rice husk as feed (Zhu 2008). Different upgrading techniques of bio-oil are being explored. Bio-oil is combusted in boilers and is being used as fuel oil or as raw material for the chemical industry (Zhang et al. 2013). Coal gasification is highly dominant in China compared to biomass gasification due to its abundance and low cost (Higman 2014). The need for clean renewable energy source has promoted the commercial application of biomass gasification in China. Current biomass gasification systems include small and medium-scale plants based on simple gas engine systems with a capacity up to 3 MW and large-scale plants operating on IGCC with a capacity exceeding 5 MW. Power generation from biomass gasification is sustainable in China due to its adaptability and ability to operate on any scale varying from several kW to 10 MW. Also, it is proving to be an excellent means of meeting the energy needs of electricity scarce and rural regions (Zhou et al. 2012). Biomass torrefaction for the pre-treatment of biomass prior to gasification is highly recommended. Additionally, the use of torrefaction gases to support efficient gasification of biomass renders the overall process more efficient. Research on biomass torrefaction as a pre-treatment of biomass feedstock on large scale is ongoing (Chen et al. 2011).

6. Thailand

Sugarcane, palm oil, rice, wood wastes and coconut residues constitute the main source of renewable biomass energy in Thailand (Papong et al. 2004). Due to their abundance, biomass fuels are home trade and not imported. Several technologies are being utilised for the successful conversion of biomass to energy in Thailand: gasification, combustion, pyrolysis and biogas. Cogeneration is well established in several industries, namely, rice, sugar, palm oil mills and sawmills; heat and electricity are produced by burning rice husk, sugarcane bagasse, fruit brunches and wood residues, respectively. Excess electricity from these small rural industries is sold to the grid. Typical energy conversion systems that are employed in Thailand include mass burn stoker boiler, stoker boiler, fluidized bed boiler, gasification with combustion in a closed-coupled boiler and pulverised fuel suspension fired boiler (Papong et al. 2004). So far, there exists only one large-scale MSW incineration plant in Phuket Province, with an installed capacity of 14 MW. Moving grate stoker combustion is applied to a feed of 700 tonnes/day. Power generated is sold to the grid. Pyrolysis is still in its infancy in Thailand; pilot lab-scale reactors are being investigated in Universities and research centres to promote production of transportation fuels from biomass. Catalytic fast pyrolysis is yet to be assessed to prove its feasibility regarding large-scale applications (Bioenergy Science Technology and Innovation Policy Group 2013). Gasification is used for small-scale power generation systems (<1 MW) or for industrial heat applications only. The fixed bed downdraft type is adopted for small capacities, whilst fluidized beds are more suitable for larger capacities (Bioenergy Science Technology and Innovation Policy Group 2013). Presently, some 26 gasification plants are operational in Thailand, out of which 80% have thermal applications and the rest are used for electricity generation. Most commercially developed demonstration plants have failed after few months of operation due to inadequate pre-treatment of biomass and lack of control of process parameters. The electricity produced is mainly fed to the grid or used in rice mills (Salam et al. 2010). Despite the great potentials of biomass torrefaction, it is still under research (Martinot et al. 2002).

7. Tanzania

Tanzania is endowed with a wide range of biomass resources. The only thermochemical technology at advanced stage of implementation is cogeneration. Sugarcane bagasse is combusted to generate heat and power, whereby excess electricity is sold to the grid. Direct combustion is common in Tanzania mainly for heating purposes. Although biomass gasification is highly appealing and is appropriate for the Tanzanian environment, drawbacks in terms of system reliability, high operation and maintenance costs are impeding its commercial application. Moreover, gasification is demanding, especially in small-scale projects where special feedstock is required. Pyrolysis is another attractive technology for effective production of transportation fuel from biomass. Yet, it has been considered inefficient in Tanzania due to water dilution of the bio-oil produced. Torrefaction is under review. To facilitate implementation of these emerging technologies in Tanzania, sensitization of its people and training sessions is necessary along with the help of academic and research institutions to promote research and development in the field (Kusekwa 2013).

8. Brazil

As one of the largest sugarcane producers in the world, Brazil has a rich stock of bagasse and field residues for combustion and cogeneration to produce heat and power. Most of the bagasse from the sugar industry is burned to power sugar and ethanol plants. Presently, 1900 MW is used internally in mills, whilst a surplus of 600 MW is sold to the grid (Bradley 2006). Several research and development projects are being promoted and funded for successful conversion of biomass to biofuels. The UCLA thermal power plant with an installed capacity of 28 MW and 25 tonnes/h boiler steam generation uses wood residues as fuel. The thermal power plant in Piratini has an installed capacity of 10 MW using about 142 thousand tonnes of wood residues annually to generate around 71,000 MWh/year (Lora and Andrade 2009). Biomass gasification is highly prospective. The GASEIFAMAZ project “Comparison between the existing technologies for biomass gasification” assesses the Indian technology of small-scaled fixed bed gasification to meet the energy demand of isolated communities in the Amazon region. A 20 kW gasification system was imported from the Indian Institute of Technology (Lora and Andrade 2009). Similarly, fast pyrolysis of sugarcane bagasse for the production of bio-oil and torrefaction of biomass are being broadly researched. Significant progress has been achieved with regards to bagasse fast pyrolysis which is economically appealing. A pilot-scale fast pyrolysis plant using sugarcane bagasse was investigated by the State University of Campinas (UNICAMP) in São Paulo. Further research works are, nevertheless, necessary to favour the commercial application of pyrolysis in Brazil (Alonso-Pippo et al. 2004).

9. Mauritius

As a legacy of the Dutch, the sugar industry has dominated the Mauritian economy over years. Consequently, sugarcane bagasse comprises the principal biomass employed for electricity generation in Mauritius. In fact, bagasse cogeneration was pioneered in Mauritius (Mohee and Mudhoo 2012; Jolly 2003). Bagasse contributes to around 92% of the local renewable sources and has an estimated share of 13–17% in electricity production (CSO 2013). Continuous research and adoption of advanced technological practices have favoured the large-scale application of bagasse cogeneration in the country. Cogeneration sustains operations in sugar mills and surplus electricity is sold to the grid. However, coal is utilised as complementary fuel during the sugarcane off-crop season. Around 597 GWh of electricity was generated from sugarcane bagasse in 2013 (CSO 2013). At present, no large-scale biomass combustion facilities are operational in Mauritius. It was proposed to establish a waste-to-energy facility at the La Chaumiere for power generation, but the project was eventually halted due to social unacceptance. Biomass pyrolysis, gasification and torrefaction are currently at infancy though continuous studies on the techniques are underway. Bagasse gasification for power production and generation of fuels is in the pipeline (Kong Win Chang et al. 1999). Deenapanray (2006) stated that initiatives are to be taken to foster bagasse gasification by establishing a demonstration Biomass IGCC plant in Mauritius. This project would play a fundamental role in displacing coal for electricity generation in the long run. Surroop and Mooloo (2013) investigated the potential of biomass torrefaction in Mauritius as a pre-treatment technology for valorising biomass as fuel. Nonetheless, more efforts are required to favour the commercial application of the different thermochemical conversion techniques in Mauritius.

8 Comparative Analysis

It can be inferred that at present biomass combustion and cogeneration have foothold at commercial scale for CHP owing to their relatively simple operation and working principle (Verma et al. 2012). Electricity generation from biomass by cogeneration and combustion technologies has established a growing market not only in developing countries but also in industrialised nations. The maturity, reliability and efficiency of these techniques justify their rapid adoption worldwide (Faaij 2006). On the other hand, biomass gasification has been clearly demarcated as a thriving technology in contrast to biomass pyrolysis and torrefaction. India and China are presently leading in biomass gasification. For every kWh of coal-based electricity that is substituted by biomass gasification-based electricity, an estimated 1 kg of CO_2 is likely to be reduced. Biomass gasification is preferred over traditional combustion as it is economically more favourable and mitigates GHG emissions and environmental pollutions (Larson 2008; Buragohain et al. 2010). Although pyrolysis offers bright prospects for the production of bio-oil, it is underdeveloped compared to gasification. Gasification is being fully explored, demonstrated and developed, allowing flexible installation and operation. Despite innumerable studies on biomass pyrolysis, very few demonstration units have managed to operate successfully so far. Market implementation is yet to be developed (Faaij 2006). Nevertheless, perpetual efforts are being deployed to foster the commercialisation of biomass pyrolysis. The BTG project in Malaysia, for instance, corroborates this fact. Torrefaction is an attractive pre-treatment for augmenting the value of biomass prior to advanced thermal treatment for energy recovery.

The resistance to openly embrace thermochemical technologies in developing countries is accounted by economic and technical barriers. Technology transfer and the maturity of evolving techniques such as pyrolysis, gasification and torrefaction have high impacts on developing nations. Moreover, high costs associated with the collection, transport and processing of biomass are major deterrents compared to the use of fossil fuels (Girard and Fallot 2006). Even though gasification is growing rapidly, biomass and waste gasification is still lagging behind when compared to coal gasification (Higman 2014). Lack of awareness about the technologies is a serious shortcoming. Governments in developing countries have to come up with financial incentives and sensitization campaigns to educate people and promote the development of necessary skills. Developing countries should learn to effectively tap their rich biomass resources instead of exporting them to industrialised nations. The emerging thermochemical technologies, especially gasification, offer great opportunities to encourage sustainable energy projects via clean development mechanism (CDM) in developing countries. Job creation, development of rural areas and energy security can easily be promoted provided the thermochemical conversion of biomass to biofuel is intelligently planned, designed and executed.

9 Conclusion

Biomass constitutes a fundamental source of renewable energy that can be appropriately tapped in developing countries to curb dependence on fossil fuels. Out of the many technologies available for conversion of biomass to biofuel, thermochemical techniques have particularly captured the attention of many researchers. The review proves that biomass combustion and cogeneration have already carved themselves a niche on the international market for the production of heat and electricity. Many developing countries have already adopted cogeneration and combustion on a large scale to meet their energy needs. However, emerging technologies such as biomass gasification, pyrolysis and torrefaction are gradually becoming more prominent on the energy market. Biomass gasification is highly sustainable compared to conventional combustion due to its effectiveness, reliability and flexibility. China, India and South African countries have already embarked on biomass gasification projects on small to medium scale. Although pyrolysis is still being thoroughly investigated, various pilot-scale demonstration plants are being set-up in developing countries to refine the technology and overcome barriers. The production of bio-oil through biomass pyrolysis to eventually generate transportation fuels will be highly lucrative in displacing the conventional fossils in the near future. Torrefaction is an important pre-treatment for valourising biomass as biofuels. Yet, biomass gasification, pyrolysis and torrefaction are continuously being examined to facilitate their large-scale application worldwide. The implementation of thermochemical technologies for conversion of biomass to energy through CDM projects will boost social and economic development in emerging nations while containing environmental impacts in the long run.

Acknowledgements The authors express their thankfulness to every individual whose contribution, through valuable scientific data and inferences, has been of great assistance in adding consistency to this paper.

References

Ahiataku-Togobo, W., & Ofosu-Ahenkorah, A. (2009) Bioenergy policy implementation in Ghana. In *COMPETE international conference,* online. Available at: http://www.compete-bioafrica.net/events/events2/zambia/Session-2/2-2-COMPETE-Conference-Lusaka-Togobo-Ghana.pdf.

Alonso-Pippo, W., Rocha, J. D., Mesa-Pérez, J. M., Olivares-Gómez, E., & Cortez, L. A. B. (2004). Emergy evaluation of bio-oil production using sugarcane biomass residues at fast pyrolysis plant in Brazil. In E. Ortega & S. Ulgiati (Eds.), *Proceedings of IV biennial international workshop "Advances in Energy Studies"*, Unicamp, Campinas, SP, Brazil.

Arena, U. (2012). Process and technological aspects of municipal solid waste gasification. A review. *Waste Management, 32,* 625–639.

Atakora, S.-B. (2000). Biomass technologies in Ghana. In *The ninth biennial bioenergy conference 2000*. Available at: http://www.nrbp.org/papers/046.pdf.

Balat, M., Balat, M., Kırtay, E., & Balat, H. (2009). Main routes for the thermo-conversion of biomass into fuels and chemicals. Part 1: Pyrolysis systems. *Energy Conversion and Management, 50,* 3147–3157.

Belgiorno, V., De Feo, G., Della Rocca, C., & Napoli, R. M. A. (2003). Energy from gasification of solid wastes. *Waste Management, 23,* 1–15.

Bioenergy Science Technology and Innovation Policy Group. (2013). *Thailand bioenergy technology status report 2013*. Available at: http://www.sti.or.th/th/images/stories/files/1_Thailand_Bioenergy_Technology_Status_Report_2013%20-%20July%2029_Final%20(2).pdf

Birol, F. (2006). World energy prospects and challenges. *Australian Economic Review, 39,* 190–195.

Bole-Rentel, T., & Bruinsma, D. (2013) *The bioenergy sector in South Africa, market entry study for technology providers from the Netherlands*. Available at: http://worldviewmission.nl/wp-content/uploads/2013/09/WM-Agendschap-nl-1339-Presentatie-biomass-study-Tjasa-Bole.pdf.

Bosmans, A., Vanderreydt, I., Geysen, D., & Helsen, L. (2013). The crucial role of waste-to-energy technologies in enhanced landfill mining: A technology review. *Journal of Cleaner Production, 55,* 10–23.

Bradley, D. (2006). European market study for BioOil (Pyrolysis Oil). In *Climate change solutions*.Available at:http://www.bioenergytrade.org/downloads/bradleyeuropeanbiooilmarketstudyfinaldec15.pdf.

Bridgwater, A. V. (2004). Biomass fast pyrolysis. *Thermal Science, 8*(2), 21–50.

Bridgwater, A. V. (2012). Review of fast pyrolysis of biomass and product upgrading. *Biomass and Bioenergy, 38,* 68–94.

Buragohain, B., Mahanta, P., & Moholkar, V. S. (2010). Biomass gasification for decentralized power generation: The Indian perspective. *Renewable and Sustainable Energy Reviews, 14,* 73–92.

Canabarro, N., Soares, J. F., Anchieta, C. G., Kelling, C. S., & Mazutti, M. A. (2013). Thermochemical processes for biofuels production from biomass. *Sustainable Chemical Processes,* 1–22. doi:https://doi.org/10.1186/2043-7129-1-22

Chen, Q., Zhou, J. S., Liu, B. J., Mei, Q. F., & Luo, Z. Y. (2011). Influence of torrefaction pretreatment on biomass gasification technology. *Chinese Science Bulletin, 56*(14), 1449–1456.

Chen, W. H., & Kuo, P. C. (2010). A study on torrefaction of various biomass materials and its impact on lignocellulosic structure simulated by a thermogravimetry. *Energy, 35,* 2580–2586.

Chen, W. H., Lu, K. M., & Tsai, C. M. (2012). An experimental analysis on property and structure variations of agricultural wastes undergoing torrefaction. *Applied Energy, 100,* 318–325.

Chew, J. J., & Doshi, V. (2011). Recent advances in biomass pretreatment—Torrefaction fundamentals and technology. *Renewable and Sustainable Energy Reviews, 15,* 4212–4222.

CSO. (2013). *Digest of energy and water statistics*. Available at: http://statsmauritius.govmu.org/English/Publications/Documents/Regular%20Reports/energy%20and%20water/Energy2013.pdf.

Deenapanray, S. (2006). *Synthesis report on renewable energy*. MRC, online. Available at: http://www.mrc.org.mu/Projects/REReport.pdf.

Deng, J., Wang, G. J., Kuang, J. H., Zhang, Y. L., & Luo, Y. H. (2009). Pretreatment of agricultural residues for co-gasification via torrefaction. *Journal of Analytical and Applied Pyrolysis, 86,* 331–337.

Dermibas, A. (2009). Biofuels securing the planet's future energy needs. *Energy Conversion and Management,50,* 2239–2249.

Duku, M. H., Gu, S., & Hagan, E. B. (2011). A comprehensive review of biomass resources and biofuels potential in Ghana. *Renewable and Sustainable Energy Reviews, 15,* 404–415.

Ericsson, K. (2007). Co-firing—A strategy for bioenergy in Poland? *Energy, 32*(10), 1838–1847.

Faaij, A. (2006). Modern biomass conversion technologies. *Mitigation and Adaptation Strategies for Global Change, 11,* 343–375.

Fei, L. (2012) *Case-based study on biomass combustion power generation in China.* Thesis (Masters), Chalmers University of Technology, Sweden.

Girard, P., & Fallot, A. (2006). Review of existing and emerging technologies for the production of biofuels in developing countries. *Energy for Sustainable Development, 2,* 92–108.

Han, J., & Kim, H. (2008). The reduction and control technology of tar during biomass gasification/pyrolysis: An overview. *Renewable and Sustainable Energy Reviews, 12*(2), 397–416.

Higman, C. (2014). State of the gasification industry: Worldwide gasification database 2014 update. In *Gasification technologies conference, Washington, DC*. Available at: http://www.gasification.org/uploads/downloads/GTC_Database_2014.pdf.

Huber, G. W., Iborra, S., & Corma, A. (2006). Synthesis of transportation fuels from biomass: Chemistry, catalysts, and engineering. *Chemical Reviews, 106*, 4044–4098.

Jahirul, M. I., Rasul, M. J., Chowdhury, A. A., & Ashwath, N. (2012). Biofuels production through biomass pyrolysis—A technological review. *Energies, 5*(12), 4952–5001.

Jolly, L. (2003). *Agriculture: A source of green energy*, online. Available at: http://www.mchagric.org/images/pdf/l_jolly.pdf.

Jones, S. B., Holladay, J. E., Valkenburg, C., Stevens, D. J., Walton, C. W., Kinchin, C., et al. (2009). *Production of gasoline and diesel from biomass via fast pyrolysis, hydrotreating and hydrocracking: A design case*. U.S. Department of Energy, USA. Available at: http://www.pnl.gov/main/publications/external/technical_reports/pnnl-18284.pdf.

Junker, H. (2005). *Cofiring of 500 MW coal-fired power plant with 10% EFB bales or 5% shells and as a 2015 scenario 10% cofiring of POFF*. Available at: http://cdm.unfccc.int/Projects/DB/DNV-CUK1174646628.13/ReviewInitialComments/L098VJ5PJ700XB67UXNM4BFJMQEG2W.

Kleinschmidt, C. P. (2011). *Overview of international developments in torrefaction*. Arnhem: KEMA.

Kong Win Chang, K. T. K. F., Lau Ah Wing, A. F., & Wong Sak Hoi, L. (1999). Bagasse gasification technologies for electricity production in the sugar industry. In *Proceedings of South African Sugar Technologists' Association* (Vol. 73, pp. 247–250).

Kusekwa, A. M. (2013). Biomass conversion to energy in Tanzania: A critique. In Hasan A. and Ibrahim Y. (Eds.), *New development in renewable energy* (pp. 240–270). ISBN 978-953-51-1040-8.

Larson, E. D. (2008). Biofuel production technologies: Status, prospects and implications for trade and development. In *United Nations conference on trade and development*. Available at: http://unctad.org/en/Docs/ditcted200710_en.pdf.

Lora, E. S., & Andrade, R. V. (2009). Biomass as energy source in Brazil. *Renewable and Sustainable Energy Reviews, 13*, 777–788.

Mamphweli, N. S., & Meyer, E. L. (2009). Implementation of the biomass gasification project for community empowerment at Melani village, Eastern Cape, South Africa. *Renewable Energy, 34*, 2923–2927.

Martinot, E., Chaurey, A., Lew, D., Moreira, J. R., & Wamukonya, N. (2002). Renewable energy markets in developing countries. *Annual Review of Energy and the Environment, 27*, 309–348.

McKendry, P. (2002). Energy production from biomass (part 2): Conversion technologies. *Bioresource Technology, 83*, 47–54.

Mohan, D., Pittman, C. U., & Steele, P. H. (2006). Pyrolysis of wood/biomass for bio-oil: A critical review. *Energy & Fuels, 20*, 848–889.

Mohanty, P., Pant, K. K., Naik, S. N., Das, L. M., & Vasudevan, P. (2011). Fuel production from biomass: Indian perspective for pyrolysis oil. *Journal of Scientific & Industrial Research, 70*, 668–674.

Mohee, R., & Mudhoo, A. (2012). Energy from biomass in Mauritius: Overview of research and applications. In A. Karagiannidis (Ed.), *Waste to energy: Opportunities and challenges for developing and transition economies*. London.

Naik, S. N., Goud, V. V., Rout, P. K., & Dalai, A. K. (2010). Production of first and second generation biofuels: A comprehensive review. *Renewable and Sustainable Energy Reviews, 14*, 578–597.

Papong, S., Yuvaniyama, C., Lohsomboon, P., & Malakul, P. (2004). *Overview of biomass utilisation in Thailand*. Available at: http://www.aist-riss.jp/old/lca/ci/activity/project/biomass/report/041028_paper/thailand_paper.pdf.

Patel, B., Gami, B., & Bhimani, H. (2011). Improved fuel characteristics of cotton stalk, prosopis and sugarcane bagasse through torrefaction. *Energy for Sustainable Development, 15*, 372–375.

Prins, M. J., Ptasinski, K. J., & Janssen, F. J. J. G. (2006). More efficient biomass gasification via torrefaction. *Energy, 31*(15), 3458–3470.

Salam, P. A., Kumar, S., & Siriwardhana, M. (2010). *Report on the status of biomass gasification in Thailand and Cambodia* (pp. 1–95). Khlong Luang: Asian Institute of Technology.

Senneca, O. (2007). Kinetics of pyrolysis, combustion and gasification of three biomass fuels. *Fuel Processing Technology, 88,* 87–97.

Sharholy, M., Ahmad, K., Mahmood, G., & Trivedi, R. C. (2008). Municipal solid waste management in Indian cities—A review. *Waste Management, 28,* 459–467.

Sharma, A. K (2011). Experimental investigations on a 20 kWe, solid biomass gasification system. *Biomass and Bioenergy, 35,* 421–428.

Shuit, S. H., Tan, K. T., Lee, K. T., & Kamaruddin, A. H. (2009). Oil palm biomass as a sustainable energy source: A Malaysian case study. *Energy, 34,* 1225–1235.

Shukla, P. R. (2001). *Biomass energy in India: Transition from traditional to modern.* Available at: http://www.decisioncraft.com/energy/papers/ecc/re/biomass/bti.pdf.

Surroop, D., & Mooloo, D. (2013). Assessing the potential of torrefaction for locally available biomass in Mauritius. In: W. Leal Filho, F. Mannke, R. Mohee, V. Schulte, & D. Surroop (Eds.), *Climate-smart technologies: Integrating renewable energy and energy efficiency in mitigation and adaptation responses.* Springer, The Netherlands.

U.S. Department of Energy. (2006). *Biomass program—Pyrolysis oil upgrading.* Available at: http://www1.eere.energy.gov/bioenergy/pdfs/pyrolysis_upgrading.pdf.

Van der Stelt, M. J. C., Gerhauser, H., Kiel, J. H. A., & Ptasinski, K. J. (2011). Biomass upgrading by torrefaction for the production of biofuels: A review. *Biomass and Bioenergy, 35,* 3748–3762.

Venderbosch, R. H., & Prins, W. (2010). Review: Fast pyrolysis technology development. *Biofuels, Bioproducts and Biorefining, 4,* 178–208.

Verma, M., Godbout, S., Brar, S. K., Solomatnikova, O., Lemay, S. P., & Larouche, J. P. (2012). Biofuels production from biomass by thermochemical conversion technologies. *International Journal of Chemical Engineering, 2012.* Available at: http://www.hindawi.com/journals/ijce/2012/542426/.

Zhang, L., Liu, R., Yin, R., & Mei, Y. (2013). Upgrading of bio-oil from biomass fast pyrolysis in China: A review. *Renewable and Sustainable Energy Reviews, 24,* 66–72.

Zhou, Z., Yin, X., Xu, J., & Ma, L. (2012). The development situation of biomass gasification power generation in China. *Energy Policy, 51,* 52–57.

Zhu, X. (2008) *Biomass fast pyrolysis for bio-oil.* University of Science and Technology of China. Available at: http://www.biomass-asia-workshop.jp/biomassws/05workshop/program/18_Zhu.pdf.

An Energy Balance Analysis of Municipal Solid Wastes (MSW) for Power Production in Fiji

Shirleen Swapna and Anirudh Singh

Abstract Waste management is a major challenge faced by Fiji and other SIDS. This paper provides a detailed analysis of the Net Energy Balance of using municipal solid waste (MSW) for producing power on the island of Viti Levu in Fiji. The research evaluates the Net Energy Balance associated with the conversion of MSW into useful power through mass incineration technique using a suitably—located thermal incineration power plant. The total mass of MSW generated in Fiji in 2012 was 48,741 tonnes. Using this feedstock, the Net Energy Balance was determined for MSW originating from each town and city in Fiji, and transported to a central incineration plant at Naboro Landfill site. The mean weighted NEB was found to be 16.76 MJ/kg.

Keywords Waste · Energy · MSW · Net Energy Balance · Thermal incineration power plant · Feedstock

1 Introduction

This paper investigates the Municipal Solid Wastes (MSW) generated in Fiji for its potential for generating power. As MSW is a diffusely distributed feedstock, there will be energy input and costs involved in the collection and transportation of the MSW from the various towns and cities of Viti Levu to the central incineration power plant proposed for installation at the Naboro Landfill site. For the assessment of producing power through incineration technology, the total mass of MSW available on Viti Levu was used as feedstock to the power plant located at the Naboro landfill site.

This research focused on determining the viability of using the municipal solid wastes (MSW) available from towns and cities on Viti Levu to generate electricity.

S. Swapna (✉) · A. Singh
Faculty of Science, Technology and Environment, School of Engineering and Physics, The University of the South Pacific, Suva, Fiji
e-mail: shirleen.swapna@gmail.com

W. Leal Filho and D. Surroop (eds.), *The Nexus: Energy, Environment and Climate Change*, Green Energy and Technology, https://doi.org/10.1007/978-3-319-63612-2_9

The aim of the study was to determine whether the amount of end-use energy produced per kilogram of the feedstock would be greater than the amount used to collect and transport the MSW to Naboro Landfill site. Essentially, this is an energy investigation between the energy contained in the MSW (input) and the energy that could be recovered from the feedstock using the mass incineration techniques in a thermal incineration power plant. However, a comprehensive economic analysis is necessary to consider the true viability of producing power from MSW and this paper is not concerned with the economic analysis.

2 Methodology

As a first step, the mass of MSW produced per annum in various towns and cities of Fiji was measured from several weeks of field sampling. Samples of bags containing garbage were collected randomly from households around Viti Levu and the composition of the MSW in the bags was determined through classification and separation at source. The contents of the MSW were characterized into organic matter, plastics, paper, and others. Thus, the mass of each of these constituents of MSW were evaluated to a fraction of the total mass of MSW contained in the bag.

The average organic fraction of the MSW was then obtained experimentally from the samples collected from each town and city of Fiji. Thus, the mass of MSW produced in Fiji per annum was estimated. Following the waste characterization, the moisture analysis for the organic fraction of the MSW was carried out in the laboratory in an oven using standard gravimetric techniques.

Further, the calorific values of the fractions of organic matter, plastic and paper were also measured using the bomb calorimeter, and the average calorific values for organic fraction, plastic fraction and paper fraction were determined in the laboratory. The energy contained in the sample was determined experimentally in the bomb calorimeter.

A thorough energy balance analysis was performed for the generation of power from MSW from a hypothetical incineration power plant proposed for installation at the Naboro Landfill site. Presently, Naboro landfill accepts approximately 25,000 tonnes of MSW annually from areas of greater Suva, Nasinu, Nausori, Lami and Navua.

The Net Energy Balance (NEB) was calculated per kilogram of MSW generated for individual towns and cities on Viti Levu. The NEB is a measure of the difference between the energy contained in MSW as determined from the bomb calorimeter and the energy consumed in handling and transportation of MSW from source to the incineration power plant proposed to be installed at the Naboro Landfill site. Calculations for net energy balances were performed for each town/city to evaluate the energetic feasibility of producing energy from MSW. Hence, the energy balance for producing electrical power from all the MSW available in Viti Levu was estimated.

3 Results

3.1 MSW Supply on Viti Levu

Table 1 outlines the mass of MSW generated at different towns and cities of Viti Levu.

The total MSW generated in the towns and cities in Viti Levu, Fiji was 48,741 tonnes per annum. Municipal Solid wastes (MSW) can be used as a source of energy to produce Refuse Derived Fuels (RDF) (European Commission 2003).

In recent years, energy from wastes, particularly MSW has advanced from the traditional composting to higher technologies such as incineration, pyrolysis, gasification and landfills. In Europe, in 2009, landfill gas to energy conversion was the most common technology for producing power from wastes (JICA 2004). In 2009 alone, 50.7 million tonnes of wastes were treated.

3.2 MSW Generation Rate per town/city

Table 2 highlights the MSW generation rate/population for each town and city in Viti Levu.

Table 1 MSW Supply from various towns and cities on Viti Levu

Town	Annual MSW generation (tonnes/annum)	% annual MSW generation (tonnes/annum)	Source of data
Suva	11,108	22.79	As per Suva City Council
Nasinu	10,890	22.34	As per Suva City Council
Navua	530	1.09	Estimated from population
Sigatoka	2246	4.61	Estimated from population
Nadi	5772	11.84	JICA Technical Report
Lautoka	8234	16.89	JICA Technical Report
Ba	2921	5.99	Estimated from population
Tavua	280	0.57	As per survey
Rakiraki	520	1.07	As per survey
Nausori	4320	8.86	Department of Environment
Lami	1920	3.94	Department of Environment
SUM	48,741	100.00	

Table 2 MSW Generation Rate/Population for various towns and cities on Viti Levu

Town	Annual MSW generation (tonnes/annum)	Population	MSW generation rate/pop (kg/person/annum)	JICA report 2011 MSW generation rate/pop (kg/person/annum)
Suva	11,108	85,711	129.6	146
Nasinu	10,890	98,828	110.2	149.7
Navua	530	5048	105	N/A
Sigatoka	2246	9622	233.4	127.8
Nadi	5772	42,284	136.5	284.7
Lautoka	8234	52,220	157.7	142.4
Ba	2921	18,526	157.7	135.0
Tavua	280	2388	117.2	109.5
Rakiraki	520	4952	105.0	N/A
Nausori	4320	47,604	90.8	135.1
Lami	1920	20,529	93.5	120.5
SUM	48,741	387,692	1436.6	

A report published by JICA in 2011 (Howes 2012) described a similar study which investigated the rate of MSW generated per person per annum. For comparison, the last column of Table 2 outlines the statistics gathered from its survey.

It is clear from Table 2 that the MSW generation rate per population ratio is largest for Sigatoka town. This is especially due to a significant tonnage of 9622 tonnes of MSW distributed over a relatively small population of 2246 individuals, resulting in a generation rate of 233.4 kg/person/annum. The average generation rate for Viti Levu was determined to be 130.6 kg/person/annum, which compares well with MSW generated from the towns of Nadi, Tavua, and Suva city.

3.3 Determination of Organic Fraction of MSW

1. Organic fraction gives a measure of the bio-degradable component of the MSW collected.

The Organic Fraction of MSW (OFMSW) was computed for samples originating from each town and city covered in this survey in accordance with Eq. 1.

$$\mathbf{OFMSW} = \frac{\text{Organic Mass of Sample contained in a bag of MSW}}{\text{Total Mass of Bag containing MSW}} \tag{1}$$

The organic fractions were determined from sampling and the average O/F for each town/city was used for analysis. Table 3 outlines the O/F values obtained from this survey.

Table 3 O/F of MSW

Town	Organic fraction	Source of O/F
Suva	0.65	As per survey
Nasinu	0.70	As per survey
Navua	0.68	As per survey
Sigatoka	0.80	Sigatoka Town Council
Nadi	0.70	Lautoka City Council
Lautoka	0.70	Lautoka City Council
Ba	0.71	Ba Town Council
Tavua	0.64	As per survey
Rakiraki	0.65	As per survey
Nausori	0.68	As per survey
Lami	0.70	Lami Town Council

3.4 Moisture Analysis

Several samples of organic matter of MSW were analysed for moisture composition and the average moisture content was recorded at 40%.

3.5 Determining the Organic Mass of MSW

The total organic mass of MSW on Viti Levu was computed as the sum of organic masses of MSW resulting from each town/city of Viti Levu, where organic mass of MSW for each town was determined as per Eq. 2;

$$\textbf{Organic Mass of MSW} = \text{Average } O/F \text{ of town} \times \text{Mass of MSW produced per annum from town} \quad (2)$$

The annual organic mass of MSW generated was evaluated for each town and city of Viti Levu. Table 4 shows the distribution of organic mass for various towns and cities.

According to Eq. 1, the organic fraction was defined as the average ratio of the mass of organic component of a bag of MSW to its total mass. It is of interest to note that an average of 69.2% O/F has been identified for Viti Levu.

The average organic fractions were determined from sampling, except for towns that partnered with JICA in 2011 for the Waste Minimisation Project. The study undertaken by JICA experimentally assessed the organic fractions for MSW generated in towns of Ba, Nadi, Sigatoka, and Lautoka City.

The total organic mass of MSW produced on Viti Levu per annum was computed as the sum of organic masses of MSW resulting from each town/city on Viti Levu. Hence, Table 4 outlines that a maximum of 33,759 tonnes of organic mass per annum is available for energy production.

Table 4 Determining the Mass of Organic Sample on Viti Levu

Town	Annual MSW generation (tonnes/annum)	O/F	Source of O/F	Organic mass (tonne/annum)
Suva	11,108	0.66	As per survey	7331
Nasinu	10,890	0.70	As per survey	7623
Navua	530	0.68	As per survey	360
Sigatoka	2246	0.80	Sigatoka Town Council	1797
Nadi	5772	0.70	Lautoka City Council	4040
Lautoka	8234	0.70	Lautoka City Council	5764
Ba	2921	0.71	Ba Town Council	2045
Tavua	280	0.64	As per survey	179
Rakiraki	520	0.65	As per survey	338
Nausori	4320	0.68	As per survey	2938
Lami	1920	0.70	As per survey	1344
SUM	48,741			33,759

3.6 *Determining the Net: Energy Content of Organic Sample of MSW*

The net energy content of the O/F of the MSW sample collected was measured in the bomb calorimeter. The average calorific value of the organic sample was recorded at 17.42 MJ/kg, which offers significant potential for energy recovery from the organic sample of MSW obtained from Viti Levu. Table 5 discusses the availability of energy from organic masses of MSW specific to its origin.

The net energy content of the organic sample as measured from the bomb calorimeter was 17.42 MJ/kg. In total, 33,759 tonnes of organic MSW was produced per annum. In accordance with Eq. 2, the net energy content of organic masses was determined. The total energy contained in 33,759 tonnes of MSW was evaluated as per Eq. 3;

$$\textbf{Total Energy Content of MSW} = \text{Total Organic Mass on Viti - Levu} \times 17.42\,\text{MJ/kg} \quad (3)$$

where

The total energy is calculated in MJ

The net energy content of the organic sample was predetermined at 17.42 MJ/kg, from bomb calorimetry

Hence, the total energy available from all MSW generated on Viti Levu over one year is 5.89×10^8 MJ.

Table 5 Calculation of the energy contained in organic component of MSW

Town	Annual MSW generation (tonnes/annum)	Organic fraction	Organic mass (tonnes/annum)	Energy available from organic mass (MJ/annum)
Suva	11,108	0.66	7297.96	1.27E+08
Nasinu	10,890	0.70	7623	1.33E+08
Navua	530	0.68	360	6.28E+06
Sigatoka	2246	0.80	1797	3.13E+07
Nadi	5772	0.70	4040	7.04E+07
Lautoka	8234	0.70	5764	1.00E+08
Ba	2921	0.71	2045	3.61E+07
Tavua	280	0.64	179	3.11E+06
Rakiraki	520	0.65	338	5.89E+06
Nausori	4320	0.68	2938	5.12E+07
Lami	1920	0.70	1344.00	2.34E+07
SUM	48,741		33,759	5.89E+08

3.7 *Energy Input from Fuel*

The calorific value of diesel was used to determine the energy input from fuel as shown in Eq. 4.

$$\text{Energy Input from Fuel} = \text{Calorific Value of Diesel} \times \text{Volume of diesel used in town per annum} \quad (4)$$

where

The calorific value of diesel is 37.4 MJ/L

Energy Input from fuel was calculated in MJ

The distances covered under the transportation of MSW were computed. Each litre of diesel spent on MSW collection or transportation inputs 37.4 MJ into the energy balance equation.

Table 6 summarizes the fuel consumed in the collection and transportation of MSW from each town and city to the centralized thermal incineration power plant proposed to be installed at the Naboro Landfill site.

Table 6 shows that a total of 38.2×10^6 MJ of energy is required per annum to transport all MSW generated on Viti Levu to the hypothetical incineration plant to be built at the Naboro Landfill site. Of this, 23.2×10^6 MJ of energy is spent on cartage of MSW from the towns of Sigatoka, Nadi, Ba, Tavua, Rakiraki and Lautoka City to Naboro.

The energy input (originating from fuel input) from each town and city was calculated as a function of the mass of MSW generated per town/city in accordance with Eq. 4. On the whole, 933.69×10^3 L of diesel is consumed per annum in

Table 6 Energy consumption in collection and transportation of MSW from source to destination

Town	Energy consumption		
	To local dump (MJ/year)	Local dump to Naboro (MJ/year)	From source to Naboro landfill (MJ/year)
Suva	1.89E+06	0	1.94E+06
Nasinu	3.88E+06	0	1.86E+06
Navua	4.21E+04	0	4.21E+04
Sigatoka	5.68E+05	2.30E+06	2.87E+06
Nadi	9.79E+05	5.56E+06	6.54E+06
Lautoka	3.16E+05	2.97E+06	3.28E+06
Ba	1.14E+05	3.95E+06	4.06E+06
Tavua	1.74E+05	3.79E+06	3.96E+06
Rakiraki	9.97E+04	4.59E+06	4.69E+06
Nausori	2.53E+06	0	2.53E+06
Lami	2.40E+05	0	0.24E+06
SUM	10.8E+06	23.2E+06	38.2E+06

transporting all MSW generated on Viti Levu to the incineration power plant proposed for installation at the Naboro Landfill site.

Further, the latest report released by Asian Development Bank (ADB 2014) highlighted that there were no large-scale facilities for the high organic wastes transported to Suva and Naboro. The report emphasized that there were "no specific guidelines for collection, treatment and disposal of wastes from electrical and electronic equipment" and also maintained that the illegal dumping and burning of wastes continued to cause problems, due to "inadequate enforcement of laws."

3.8 Estimating the Energy Input per kg of MSW

For comparative analysis of the energy invested in each kilogram of MSW, Eq. 5 was developed.

The energy input (originating from fuel input) from each town and city was calculated as a function of the mass of MSW generated per town/city by Eq. 5.

$$\textbf{Energy Input per kilogram of MSW}\ [\mathrm{MJ}/L] = \frac{\text{Litres of Diesel consumed per annum} \times \text{calorific value of Diesel}\ [\mathrm{MJ}/L]}{\text{Total mass of MSW}\ [\mathrm{kg}]} \tag{5}$$

where

The calorific value of diesel is 37.4 MJ/L

Strategies for energy recovery have been well elaborated in 2004 for small Waste Management Systems (WMS) serving 200,000 people as well as large WMS

Table 7 Determination of energy input per kilogram of MSW

Town	Energy consumption			Annual MSW generation (tonne/year)	Energy input per kg of MSW (MJ/kg)
	To local dump (MJ/year)	Local dump to Naboro (MJ/year)	To Naboro landfill (MJ/year)		
Suva	1.89E+06	0	1.89E+06	11,108	0.17
Nasinu	3.88E+06	0	1.86E+06	10,890	0.17
Navua	4.21E+04	0	4.21E+04	530	0.08
Sigatoka	5.68E+05	2.30E+06	2.87E+06	2246	1.28
Nadi	9.79E+05	5.56E+06	6.54E+06	5772	1.13
Lautoka	3.16E+05	2.97E+06	3.28E+06	8234	0.40
Ba	1.14E+05	3.95E+06	4.06E+06	2921	1.39
Tavua	1.74E+05	3.79E+06	3.96E+06	280	14.15
Rakiraki	9.97E+04	4.59E+06	4.69E+06	520	9.02
Nausori	2.53E+06	0	2.53E+06	4320	0.58
Lami	2.40E+05	0	2.40E+05	1920	0.12
SUM	10.8E+06	23.2E+06	38.2E+06	48,741	28.50

serving up to 1.2 billion people (JICA 2004). In this study, an organic fraction content of 31.5% in Materials Residue (MR) was determined (Consonni et al. 2004). Calculations for energy and environmental balances with and without RDF have also been discussed (Consonni et al. 2006).

Table 7 illustrates the energy input values per kg of MSW incinerated at Naboro thermal power plant.

Table 7 reveals that the energy invested per kilogram of MSW is extremely high for towns of Tavua and Rakiraki, attributing explicitly to its distribution of its MSW generated over a relatively low population. In addition, the large energy input per kilogram of MSW generated in Tavua and Rakiraki is more pronounced due to energy input in the transportation of a total MSW mass of 800 tonnes per year, consuming additional diesel amounting to 230.14×10^3 litres per annum, resulting in 8.65×10^3 MJ per annum.

It is evident from Table 7 that Suva, Nasinu, Nausori, Navua and Lami towns had minimum energy input, ranging from 0.08 MJ to 0.17 MJ/kg of MSW.

The energy input per kilogram of MSW generated for the majority of towns and cities of Viti Levu were especially low, indicating a promising potential of waste-to-energy conversion from MSW.

3.9 Calculation of the NEB for MSW for Incineration

The Net Energy Balance (NEB) was calculated per kilogram of MSW generated for individual towns and cities on Viti Levu. The NEB is a measure of the difference between the energy contained in MSW as determined from the bomb calorimeter and the energy consumed in handling and transportation of MSW.

Energy per kilogram of MSW was determined from bomb calorimetry as 17.42 MJ/kg. Further, for comparative analysis, the weighted average of NEB for each town/city was computed in accordance with Eq. 6.

$$\textbf{Weighted Average of NEB}\ [\text{MJ/kg}] = \frac{\text{Mass of MSW generated ine each town}}{\text{Mass of MSW generated on Viti - Levu}} \times 17.25\,\text{MJ per kg} \quad (6)$$

The weighted averages of NEB for each town and city are represented in Table 8.

The weighted average of NEB is a reflection of the density of population of a town to the total population of Viti Levu with regards its NEB. Table 8 outlines that the weighted average for NEB is highly pronounced for Tavua, Rakiraki and Navua, specially attributed to its low population density related to its relatively low generation of MSW.

On the contrary, it was of interest to note that the high energy input in transportation of MSW produced in Lautoka and Nadia was not highly pronounced in its weighted average of NEB. It is clear from Table 8 that Lautoka and Nadi played major contributors (third and fourth largest respectively) of MSW, based on its dense populations, accounting to 28.7% of that generated on Viti Levu.

The mean weighted average of the NEB per kilogram for Viti Levu was calculated at 16.76 MJ.

Table 8 Weighted averages of NEB

Town	Net energy balance for MSW (MJ/kg)	Weighted average of net energy balance (MJ/kg)
Suva	17.25	3.93
Nasinu	17.25	3.85
Navua	17.34	0.19
Sigatoka	16.14	0.74
Nadi	16.29	1.93
Lautoka	17.02	2.88
Ba	16.03	0.96
Tavua	3.27	0.02
Rakiraki	8.40	0.09
Nausori	16.84	1.49
Lami	17.30	0.68
SUM	163.12	16.76

3.10 NEB for Producing Energy from MSW via Incineration Technology

Hence, in accordance with Table 8, the NEB is proven to be positive for energy production at Naboro from MSW. It has therefore been demonstrated that, it is worthwhile to consider generating energy from all MSW produced on Viti Levu via mass incineration technique using a thermal incineration power plant installed at Naboro.

The Journal of Municipal Solid Wastes Professionals, MSW Management (2014) emphasizes that "direct combustion of MSW with energy recovery (mass burn) is a mature technology and the most commonly used MSW conversion technology in the US. At mass burn facilities, waste is combusted minimally or with no front-end processing, and heat generated from combustion creates steam. This steam is used in district heating networks or industrial applications, or to power turbine generators for electricity production." (Nan 2004). Moreover, In Taiwan, 64.2% (about 43,232 thousand metric tonnes) of MSW is treated via incineration WTE technology in a thermal power plant.

Gerber et al. (2008) identified 47 landfill sites the United States that processed 3300 short tonnes per day or more of MSW. "Together those sites could produce more than 310,000 barrels of oil per day of liquid fuels as fuel grade ethanol, which is equal to about 1.4% of current U.S. transportation fuels demand."

Further, Liss et al. (2013) outlined that "between approximately 0.4 and 0.7 MWh of electrical energy can be generated from a tonne of MSW through incineration. Incineration of 1000 short tonnes per day of waste will produce about 650 MWh of electrical energy per day (27 MW of electrical power continuously for 24 h) or approximately 2000 MWh of district heating energy each day."

4 Summary and Conclusion

MSW feedstock is a reliable source of energy due to consistency in its supply. The total mass of MSW produced on Viti Levu per annum is 48,741 tonnes. The average O/F for Viti Levu as obtained from this survey was 69.27%. On average, an individual on Viti Levu generates 130.6 kg of MSW per annum. The average calorific value of the organic sample was recorded at 17.42 MJ/kg, which offers significant potential for energy recovery from the organic sample of MSW obtained from Viti Levu. The total energy available from all MSW generated on Viti Levu over one year is 5.89×10^8 MJ. The mean weighted average of the NEB for Viti Levu was calculated at 16.76 MJ/kg. A total of 38.2×10^6 MJ of energy is required per annum to transport all MSW generated on Viti Levu to the hypothetical incineration plant to be built at the Naboro Landfill site. Of this, 23.2×10^6 MJ of energy is spent on cartage of MSW from the towns of Sigatoka, Nadi, Ba, Tavua,

Rakiraki and Lautoka City to Naboro. On the whole, it is highlighted that NEB is positive for all MSW generated in Fiji.

It can be concluded from this study that it is worthwhile to consider producing energy from MSW via mass incineration technique, in support of developing renewable energy in Fiji. What is required next is an analysis of the economic viability of such a proposition.

References

Consonni, S., Giulgiano, M., & Grosso, M. (2004). Alternative strategies for energy recovery from municipal solid wastes; part A: Mass and energy balances. *Waste Management, 25,* 123–135.

Consonni, S., Giulgiano, M., & Grosso, M., & Rigaminti, L. (2006). Energy and environmental balances for energy recovery from municipal solid wastes with and without RDF production. In *Proceedings: Venice 2006 biomass and waste to energy symposium*, Italy 29 November–1 December 2006.

European Commission. (2003). *Refuse derived fuel, current practice and perspectives.*

Gerber, M. A., Jones, S. B., Stevens, D. J, Valkenburg, C., Walton, C. W., Thompson, B. L. (2008). *Municipal solid wastes (MSW) to liquid fuel synthesis, volume 1: Availability of feedstock and technology.*

Howes, P. (2012). Integration of energy recovery from solid waste. In *IEA bioenergy conference*, Vienna, November 13–15, 2012.

Liss, B., Williams, N., Wison, B., & Wilson, B. (2013). *A comparative assessment of commercial technologies for conversion of solid waste to energy.*

Nan, C. (2004). An overview of renewable energy utilisation from municipal solid wastes (MSW) incineration in Taiwan. *Renewable and Sustainable Energy Reviews, 10,* 491–502.

Solid Waste Management and Recycling in the Fiji Islands. (2004). *Preliminary analysis of the situation in the Western region of Fiji and the handling of some difficult Solid Waste.* Final Report to JICA prepared by University of the South Pacific Centre for Environment and Sustainable Development with contributions from Dr Fabrice Mathiuex, Melchior Mataki and Professor Kanayathu Koshy.

Solid Waste Management in the Pacific. (2014, June). *Fiji country snapshot released by ADB.*

Potential of Small-Scale Power Generation from the MSW Incineration Techniques in Fiji

Arvind Kumar and Anirudh Singh

Abstract Fiji like other Small Island Developing States (SIDS) is sharing similar sustainable development challenges, including small population, lack of resources, remoteness, susceptible to natural disasters, highly dependent on international trade and vulnerability to global developments. In addition, they suffer from lack of economies of scale, high transportation and communication costs, and costly public administration and infrastructure. Management of municipal solid waste presents one such challenge where preventative strategies like reuse, recycling, and recovery (3Rs) are being emphasized. One of the best ways of these 3Rs is converting the waste to energy. This study will investigate the potentials of setting up small-scale *msw* incineration power generation in major town and cities around Fiji. The aim of this study was to estimate the potential electricity that could be produced by MSW incineration technique in Fiji. The *msw* data used in this study is was obtained from the Fiji Solid Waste Management Strategy Report (2011–2014), compiled by the Japanese International Co-operation Agency (JICA). Also further data has been retrieved from the Department of Energy and from the Town Council Waste Management reports. The simple economic analysis model reveals Fiji has potentials to generate electricity from *msw*. For instance, 50,000 tonnes of *msw* are collected and disposed annually in the Central division which has potential of generating 2.2 MW electricity via *msw* incineration technique. Similarly, Western division has a similar electricity potential as they collect and dispose similar quantities of *msw*. For a 20-year lifespan project, the economic model reveals that the cost of energy of producing power using *msw* for a 5 MW plant with an investment cost of \$200 M (FJD), the cost of energy is \$0.228 per kWh. The payback period for this 5 MW plant is approximately 14 years. The power output for a 5 MW plant suggests that the concept of producing power via *msw* incineration technique may become viable for Fiji. This could be an attractive technology for electricity production and waste management in near future as Fiji has initiated the Green Growth Framework. This Green Growth Framework is an innovative tool

A. Kumar (✉) · A. Singh
Faculty of Science and Technology, School of Engineering and Physics, University of the South Pacific, Suva, Fiji
e-mail: kumararvin880@gmail.com

W. Leal Filho and D. Surroop (eds.), *The Nexus: Energy, Environment and Climate Change*, Green Energy and Technology, https://doi.org/10.1007/978-3-319-63612-2_10

to accelerate integrated and inclusive sustainable development that inspires action at all levels in the country to build environmental resilience, build social improvement and reduce poverty, build economic growth, and build resilience to the anticipated adverse affects of climate change.

Keywords Energy · Environment · Climate change · Sustainable development Municipal solid waste (*msw*)

1 Introduction

Fiji with 332 islands, land area of 18,270 km^2 has two major islands, Viti-levu and Vanua-levu accounting for 87% of the population of almost 837,231 (2007 Census). For small island countries in the Pacific region, including Fiji, solid waste management has become a major concern with the potential to cause negative impacts on national development activities, including tourism and trade, food supplies, public health, and the environment. Waste materials represent a loss of money and opportunity, in terms of added cost of management and lost potential as a reusable resource. Poorly managed waste can have negative effects on tourism, by jeopardizing the image of "Pacific Paradise" promoted by Fiji. As the change of lifestyle of the people, Fiji has been heavily dependent on goods and materials imported from industrialized countries. It is however; very difficult to recycle those materials and goods because of the geographical isolation and uneconomically small size of the recycle market in the country. According to Fiji Solid Waste Management report, most of the solid waste is still currently being thrown in open dumpsites, illegally disposed off in the sea on unused land, or in the streets. Burning of municipal waste is also quite common, with the population being continuously exposed to the harmful health effects of poisonous gases.

Solid waste minimization and recycling remains priority strategies for Fiji, and several ways are considered for the recycling of this. Currently, the *msw* management is governed by departments such as, Department of Environment (responsible for the implementation, monitoring, and enforcement of the environment and waste management acts and regulations, and the development of policies on waste management), Ministry of Health (monitors the disposal of garbage making them one of the key players in this issue), Ministry of Fijian Affairs (monitoring of the solid waste management in villages through the Fijian Affairs Act), Ministry of Labour (OHS Department ensuring that the handling of all hazardous waste complies the OHS regulations), and the Department of Local Government which ensures the health, welfare of the towns and city environment, governs the collection and disposal of garbage [Country Report (Fiji), Asia, Ha Noi, Viet Nam, March 18–20, 2013].

Fiji is heavily dependent on the fossil fuels which accounts for billions of dollars of imports every year. Renewable like solar, wind turbines, and hydroelectricity are advancing technologies in Fiji. Solving issues like the disposal of Municipal Solid Wastes (*msw*) has cost the country millions of dollars which could be saved if this could be converted to renewable energy. One of the possible solutions to these would be, producing power via *msw* incineration techniques. This is one of the techniques where MSW will be recycled by incineration process to generate renewable power. An incineration technology is the controlled combustion of waste with the recovery of heat to produce steam that in turn produces power through steam turbines. MSW after pretreatment is fed to the boiler of suitable choice wherein high pressure steam is used to produce power through a steam turbine.

Converting *msws*' to energy could help combat the greenhouse effects (GHGs) since there would be less dependency on the fossil fuels and a healthier green environment. For instance, in 2005, waste incineration produced 4.8% of the electricity consumption and 13.7% of the total domestic heat consumption in Denmark (Danish Energy Statistics 2005). In Switzerland and Luxembourg, over 75% of the municipal waste stream is treated by combustion with energy recovery. In Sweden, it is over 60%, in France 43% and in USA 17%. Japan uses waste combustion to treat over 75% of the waste remaining after recycling (Malua 2001). Other *msw* plants operating globally are illustrated in the Table 1.

In addition to GHG mitigation, these types of projects would contribute to the sustainable development of the local area as the residents will have a healthier environment in general because of improved local air quality, as well as the abatement of water and soil pollution. It prevents the generation and release of air pollutants in the landfill gas by reducing the amount of waste that is to be land filled and would create job opportunities over the project construction and operation period.

This paper reveals the techniques that would be used to access the potentials of power generation via the incineration of *msw* in Fiji. It accounts for the total municipal solid waste collected by the various municipalities around Fiji and the

Table 1 MSW facilities available globally (World Energy Resources 2013)

Country	Region	Quantity of *msw* production (tonnes/year)	Electricity generation (MW)	Thermal energy recovered (MW)
England	Sheffield	225,000	17.0	39.0
	Nottingham	160,000	14.4	44.2
	Coventry	250,000	18.0	7.5
	Grimsby	56,000	3.2	3.3
	Slough Heat and Power	110,000	104	22.0
Japan	Tokyo	400 M	5.3	N/A
Netherlands	Amsterdam	1.5 M	114.2	N/A
Thailand	Phuket	180,000	14.0	N/A

potential of setting up small *msw* incineration power plants technology around various towns and cities. In addition, the basic working principles of these plants are determined with the given considerations to the hazardous emissions and its controls. The power output for the proposed incineration power plants using waste such as paper, cardboard, plastics, and other organic materials is also determined. An economic analysis involving the cost of installation, operating and maintenance cost, and the payback period for the projects is done to investigate whether such projects are viable for Fiji.

2 Methodology

The main aim of this research is to estimate the potential energy that could be produced by the incineration of *msw* around the various towns and cities in Fiji. The following procedure was carried out to investigate the potentials of the power production from MSW incineration. The amount of the *msw* collected in Fiji was taken from the Fiji Solid Waste Management Strategy Report (2011–2014). For the calculations of power production, the calorific values of the *msw* were obtained from the literature. The following formula was used to determine the total energy derived via *msw* incineration:

$$\begin{aligned}&\text{Total Energy of MSW (MJ)}\\&\quad = \text{Total MSW(kg) per year} \times \text{Calorific Value} \times \text{Organic Fraction of MSW (MJ)}.\end{aligned}$$

The literature value for the conversion efficiency was used to determine the total power output from *msw*. This was calculated using the formula

$$\text{Energy Output (kWh)} = \text{Conversion Efficiency} \times \text{Energy Input (kWh)}.$$

The power output for the incineration power plant determined using the equation

$$\text{Power Output (MW)} = \frac{\text{Total energy Output (kWh)}}{8760\,\text{h}},$$

where 8760 h represents total hours per annum.

A simple economic analysis model was used to assess the potentials for power generation via *msw* incineration technique for Fiji. The Cost of Energy (COE) and the Payback period was determined using the following formulae:

$$\text{COE (\$)} = \text{Total Costs per Annum/kWh of Energy Produced},$$

$$\text{Payback period} = \text{Total Capital Cost/Savings per year}.$$

3 Results and Discussion

The *msw* generated rate for Fiji is 0.4 kg/person/day (Table 2) based on the data gathered from Naboro landfill. Table 2 given below shows the waste generation statistics for the Naboro Sanitary Landfill.

From the Naboro landfill waste data, a total of 122,240 tonne/year of *msw* was collected. These *msw* was collected from the four divisions around Fiji.

3.1 *The Energy Content of MSW in Fiji*

Calorific value is considered as the main parameter of any combustible matter. Not all waste is suitable for combustion. The waste can be humid or of low calorific value due to the low concentration of organic fractions. Table 3 given below gives the fractions (for dry waste) and the energy contents of *msw* components.

3.2 *Total Refuse-Derived Fuel (RDF)*

The feedstock would be the Refuse-Derived Fuel (RDF). This fuel component of municipal solid waste is the by-product of shredding *msw* to a uniform size,

Table 2 Waste generation statistics for Naboro Sanitary Landfill

Division	Population	MSW generated kg/person/day	MSW Generated tonne/year	Cost of collection and disposal $/tonne	Total cost FJD$ million
Central	342,386	0.4	49,988	115	5.748
Western	319,611	0.4	46,663	130	6.066
Northern	135,961	0.4	19,850	150	2.977
Eastern	39,313	0.4	5739	150	0.861
Totals	837,271		122,240		16.652

Source Fiji Solid Waste Management Strategy Report (2011–2014)

Table 3 Fractions and energy contents of components of MSW

Component of MSW	Fraction (%)	Gross Calorific Value (MJ/kg)	Net Calorific Value (dry) (MJ/kg)	Total Energy Content of Component (TJ)
Paper	13	16.44	2.14	104.17
Plastic	14	41.50	5.81	283.19
Organic	70	17.42	12.19	594.35
Miscellaneous	3	N/A	N/A	N/A

Shirleen (2012)

screening out oversized materials, and isolating ferrous material in magnetic separation. The resulting RDF can be burned as a fuel source as it will have higher energy content. The assumption is that RDF quantity that will be 30 percent of the total MSW collected. (Zafar et al. 2015). A sample calculation for the power output of a possible MSW incineration power plant is done using *msw* quantity deposited at the Naboro Sanitary Landfill in the Central division. The Power Output is determined from the total energy contents of the *msw* components mainly, paper (net calorific value 2.14 MJ/kg), plastics (net calorific value 5.81 MJ/kg), and organic materials (net calorific value 12.19 MJ/kg) (Shirleen 2012). The total *msw* deposited at the Naboro Sanitary Landfill is given by

$$\underline{\text{MSW (Suva, Nausori, Nasinu, Lami} - \text{Navua at Naboro Sanitary Landfill)}}$$
$$= \text{Total population} \times \text{Average Geneartion Rate (kg/person/day)} \times 365\,\text{days}$$
$$= 342{,}386 * 0.40 * 365 \approx \underline{50{,}000}\ \text{tonnes/year}$$

3.3 *Total Energy Content of MSW*

$$\text{Total MSW produced (Central Division)} = 50{,}000\ \text{tonnes/year}$$
$$\text{RDF} = \underline{15{,}000}\ \text{tonnes/year}$$

Considering the major components of *msw* collected are paper, plastic, and organics

The Total Energy Content of MSW

= energy content of Organic + energy content of paper + energy content of plastics
= [15,000 × 10^3 kg × 12.19 MJ/kg] + [15,000 × 10^3 kg × 2.14 MJ/kg]+ [15,000 × 10^3 kg × 5.81 MJ/kg]
= $\underline{\mathbf{3.02}} \times \underline{\mathbf{10^8}}$ MJ per annum.

MSW quantity depends on the population size. For a larger the population, the greater quantity of *msw* would be generated. The energy content of the MSW could be greater if the *msw* are sorted and the moisture contents are reduced before being used as a feedstock.

3.4 *MSW Incineration Power Plants: Energy Output*

The energy input to a power plant is given by

$$\text{Energy Input (kWh)} = \frac{\text{Total Energy per year (MJ)}}{3.6\ \text{MJ per kWh}},$$

where

One kilowatt-hour (kWh) is equivalent to 3.6 MJ of energy

$$\text{Energy Input (kWh)} = \frac{3.02 \times 10^8\ (\text{MJ})}{3.6\,\text{MJ per kWh}}$$
$$= \underline{83.89 \times 10^6}\ \text{kWh per Annum}$$

The energy output of a power plant is given by;

$$\text{Energy Output (kWh)} = \text{Conversion Efficiency} \times \text{Energy Input (kWh)}$$

Assuming that the energy efficiencies of the furnace, steam turbine, and generator are as 80, 30, and 95% respectively

$$\underline{\text{Energy Output}}\ (\text{kWh})$$
$$= 0.80 * 0.30 * 0.95 * 83.89 \times 10^6\ \text{kWh/year}$$
$$= \underline{19.13 \times 10^6}\ \text{kWh}$$

3.5 *Power Output*

$$\text{Power Output (MW)} = \frac{\text{Energy Output (kWh)}}{8760\,\text{h}} = \frac{19.13 \times 10^6\,\text{kWh}}{8760\,\text{h}} = \underline{2.2}\,\text{MW}$$

The energy output depends on the conversion efficiency of MSW power plants. Higher energy from the *msw* feedstock could be retrieved with better efficiencies. Assuming that the energy efficiencies of the furnace, steam turbine and generator are as 80, 30, and 95%, respectively, the energy output of the *msw* for Central division is found to be $\underline{19.13 \times 10^6}$ kWh and the *msw* Incineration power plant capacity is approximately $\underline{2.2}$ MW.

3.6 *Potential for Setting Up MSW Incineration Power Plants for Various Divisions in Fiji*

Table 4 given below refers to the potential of electricity generation via *msw* incineration for the four divisions in Fiji.

Based on the latest population based on 2007 Census Report and assuming the MSW generation rate for each town is approximately 0.4 kg/person/day (Table 2). Each town's potential to produce power via *msw* incineration are shown in Table 5 given below.

Table 4 Electricity generation by MSW incineration for each division

Division	Population	MSW generated (tonnes)	RDF (tonnes)	Total energy content MSW ($\times 10^8$ MJ)	Energy input $\times 10^6$ (kWh)	Energy output (kWh) $\times 10^6$	Power output (MW)
Central	342,386	49,988	14,997	3.02	84	19	2.2
Western	319,611	46,663	13,999	2.82	78	18	2.0
Northern	135,961	19,850	5955	1.20	33	7.6	0.9
Eastern	39,313	57,340	1722	0.34	9.4	2.1	0.24

Table 5 Potential power production via MSW incineration power plants

Major towns	Population	MSW generated (tonnes)	RDF (tonnes)	Total energy content $\times 10^8$ (MJ)	Energy input (kWh) $\times 10^6$	Energy output (kWh) $\times 10^6$	Power output (kW)
Nasinu	76,064	11,105.34	3331.6	0.67	18.6	4.2	484
Suva	74,481	10,874.23	3262.3	0.65	18.3	4.2	470
Lautoka	43,473	6347.058	1904.1	0.38	10.7	2.4	278
Nausori	24,919	3638.174	1091.5	0.22	6.1	1.4	159
Nadi	11,685	1706.01	511.8	0.10	2.9	0.7	75
Lami	10,752	1569.792	470.9	0.095	2.6	0.6	69
Labasa	7706	1125.076	337.5	0.068	1.9	0.4	49
Ba	6826	996.596	299.0	0.060	1.7	0.4	44
Savusavu	3285	479.61	143.9	0.029	0.8	0.2	21

3.7 *Economics of Power Production from MSW Incineration Power Plant*

The cost analysis matrix is determined using the following assumptions:

- Project lifetime is 20 years
- Cost of Power Plant is 20% of the Total Cost of Installation
- Operations and Maintenance Cost is 10% of the Total Cost of Installation
- Other Cost at 1% of the Total Cost of Installation
- FEA current Tariff of $0.331 per kWh.

The above table shows the cost of energy ($/kWh) for setting up different sizes of power plants with an estimated cost. Analysis shows that with from the amount of *msw* collected in Fiji, there is a potential of setting upto 5 MW plant with an estimated cost of $300 Million dollars (FJD). The cost of energy is $0.228 (FJD) per kW (Table 6) which is still lower than the Fiji Electricity Authority tariff of $0.331 per kW.

Table 6 Cost analysis matrix of MSW incineration power plants

Total cost FJD $M	Cost of energy ($/kWh) for various power plant sizes									
	2 kW	10 kW	50 kW	100 kW	200 kW	500 kW	1 MW	2 MW	3 MW	5 MW
0.01	0.03	0.01	0.001	0.001	0.000	0.000	0.000	0.000	0.000	0.000
0.05	0.14	0.03	0.006	0.003	0.001	0.001	0.000	0.000	0.000	0.000
0.1	0.29	0.06	0.011	0.006	0.003	0.001	0.001	0.000	0.000	0.000
0.5	1.43	0.29	0.057	0.029	0.014	0.006	0.003	0.001	0.001	0.001
1	2.85	0.57	0.114	0.057	0.029	0.011	0.006	0.003	0.002	0.001
5	14.27	2.85	0.571	0.285	0.143	0.057	0.029	0.014	0.010	0.006
10	28.54	5.71	1.142	0.571	0.285	0.114	0.057	0.029	0.019	0.011
25	71.35	14.27	2.854	1.427	0.713	0.285	0.143	0.071	0.048	0.029
50	142.69	28.54	5.708	2.854	1.427	0.571	0.285	0.143	0.095	0.057
100	285.39	57.08	11.416	5.708	2.854	1.142	0.571	0.285	0.190	0.114
150	428.08	85.62	17.123	8.562	4.281	1.712	0.856	0.428	0.285	0.171
200	570.78	114.16	22.83	11.42	5.708	2.283	1.142	0.571	0.381	0.228
300	856.16	171.23	34.25	17.12	8.562	3.425	1.712	0.856	0.571	0.342

3.8 Payback Period

Form the cost analysis matrix as stated in Tables 6 and 7; the cost of energy remains low. For example for a 5 MW plant with investment cost of $200 M (FJD) the cost of energy is $0.228 per kWh. The payback period for setting up to 5 MW plant at a cost of $200 M (FJD) is approximately 14 years of the project life of 20 years. The Fiji Electricity Authority sells electricity to the grids at a rate of $0.331 per kWh. These figures indicate that there are potential of setting up small-scale incineration plants at a lower investment costs in Fiji. For instance, a 2 kW plant at a cost of $100,000 is considered viable for Fiji.

3.9 Incineration Power Plant: Suitable for Fiji

Based on the Population Census (2007) and Fiji Solid Waste Strategy Report information, each of the four divisions in Fiji has potentials for small scale power generation. These four divisions have well developed Sanitary Landfills. For example; Naboro Sanitary Landfill (in Central Division) and Lautoka—second largest municipality in Western Division are built to cater about 100,000 tonnes of MSW per year. Their land size is approximately 10 ha and can accommodate medium to large-sized Moving grate Technology or Small Oscillating Kiln technology plants as stated below in Fig. 1.

The basic working principle of such technologies will be as follows: the *msw* will be lifted with the crane and dumped into a funnel that will lead to the furnace where incineration will takes place. The heat in the flue gases would rise through the boiler where heat exchange will take place with the water running through the tubs that line the boiler. This heat exchange continues through the economizer until the water changes phase to wet steam. The over heater will heat the steam further making it dry so that it can drive the turbine to generate power. The steam then condenses to start the process again. The rest products from incineration will be the bottom ash and fly ash. The bottom ash is what is left at the end of the stoke grate after incineration has taken place. It is usually harmless and can thus be used for road constructions. The flue gases however need to go through some treatment steps before being released to the atmosphere. Ammonia present in the boiler can be injected to minimize NOx production.

The flue gases will then pass through a scrubber system to separate SOx and heavy metals from the gases. This is done with the help of lime slurry. Active carbon is injected to bind the last of the dust particles that separates from the flue gases in the bag filters. Then the clean flue gases are released through the smoke stack (Chimney) to the surrounding atmosphere. The fly ash is toxic and is placed on landfills. The final product is the electrical energy generation which will be sent to the grids upon agreement with the Fiji Electricity Authority.

Table 7 Payback period of MSW incineration power plants

Total cost FJD $M	2 kW	10 kW	50 kW	100 kW	200 kW	500 kW	1 MW	2 MW	3 MW	5 MW
0.01	1.72	0.34	0.07	0.03	0.02	0.01	0.00	0.00	0.00	0.00
0.05	8.62	1.72	0.34	0.17	0.09	0.03	0.02	0.01	0.01	0.00
0.10	17.24	3.45	0.69	0.34	0.17	0.07	0.03	0.02	0.01	0.01
0.50	86.22	17.24	3.45	1.72	0.86	0.34	0.17	0.09	0.06	0.03
1.00	172.44	34.49	6.90	3.45	1.72	0.69	0.34	0.17	0.11	0.07
5.00	862.20	172.44	34.49	17.24	8.62	3.45	1.72	0.86	0.57	0.34
10.00	1724.40	344.88	68.98	34.49	17.24	6.90	3.45	1.72	1.15	0.69
15.00	2586.60	517.32	103.46	51.73	25.87	10.35	5.17	2.59	1.72	1.03
20.00	3448.80	689.76	137.95	68.98	34.49	13.80	6.90	3.45	2.30	1.38
25.00	4311.00	862.20	172.44	86.22	43.11	17.24	8.62	4.31	2.87	1.72
50.00	8622.00	1724.40	344.88	172.44	86.22	34.49	17.24	8.62	5.75	3.45
70.00	12,070.80	2414.16	482.83	241.42	120.71	48.28	24.14	12.07	8.05	4.83
100.00	17,244.00	3448.80	689.76	344.88	172.44	68.98	34.49	17.24	11.50	6.90
150.00	25,865.99	5173.20	1034.64	517.32	258.66	103.46	51.73	25.87	17.24	10.35
200.00	34,488.0	6897.6	1379.5	689.8	344.9	138.0	69.0	34.5	23.0	13.8
300.00	51,732.0	10,346.4	2069.3	1034.6	517.3	206.9	103.5	51.7	34.5	20.7

TT Facility	Size, tonnes per annum	Buildings Area m²	Total Landtake Ha	Indicative Stack Height
Medium Sized Moving Grate Technology*	90,000	5850	1.7	65m
Large Moving Grate Technology*	250,000	6,600	4	70m
Large Fluidised Bed Technology#	500,000	70,000	34	80m
Small Oscillating Kiln Technology*	70,000	3,500	4	60m

Fig. 1 Shows the land area required for the entire site. *Source* Incineration of Municipal Solid Waste (2013)

4 Conclusion

From the analysis, it is evident that Fiji has potentials to build a small-scale incineration power plant that could use MSW as the feedstock. As solid waste minimization and recycling remains priority strategies for Fiji this could be one of alternative solutions. The simple economic analysis model reveals that Fiji has potential to generate electricity using *msw*. For example from the 50,000 tonnes of *msw* collected and disposed annually, in the Central division, there are potential of 2.2 MW electricity using incineration techniques. Also, other division like Western has similar 2.0 MW electricity potentials. A more comprehensive research on the solid waste management data to be accurate is required for a comprehensive report. For instance data including the quality and quantity of solid waste collected, the types of waste collected the energy and moisture contents of these wastes. Most of the data used in this study were gathered from the past few researches on the waste management conducted in Fiji.

Furthermore, the economic model shows that the cost of energy remains low. For example for a 5 MW plant with investment cost of $200 M (FJD) the cost of energy is $0.228 per kWh. In Fiji, the Fiji Electricity Authority sells electricity to the grids at a rate of $0.331 per kWh. The payback period for setting up to 5 MW plant at a cost of $200 M (FJD) is approximately 14 years of the project life of 20 years. These figures indicate that there are possibilities of setting up even small-scale incineration plant at a lower investment costs in Fiji. For example, a 2 kW plant at a cost of $100,000 is viable for Fiji.

While these projects may become viable, there are also many drawbacks. First, *msw* incineration technology has a higher capital cost (investment cost) and considering our current economic status the setting of such technologies may not be viable unless there are funding available from the donor countries. Moreover, Fiji lacks technical expertise in this area and hiring experts from overseas to do training and installation of these power plants would be costly. In addition, there are a lot of environmental concerns on incineration of MSW. For instance, the toxic emissions released from the combustion of wastes. For this a proper environmental risk assessment and a suitable technology (like moving grate or oscillating kiln) that controls all the toxic emissions are to be considered for Fiji.

Despite this, *msw* incineration for Fiji could become attractive for power generation and waste management if there are incentives given by the Fiji government for the electricity generation.

Finally, in terms of environmental sustainability and climate change mitigation, this will greatly help Fiji to reduce the GHGs and minimize the effects of climate change and will also minimize the current problems of the disposal of waste (MSW) in nearly all towns and cities. Furthermore, developing such technologies would help the economy as it would reduce some dependence on the use of fossil fuels.

Acknowledgements All the authors work that has been used in this paper are being acknowledged. Also would like to thank Dr. Anirudh Singh (Associate Professor of Physics), of the University of the South Pacific, Fiji for his guidance and contributions towards this paper.

References

Calorific values of various waste fractions. *Polish Journal of Environmental Studies, 19*(5).

Clarke, M. J. (2002). *Introduction to municipal solid waste incineration.* Lehman College, Geography Department of Bronx, NY. http://www.geo.hunter.cuny.edu/~mclarke/IntroMSWincineration.htm. Accessed February 6, 2015.

Danish Energy Statistics. (2005). http://www.ens.dk. Accessed February 5, 2015.

Database of Waste Management Technologies. (2015). http://www.epem.gr/waste-c-control/. Accessed February 5, 2015.

Fiji Solid Waste Management Strategy Report. (2011–2014). Department of Environment Level 1, P D Patel Building 90, Raojibhai Patel Street, Suva, Fiji.

General Country Information and Waste Management in Fiji. (2013). Brief country analysis paper for the 4th regional 3R forum, Department of Environment, Country Report (Fiji), Asia, Ha Noi, Viet Nam, March 18–20, 2013.

Incineration of Municipal Solid Waste. (2013). http://www.defra.gov.uk. Accessed February 5, 2015.

Malua, L. (2001). *Application of solid waste assessment to the formulation of strategic solutions for solid waste management.* http://www1.mnre.gov.ws/. Accessed February 28, 2015.

Shirleen, S. (2012). *An energy balance analysis of MSW for energy production in Fiji.* M.Sc. thesis, USP, Fiji.

World Energy Resources. (2013). *Waste to energy*. World Energy Council 2013 Report (pp. 1–14).

Zafar, A., Martinez, A., Sebastian, C., École des Mines, K. A. (2015). *Decentralized CHP from MSW-RDF for residential buildings*. http://www.academia.edu/4298040. Accessed February 27, 2015.

Replacing Fossil Fuel with PV Systems Through Technical Capacity Building in Kenya

Izael Da Silva, Ronoh Geoffrey, Teddy Nalubega and Mwaura Njogu

Abstract In this paper, it is described how Strathmore Energy Research Centre developed an outreach project funded by USAID, National Science Foundation (NSF) to empower technical institutions to offer solar courses hence creating a pool of qualified technicians spread throughout the geography of the country. Presently, the situation is that Kenya has around 1000 solar technicians working in the market with no formal solar PV training or accreditation. The National Industrial Training Authority—NITA, which regulates non-academic skills or craft-based training, was helped by SERC (with GIZ support) to develop a PV solar curriculum at three levels (T1/T2/T3) which empowers from craft-level technicians to engineers to deal with design, installation, and maintenance of PV systems from solar lantern up to utility size level. ERC, the Energy Regulatory Commission for Kenya has supported the initiative as technicians once trained can be accredited by ERC and thus further strength the industry. By November 2016, which is the end of the program (after a 6 months no cost extension), Kenya is to have more than 1000 accredited technical personnel near almost every major town in the country. The paper describes the positive and negative aspects of this venture.

Keywords Solar photovoltaic · Capacity building · Technical training institutions Solar PV penetration logistics · Phasing out kerosene for lighting · Empowering the bottom of the social pyramid

---

I. Da Silva (✉) · R. Geoffrey · T. Nalubega · M. Njogu
Strathmore Energy Research Center, SERC, Strathmore University, Nairobi, Kenya
e-mail: idasilva@strathmore.edu

R. Geoffrey
e-mail: gronoh@strathmore.edu

T. Nalubega
e-mail: tnalubega@strathmore.edu

M. Njogu
e-mail: kaiyahz@gmail.com

W. Leal Filho and D. Surroop (eds.), *The Nexus: Energy, Environment and Climate Change*, Green Energy and Technology, https://doi.org/10.1007/978-3-319-63612-2_11

1 Introduction

At the time of this project, Kenya ranked seventh globally and second in Africa, in terms of best performance on Clean Energy Enabling Framework, largely due to the country's solid policy framework (Power Africa Annual Report 2014). Although the Kenyan solar PV market is often considered as a successful commercially driven market (Target Market Analysis Kenya's Solar Energy Market GiZ/GtZ 2009), market spoilage due to poor quality service significantly hindered market growth. In December 2012, the Energy Regulatory Commission (ERC) published the "Energy Solar Photovoltaic System Regulation 2012" (The Energy (Solar Photovoltaic Systems) Regulations 2012). This regulation proposes a path for training and accreditation of Solar Photovoltaic technicians. There are approximately 800–1000 solar PV technicians working in the Kenyan market, majority of them did not attend any formal training and certification (Improving Technical Skills for Solar Systems in Kenya 2014). The regulations require that only licensed technicians are allowed to design and install solar PV systems. To be licensed, technicians are required to have undertaken a solar training course. The presence of few competent technicians to offer design, installation and maintenance services within the solar PV sector in Kenya presented an opportunity to launch into an outreach program to support the development of the solar PV technology within Kenya outside of the area around Nairobi. The number of technicians has since increased by a small percentage due to awareness creation and development of the solar curriculum (Solar Energy Licenses Growing In Popularity—Energy News Energy Regulatory Commission 2014).

1.1 The Problem

In a survey funded by JICA (2012) in which 41 higher education institutions participated, it was identified that only 50% of those institutions offered short courses in Solar PV technology. Of the 368 teaching staff interviewed, only 16% had received prior training in solar PV technology. 52% of the institutes did not have access to adequate training materials while 60% of the respondents felt that they did not have appropriate hands-on training equipment. 20% did not have any hands-on training equipment. This survey clearly uncovered the sorry state of capacity building on solar PV in Kenya and somehow explains the small market penetration. As someone commented "would you buy a car if the nearest mechanic were 200 km away?" Certainly not!·

1.2 Scope and Objective

There are 34 Technical Training Institutions (TTI) in Kenya and SERC has signed a memorandum of understanding with 18 of them. This paper shares the approach,

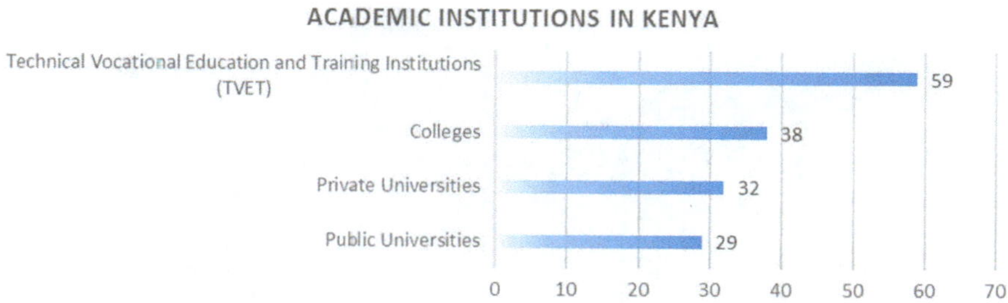

Fig. 1 Numbers of academic institutions in Kenya by category

progress, setback, methodology, and impact so far with the creation of a solar PV capacity building in Kenya. The use of Training of Trainers was the most suitable way to achieve numbers and geographical spread. For that, capacity building and empowerment of technical and vocational institutions to deliver such courses at the institutional level was required (Fig. 1).

Strathmore University, through Strathmore Energy Research Center is coordinating a USAID-funded NSF project under the Partnership in Enhanced Engagement in Research (PEER) project (http://sites.nationalacademies.org/PGA/PEER/PGA_152068). The project's objective is to train a minimum of 1000 solar technicians. It was launched in September 2014. Through the project, TTIs were capacitated by having two of their teaching staff attending training of trainer sessions and equipped with hands-on training equipment like the ones seen in Fig. 2. The memorandum of understanding signed between SERC and each TTI requires that each institute trains at least 100 technicians in a 2-year period. These trained technicians are then encouraged to approach ERC to seek accreditation. These trained/accredited technicians will add to the pool of the technical personnel within the country making it easier for any customers, irrespective of their location, to invest in solar PV installations of any scale with confidence.

The scope of the PEER project was to train at T2 level only. Nonetheless, the German Cooperation Agency—GIZ picked interest and added funds to upgrade the training of trainers to T3 level (grid-tie utility size systems and hybrid solar/diesel systems). GIZ further helped the project by sponsoring SERC and the Kenya Renewable Energy Association (KEREA) to develop a criterion to accredit institutions to carry out PV training. These regulations were gazetted and adopted by NITA.

SERC has designed and patented a specific mobile training kit to serve the purpose of this project (Fig. 2). At this time, the project has so far trained a total of 936 technicians of which 32 have been accredited, 6 of them being ladies.

1.3 *Overview of Solar Technicians Licensing Procedure*

Energy Regulatory Commission (ERC) gazetted Solar PV regulations in 2012 for regulating the solar PV sector/market as well as protecting consumers. For solar PV

Fig. 2 SERC solar training kits

technicians, licensing is done at three levels, i.e., basic (T1) level Solar PV system installation work for small systems or single-battery DC system of up to 100 Wp, intermediate (T2) level Solar PV system installation work for medium systems or multiple batteries which may include an inverter and advanced (T3) level Solar PV system installation work for advanced, including grid connected and hybrid systems. According to ERC records, in August 2014, 64 solar PV technicians were licensed and by November 2015, 300 had been licensed since the regulations were gazetted (Energy Regulatory Commission. Solar Photovoltaic Technicians Register 2015). The number of licensed technicians per category is shown in Table 1.

The analysis of these technicians per location indicates that 42 are based in Nairobi (66%) while the remaining 22 are scattered in 8 different towns across the country (i.e., an average of two technicians per location). The number of licensed technicians is therefore too low and their distribution is skewed towards Nairobi and other bigger towns (e.g., Mombasa, Kisumu, and Eldoret). This therefore implies that technicians currently installing solar systems in different parts of the country are operating in contravention with the 2012 solar PV regulations. More efforts is therefore required to be put in decentralized training through training institutions and supporting licensing of solar PV technicians in order to improve this situation.

Table 1 Number of licensed technicians in Kenya

Level	Number of licensed technicians	
	August 2014	November 2015
Basic (T1) level	2	6
Intermediate (T2) level	21	148
Advanced (T3) level	41	146
Total	64	300

Ref. ERC solar PV register, Nov 2015

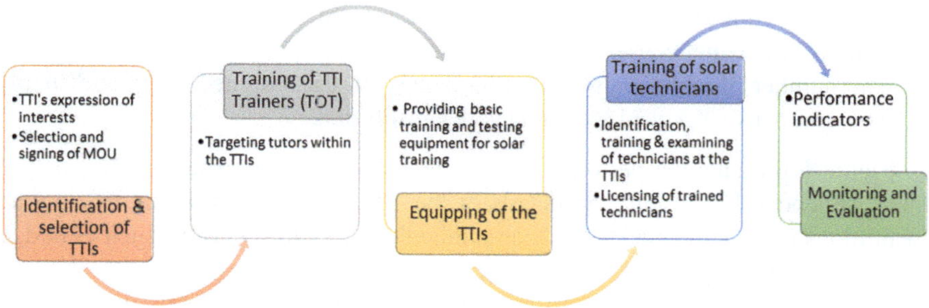

Fig. 3 TTI capacity building criteria inception to final auditing

2 Methodology

With the target of building capacity for basic and intermediate solar technical skills (T1/T2) in rural areas, SERC with other stakeholders such as GIZ-Endev (Energy for Development) and the National Commission for Science, Technology and Innovation (NACOSTI) designed the ToT curriculum and hands-on mobile training kit. Once this was done, SERC undertook the following activities shown below.

The course content for T1/T2 is shown in "Appendix 2". The duration of the course is 5 days and most of the theory is covered in the morning sessions and the afternoon session addresses the practical aspect. The course has an e-learning platform which serves the purpose of pre-evaluation of the knowledge level of the candidates. The prerequisite for this training is that the participant has a basic knowledge in electrical wiring. "Appendix 1" presents the 5-day course details.

As per the MoU, SERC is tasked to carry out auditing of the TTI to access the quality of the hands-on training kit assembly which is to be done by the beneficiary, the prowess of the trained teaching staff and finally the number of technicians trained/accredited per institution. "Appendix 1" gives the details of the selected TTIs presented in Fig. 3.

3 Results

The impact of the project is presented in two categories, namely qualitative and quantitative. Five parameters have been considered

1. The level of responsiveness measured by the speed of signing the MOU, the support received by the project coordinators at the institution and the financial support to the design and completion of the training kits.
2. Participation and preparation of the TTIS by availing two technicians for the training of trainers and the absorption of the course materials and content.

3. The fabrication of the training kits according to SERC design, which has to be done by the beneficiaries TTIs in their workshops.
4. The number of technicians trained in respect to the assigned target and finally
5. The number of licensed technicians per institution.

Figure 5 presents the performance per institution in each aspect of the analysis and "Appendix 1" presents the details of the TTIs. It is worth noting that private institutions have performed better than public due of various reasons such as budget constraint and delays due to bureaucracy (Fig. 4).

25% of the selected beneficiary institutions are private and have trained 32% of the total technicians trained under the project. The project has trained a total of 60 women technicians, 6 are now licensed bringing to 15 the number of licensed female technicians in the whole country. The female licensed technicians only account for 3% of the licensed technicians in Kenya (ERC, 2015).

Qualitatively, the project has created a lot of impact through empowering institutions to run the solar PV courses, building capacity of the youth through the institutions, creating awareness and supporting the regulations within the solar sector, empowering women, creating confidence in the end users about the solar PV technology and supporting the development of the solar PV technology within Kenya.

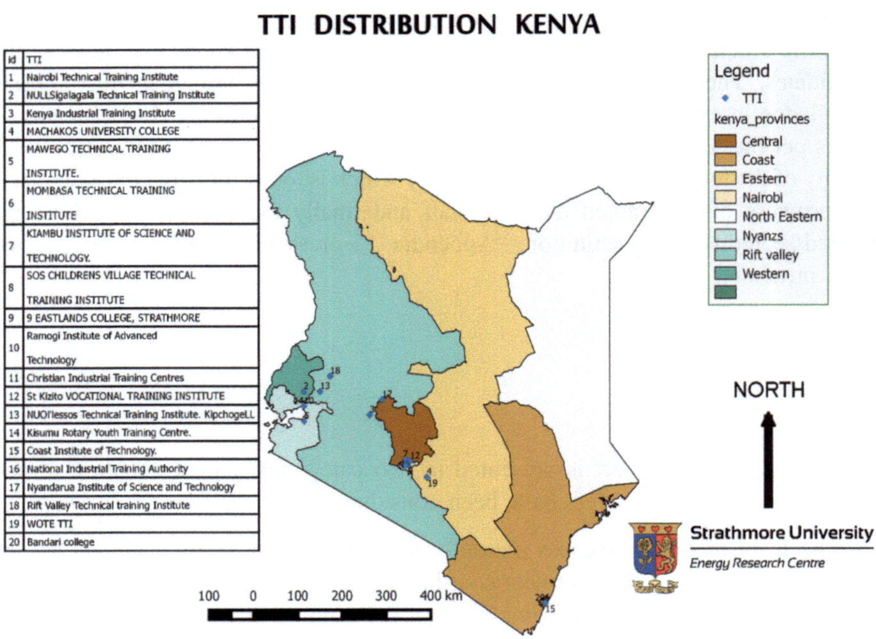

Fig. 4 Regional distribution of the selected TTIs in Kenya

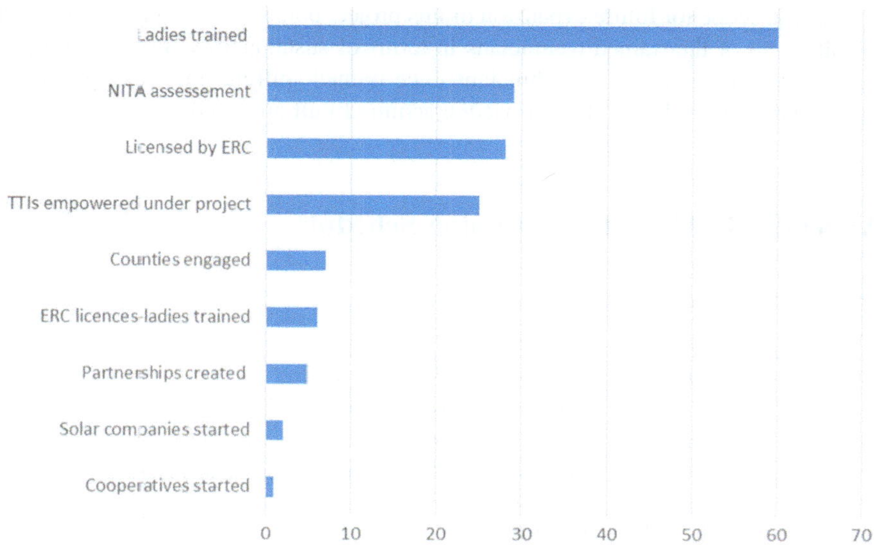

Fig. 5 Performance of the PEER beneficiary in general

4 Conclusion

The project has a target of training, through the TTIs, a minimum of 1000 technicians by the September 30, 2016. So far, 936 people have been trained within 90% of the project period. This is the cause for concern though we expect many more people being trained beyond the scope of the PEER program. The impact within the country is much more than what the quantitative figures may suggest.

SERC has recognized also that supporting the technicians to get licensed and accredited requires attention as many may fear taking the exams. Another issue is that the institutions have to ensure gender balance by training female technicians. Large numbers of licensed technicians are anticipated with regards to the impact by the project which is in line with Kenya's vision 2030.

It was noted that awareness creation of the solar PV regulations was achieved through the partnership with other stakeholders such as GIZ EnDev Kenya, ERC, NACOSTI, KEREA, TTIs, and the county governments. The awareness focused on the content and importance of complying with the regulations and the penalties for non-compliance. Many players in the market did not know that they were required to have a trained and licensed technician as a member of staff in order to carry out their business. This awareness is anticipated to increase demand for solar technician training services in the relevant areas. Finally, we envision that this project should continue on a second stage to cover the other TTIs which did not benefit from this phase.

There is scope for future expansion of this project both in geographical reach and in follow up of the trained technicians in terms of sustainability. Previous project have faced the challenge of technicians once trained shifting into more profitable trades and thus rendering the knowledge acquired null and void.

Appendix 1: The 5-Day Training Schedule

Training schedule and contents

Date/Day	Time	Content
Day 1	Morning	Introduction to the Training
		S1O1—Basic and fundamentals of electricity
	Mid-morning	S102—Solar resource and energy conversion
		Introduction of PV project development and preparation (excel doc)
	12:30–13:30	**LUNCH**
	Afternoon	SPV H01—Hands-on (solar system, tools, and measuring instruments)
Day 2	Morning	S103—Photovoltaic modules I
	Mid-morning	S103—Photovoltaic modules II
		S104—Energy Storage (Batteries)
	12:30–13:30	**LUNCH**
	Afternoon	SPV H02—Hands-on (PV modules and battery, System connection and operation)
Day 3	Morning	S105—Charge Controllers & Inverters
	Mid-morning	S106—Solar loads and appliances
	12:30–13:30	**LUNCH**
	Afternoon	S108—Policies and legal regulatory framework for solar PV in Kenya, National Industrial Training Authority Session
		S107—System design, installation, commissioning and safety
		SPV H03—Maintenance and troubleshooting hands-on session
Day 4	Morning	Review and question session
		Possible System situation analysis
		Summary of the training
	Mid-morning	Hands-on/ Project preparation and design
	12:30–13:30	**LUNCH**
	Afternoon	Hands-on/ Project preparation and design
Day 5	Morning	Project preparation and design/ Discussion of completed projects with the tutors
		Energy Regulatory Commission (ERC) session
	12:30–13:30	**LUNCH**
	Afternoon	End of training

Appendix 2: Details of the Project Beneficiaries

No.	TTI	County
1	Christian Industrial Training Centre (CITC) [Meru road-off General Waruinge Street]	Nairobi
2	St Kizito VOCATIONAL TRAINING INSTITUTE. P.O. Box 759, Githurai, Nairobi, Kenya	Kiambu
3	SOS TTI BURUBURU, Nairobi	Nairobi
4	Eastlands college of technology, Nairobi	Nairobi
5	Kiambu Institute Of Science and Technology	Kiambu
6	Nyandarua Institute of Science and Technology. Nyahururu	Nyandarua
7	Machakos University, Machakos	Machakos
8	Wote, Makueni	Makueni
9	KITI, Nakuru	Nakuru
10	Sigalagala TTI P.O. BOX 2966, Kakamega	Kakamega
11	Nairobi TTI, Nairobi Technical Training Institute	Nairobi
12	Mawego TTI. P.O. BOX 289 40222, Oyugis	Homa Bay
13	Kisumu Rotary Youth Training Centre. Nyanza Kenya	Kisumu
14	Olesscs TTI, Nandi	Nandi
15	Bandari College (KPA), Mombasa	Mombasa
16	Rift Valley TTI, Eldorate	Uasin Gishu
17	Ramogi TTI, Kisumu	Kisumu
18	Coast Institute of Technology. P.O. Box 34, 80300 Voi Coast 80300 Kenya, Voi	Kwale
19	National Industrial Training Authority (NITA) School. Commercial St, Nairobi, Kenya	Nairobi
20	Mombasa TTI	Mombasa

References

Development and Implementation of a Solar PV Outreach Training Module for Capacity Building in East Africa. (December, 2016). http://sites.nationalacademies.org/PGA/PEER/PGA_152068.

Energy Regulatory Commission. Solar Photovoltaic Technicians Register. (November, 2015) Kenya. http://erc.go.ke/index.php?option=com_docman&view=docman&Itemid=693.

Improving Technical Skills for Solar Systems in Kenya (May, 2014). http://www.jica.go.jp/kenya/english/office/topics/140512_01.html.

JICA Annual Report. (2012). Retrieved from http://www.jica.go.jp/kenya/english/office/others/c8h0vm000001pzr0-att/report2012.pdf.

Power Africa Annual Report (2014, July). Retrieved from http://www.afdb.org/fileadmin/uploads/afdb/Documents/Generic-Documents/USAID_-_Power_Africa_Annual_Report.pdf.

Solar Energy Licenses Growing In Popularity—Energy News Energy Regulatory Commission (December 2014). Quarterly Newsletter. Retrieved from http://erc.go.ke/index.php?option=com_docman&task=cat_view&gid=30&Itemid=429.

Target Market Analysis Kenya's Solar Energy Market GiZ/GtZ. (2009, November). Retrieved from https://www.giz.de/fachexpertise/downloads/gtz2009-en-targetmarketanalysis-solar-kenya.pdf.

The Energy (Solar Photovoltaic Systems) Regulations. (2012). Retrieved from http://www.erc.go.ke/index.php?option=com_docman&task=cat_view&gid=8&Itemid=429.

Design Technology for Bioenergy Conversion of Organic Fraction of Municipal Solid Waste

Anthony Njuguna Matheri, Charles Mbohwa, Mohamed Belaid and Jane Catherine Ngila

Abstract In this study, we investigated the use of a laboratory batch anaerobic digester technology, for biochemical methane potential (BMP) and the bioenergy process design from anaerobic co-digestion of different organics fraction of municipal solid waste (OFMSW) originating from the city's landfills. The carbon-to-nitrogen (C/N) ratio of OFMSW was found to be below 13. Through co-digestion, the C/N ratio settled at 14. Laboratory experimental data from 500 ml batch anaerobic digester operating at mesophilic temperature of 37 °C and pH of 7 had good productivity of methane of average 59.49% with (54–62%) recommended in the literature and was used to derive volume of digester and surface area. Via the application of the simple multi-attribute rating (SMART) technique of multiple-criteria decision analysis (MCDA) as a decision support tool, the most preferred model option for bioenergy design technology was selected from a list of potential alternatives available in the market. The geometry of the biodigester parameters was comparable with the process.

Keywords Anaerobic digestion · Co-digestion · Energy · Mesophilic temperature OFMSW · Process design

Nomenclature

AD	Anaerobic digestion
AMPTS II	Automatic methane potential test system
AHP	Analytic hierarchy process

A.N. Matheri · M. Belaid (✉)
Department of Chemical Engineering, University of Johannesburg, Johannesburg 2028, South Africa
e-mail: mbelaid@uj.ac.za

C. Mbohwa
Department of Mechanical Engineering, University of Johannesburg, Johannesburg 2028, South Africa

J.C. Ngila
Department of Applied Chemistry, University of Johannesburg, Johannesburg 2028, South Africa

W. Leal Filho and D. Surroop (eds.), *The Nexus: Energy, Environment and Climate Change*, Green Energy and Technology, https://doi.org/10.1007/978-3-319-63612-2_12

APHA	American Public Health Association
BMP	Biochemical methane potential
$CaCO_3$	Calcium carbonate
CO_2	Carbon dioxide
C/N	Carbon to nitrogen
CBR	Case-based reasoning
CSTR	Continuous stirred tank reactor
CM	Cow manure
D	Diameter of the tank
DM	Dry matter
DS	Dry solid
f	Height of dome
GHG	Greenhouse gas
H	Height of the digester
HRT	Hydraulic retention time
MCDA	Multiple-criteria decision analysis
MSW	Municipal Solid Waste
OFMSW	Organics fraction of municipal solid waste
PEETS	Process Energy and Environmental Technology Station
Q	Volumetric flow rate
r	Radius of the digester
S_1	Surface area of the digester top dome
SMART	Simple multi-attribute rating
NaOH	Sodium hydroxide
TS	Total solid
UASB	Up-flow anaerobic sludge blanket
V_1	Volume of the digester top dome
V_2	Volume of the digester cylinder
VS	Volatile solids
V_r	Volume of the digester

1 Introduction

The global energy demand has recorded exponential growth over time with a predicted 85% increase between 2010 and 2040. Approximately 85% of the world's energy supply is obtained from non-renewable fossil fuels sources such as coal, oil, and natural gas. These fuels yield high quantities of greenhouse gas (GHG) emissions, particularly carbon dioxide (CO_2). The continuous use of fossil fuels is leading to the long-term potential risk of energy insecurity and simultaneously degrading the environment with the high CO_2 emissions (Allan 1998; Latinwo and Agarry 2015; E.I.A 2013). Municipal Solid Waste (MSW), a by-product of the lifestyle of urban dwellers, comprises of wastes from household, offices, restaurants, market, and food processing industries among others. In some countries, construction wastes are also included as MSW but it excludes hazardous waste (McDougall et al. 2008; White et al. 2012). MSW management encompasses the generation, handling, storage, collection, transfer, transportation, processing, and final disposal of wastes. The management of MSW within a city is of utmost concern as the volume of waste generated continues to increase along with population and economic growth. There are several obstacles confronting MSW management within the cities. Some of such obstacles are; interrelation of economic growth and urbanization; complexity of the waste stream due to different class of citizen living within the city; lack of adequate facilities that will expedite waste separation at source; overstretching of the superannuated infrastructure; and also the waste management technologies that are handy, are very costly compared to the cost of land-filling (McDougall et al. 2008; White et al. 2012). Separation of waste at source or the use of waste transfer station have both achieved some degree of success (McDougall et al. 2008; White et al. 2012).

Biogas is a mixture of gases such as methane, carbon dioxide, ammonia, hydrogen sulphide, and trace amount of hydrogen, oxygen, and carbon monoxide. It is produced by break down of organic materials using bacteria under controlled conditions (Matheri et al. 2016). The organic source is composed of biodegradables such as municipal solid waste, animal waste, agricultural waste, and industrial waste (Sambo et al. 1995; Dangoggo et al. 1996). Energy derived from biogas is used in the form of fuel, electricity, and heat. It is desirable to create sustainable and with zero carbon emissions world-wide energy system.

Biogas production takes place in series of four fundamental steps namely: hydrolysis, acidogenesis, acetogenesis, and methanogenesis. Figure 1 shows degradation steps of anaerobic digestion process. In hydrolysis, large organic polymers such as fats, carbohydrates and proteins are broken into fatty acids, simple sugar, amino acids, respectively. This step is carried out by bactericides. Hydrolysis is followed by acidogenesis where low alcohol and organic acids are produced through fermentation process utilized by fermentative bacteria. This includes volatile fatty acids (butyric acid, acetic acid, and propionic acid), gases like carbon dioxide, ammonia, and hydrogen and aldehydes. In third fundamental step (acetogenesis), the products of acidogenesis are converted to acetic acid, carbon dioxide and hydrogen by acetogenic bacteria. Methanogenesis is the final stage whereby methanogenes bacteria converts hydrogen, acetic acid, and carbon dioxide to methane and carbon dioxide (Sundararajan et al. 1997; Angelidaki 1996).

1.1 Process Parameters

There are various parameters that control the efficiency of anaerobic digestion process. These parameters when optimized provide an appropriate environment for the growth of anaerobic microorganisms. They include: nutrient supply (carbon nitrogen ratio), constant temperature, stirring intensity, partial pressure, nature of

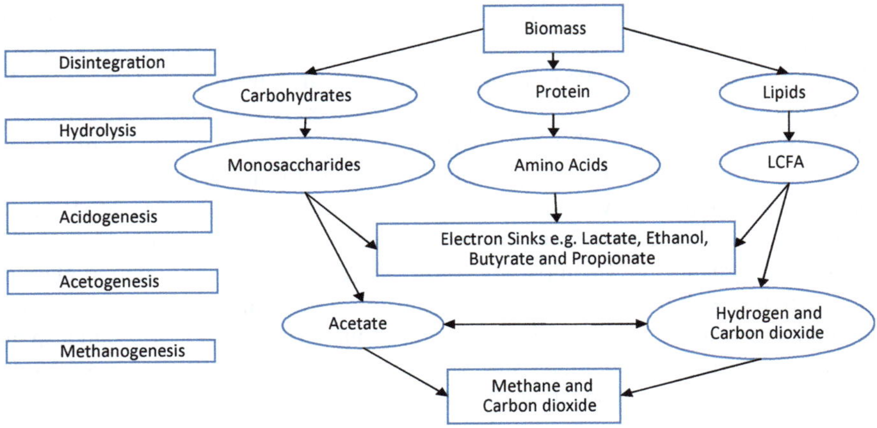

Fig. 1 Degradation steps of anaerobic digestion process (Matheri et al. 2015a, b)

substrate, optimum trace element concentration, moreover presence and amount of inhibitors (e.g., ammonia) and exclusion of oxygen. The presence of oxygen into digestion process must strictly be avoided since methane bacteria are anaerobes.

1.1.1 Temperature

The optimum temperature, i.e., the temperature at which the microorganisms grow most efficiently varies among species. Microorganisms can be divided into three groups depending on the temperature at which they can best thrive and grow: psychrophilic, mesophilic, and thermophilic (Schnurer and Jarvis 2010). The optimum temperature for a specific organism is strongly linked to the environment from which it originates. The two convectional operational temperature levels for anaerobic digesters determine the species of methanogens in the digesters (Song et al. 2004).

Psychrophilic occur at a low optimum temperature of around 10 °C, whereas mesophilic is around 20–45 °C and thermophilic with an optimum temperature above 50 °C as shown in the Fig. 2. At low temperatures of less than 10 °C, the anaerobic process is slow, taking three times more than the normal mesophilic time process (Song et al. 2004).

1.1.2 pH

pH is the measure of H^+ ions in a solution, it determine whether a solution is an acid or a base. In anaerobic digestion, it is crucial to measure the pH throughout the entire process to ensure the well growth of the methanogens (http://www.e-inst.com/bimass-to-biogas). As with living beings, methanogens require an optimum

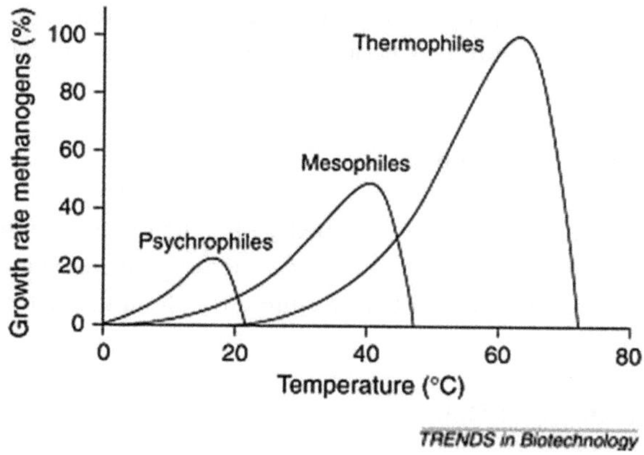

Fig. 2 Growth of microorganisms at different temperatures (Matheri et al. 2015c; Kuria and Maringa 2008)

environment for their growth. They require an environment between the pH ranges of 6.5–7.5 (Schnurer and Jarvis 2010). It was reported that there are several biogas processes in Sweden currently operating at pH values of 8 (Nordberg 2006). In the acidogenesis process, acid produced lowers the pH in the digester. It is therefore important to constantly measure the pH in the digestate to ensure continued well-being of methanogens and thus methane production.

1.1.3 Retention Time

Retention time is the time taken to replace all the materials in the digestion tank (Schnurer and Jarvis 2010). It varies with the amount and type of the substrate, configuration of the digestion system and the number of the stage process. The length of the retention time depends partly on the composition of the substrate and the digestion temperature (Schnurer and Jarvis 2010).

1.1.4 Digestion Chamber Loading

Digestion chamber loading refers to the amount of substrate feeding into the digestion chamber per day per m^3 of digestion chamber volume. Increasing the digestion chamber loading will reduce the digestion chamber volume and also reduce the percentage of volatile solids converted to gas. Studies have shown that methane yield increased with a reduction in the loading rate (Babaee and Shayegan 2011). Mesophilic digesters appear to achieve greater conversions at lower loadings while thermophilic reactors appear to achieve greater conversions at high loadings. In typical anaerobic digester, the digestion chamber loading is approximately from 1 to 5 kg/m^3 day (Schnurer and Jarvis 2010). In this case, it is important to know the dry solids (DS) and volatile solids (VS) content in the substrate in order to give the biogas process the right loading rate. Studies have shown that methane yield increased with a reduction in the loading rate (Babaee and Shayegan 2011).

1.1.5 Agitation

Digesters should be equipped with agitators to mix the substrate. Mixing facilitates contact between the substrate, microorganisms, and nutrients and provides a uniform distribution of temperature throughout the process (Schnurer and Jarvis 2010).

1.1.6 C:N Ratio

Microbes need a 10–30:1 ratio of carbon to nitrogen with the largest percentages of the carbon being readily degradable to meet the microbes' nutrients requirement (Bardiya and Gaur 1997). A methanogenic bacterium uses nitrogen to meet their protein requirements. Therefore in cases of high C/N ratios higher than the optimum ranges, the nitrogen will be depleted rapidly by the microbes. For cases of lower rations than the desired range, the excess nitrogen will result into ammonia formation hence raising the pH over the desired eight and inhibiting the microbes and ultimately lowering gas production rates (Kumar 2012).

1.1.7 Particle Size

According to EU regulation EC 208/2006, the proposed maximum particle size for adequate digestion is 12 mm. Several studies (Schnurer and Jarvis 2010) also show a clear correlation between particle size and methane yield, and for maximum digestion, particle size should preferably be just a few or less 12 mm (Schnurer and Jarvis 2010).

1.2 Anaerobic Digesters Configuration

1.2.1 Batch or Continuous Configuration

AD can be performed as a batch or a continuous process depending on the substrates being digested and the configuration of the digester (Schnurer and Jarvis 2010). In a batch process, the substrate is added to the digester at the start of the process and sealed for the duration of the retention time. After digestion, biogas is collected and digester is partially emptied. They are not emptied completely to ensure inoculation of fresh substrate batch with bacteria from previous batch (Schnurer and Jarvis 2010).

Table 1 Main characteristics of anaerobic digesters technologies (Lfu 2007)

Characteristics	Technologies
Construction of digester	Covered lagoon, plug flow, complete mix, fixed film, UASB, vertical, horizontal and etc.
Temperature in digester	Psychrophilic, mesophilic and thermophilic
Environment in digester	Wet and dry
Process stages	One-stage, two-stages and multiple stages
Loading (feeding) strategy	Batch, continuous and semi-batch

In a continuous digestion process, organic matter is constantly added in stages to the digester on a daily basis (http://www.e-inst.com/bimass-to-biogas). In this case, the end products are constantly removed resulting in constant biogas production (http://www.e-inst.com/bimass-to-biogas). A single or multiple digesters in sequence may be used.

The selection of biogas digester depends on the dry matter (DM) content of the digested substrate. There are two AD technologies systems: wet digestion which is liquid digestion; when the average DM content of the substrate is less than 15% and dry digestion which is solid digestion; when the DM content of the substrate is more than 15% (usually from 20 to 40%). Wet digestion is applied for substrate like manure and sewage sludge, while dry digestion is applied for solid municipal biowaste, solid animal manure, high straw content, household waste and green cuttings, grass from landscape maintenance or energy crops (Al Seadi 2008; Rapport et al. 2008). Table 1 shows the characteristics of anaerobic digesters technologies while Table 2 shows the comparison of various digesters types.

Table 2 Comparison of various digester types (Lfu 2007)

Technology	Digester type	Feedstock type	HRT (days)	Biogas yield	Technology level
Wet digestion	Covered lagoon	Thin manure	20–200	Poor	Low
	Plug flow	Think manure	20–40	Poor	Low
	Complete mix	Liquid and solid	20–80	Good	Medium
	Fixed film	Liquid	1–20	Good	High
	UASB	Liquid	0.5–2	Good	High
Dry digestion	Batch	Agricultural and municipal feedstock	20–30	Good	Medium
	Vertical		20–40	Good	High
	Horizontal		20–40	Good	High

1.3 Conditions Affecting the Choice of a Biogas Plant

Developing a biogas plant design is essentially the final stage of the planning process. However, it is mandatory for the designer to familiarize themselves with basic design considerations in advance. Ultimately, a successful plant design should be able to respond to quite a number of factors, and these include.

1.3.1 Climate

The design should respond to the prevailing climatic conditions of the location. Bearing in mind that biogas plants operate optimally at temperature ranges between 30 and 40 °C, in cooler regions, it is advisable for the designer to incorporate insulation and heating accessories to the design.

1.3.2 Substrate Quality and Quantity

The type and amount of substrate to be used on the plant will dictate the sizing of the digester as well as the inlet and outlet design.

1.3.3 Construction Materials Availability

If the materials required for the plant setup can be sourced locally at affordable rates so as to maintain the plant setup costs within manageable ranges, then the design is preferred to that whose materials have to be imported.

1.3.4 Ground Conditions

Preliminary geotechnical investigations can guide the designer on the nature of the subsoil. In cases where the hard pan is a frequent occurrence, the design installation plan must be done in such a way that deep excavations are avoided because this would then increase the construction costs tremendously.

1.3.5 Skills and Labour

Biogas technology is sophisticated and hence requires high levels of specialized skilled labour. The labour factor cuts across from the planner to the constructor up to the user. However, gaps can be reduced through training of the involved parties at a cost.

1.3.6 Standardization

Prior to commissioning of the design, the planner must carefully study the prevailing standards already on the market in terms of product quality and pricing especially for large scale projects.

1.4 Technology Selection Methods

Several methods have been developed to give unbiased results when it comes to decision making on a particular choice of technology. In principle, all methods are based on the steps summarized below

- Identification of the problem,
- Identification of stakeholders,
- Seeking the unbiased opinions of the stakeholders in the form of solutions to the identified problem. The identified solutions are treated as alternatives and the key performance indicators of the chosen options become the selection criteria,
- Modelling the obtained solutions so as to obtain impartial results through detailed analyses. At the modelling stage is when the decision-maker decides on which particular selection method to employ basing on the nature of the problem at hand.

In modern times, technology designs are probabilistic in nature and the evaluation criterion are multi-dimensional. This calls for complex tools that can capture all the dimensions of a decision problem. The existing technology selection methods includes the following:

1.4.1 Multi-Criteria Decision Analysis (MCDA)

MCDA is an approach employed by decision makers to make recommendations from a set of finite seemingly similar options basing on how well they score against a predefined set of criteria. MCDA techniques aim to achieve a decision goal from a set of alternatives using pre-set selection factors herein referred to as the criteria (Chai et al. 2013). The selection criteria are assigned weights by the decision-maker basing on their level of importance. Then using appropriate techniques the alternatives are awarded scores depending on how well they perform with regard to particular criteria. Finally ranks of alternatives are computed as an aggregate sum of products of the alternatives with corresponding criteria. From the ranking, a decision is then made (Dodgson et al. 2009).

There are several variations in MCDA techniques used currently employing mathematics and psychology. These include; analytic hierarchy process (AHP), simple multi-attribute rating technique (SMART) and case-based reasoning (CBR)

Table 3 Saaty's scale intensity 1–9 (Saaty 2004)

Scale Intensity	Definition	Explanation
1	Equal Importance	Two elements equally contribute to the intended objective
3	Moderate importance	Basing on judgement and experience one element is favoured over the other
5	Strong Importance	Basing on judgement and experience one element is strongly favoured over the other
7	Very Strong Importance	One element is very strongly favoured over the other and its dominance can be demonstrated in practice
9	Extreme Importance	The evidence favouring one element over another is of the highest order of affirmation

AHP aims at organizing and analysing complex decisions basing on their relative importance independent of each other (Saaty 2004; Pohekar and Ramachandran 2004). Saaty (2004) developed a scale of 1–9 to score alternatives basing on their relative importance as shown in Table 3. However, the major drawback of the AHP is the alteration of ranks in cases where new alternatives are introduced into an already analysed problem (Saaty 2004; Pohekar and Ramachandran 2004).

By applying the SMART technique, alternatives are ranked basing on ratings that are assigned directly from their natural scales (Barron and Barrett 1996; Belton 1986). The advantage of the SMART technique over AHP is the fact that the decision-making model is developed independent of the alternatives. Therefore the scoring of the alternatives is not relative and therefore introduction of new alternatives does not affect the ratings of the original ones making it a more flexible and simpler technique (Belton 1986). In CBR, problem solving is done basing judgement on similar past problems and experiences. Basically, the decision is made basing on what has happened before (Leake 1996).

The main objective of this study was to investigate the use of laboratory batch anaerobic digester for biochemical methane potential (BMP) and the bioenergy process design from anaerobic co-digestion of different organic fraction of municipal solid waste (OFMSW) originating from the city's landfill to be used as biofuel for the city buses and electricity generation.

2 Materials and Methods

2.1 Waste Quantification and Characterization

Waste generated in this study was quantified (Fig. 3) at the City of Johannesburg damping site, Gauteng Province. This involved measurement of the waste at the

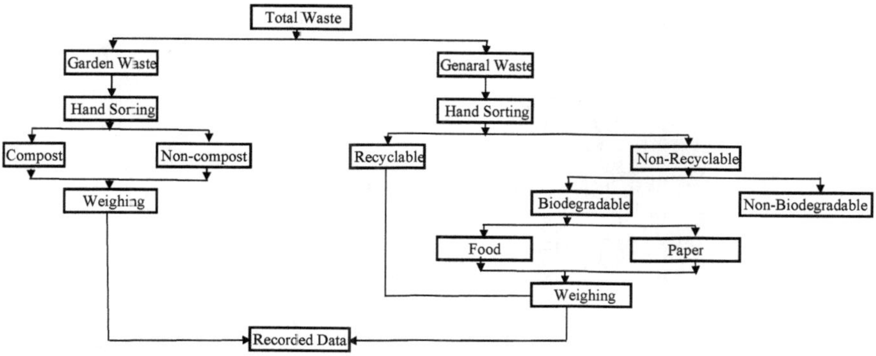

Fig. 3 Feedstock quantification flow diagram

point of generation to obtain the total amount of waste is generated. Waste characterization was done to ascertain the composition. This included physical and chemical composition with regards to C/N ratio, volatile solids, total solids, and elemental analysis for carbon, nitrogen, sulphur, and hydrogen in accordance with the standard methods (APHA 1995) (Rice et al. 2012).

2.2 *Methane Production from OFMSW*

Bioprocess control AMPTS II was used to perform BMP for OFSMW. The AMPTS II consist of a digester, CO_2 fixing unit, and gas collection unit. The setup is batch process. A 500 mL digester, with effective volume of 400 mL, was used for biogas production which had head space of 100 ml. Sodium hydroxide (NaOH), obtained from Sigma-Aldrich, South Africa, was used for CO_2 removal. A 3 M NaOH solution was prepared by mixing 240 g pure NaOH with distilled water up to 2 l. The solution was used as the scrubbing solution to absorb the impurities. A pH indicator solution was added to NaOH solution with 0.4% thymolphthalein pH indicator solution (40 mg in 9 ml ethanol 99.5% and 1 ml water). The prepared NaOH with pH indicator was used to determine the saturation point for the cleaning solution to be replaced. The substrate was prepared and fed into the digester. The digester was purged with nitrogen to remove the oxygen and create an anaerobic condition. The digester was connected to a 100 ml bottle containing 80 ml NaOH and pH indicator solution, which was used as scrubber. The gas exiting the CO_2 fixing unit was sent to the flow cell (gas collection) where the volume of biomethane is determined using the buoyancy principle. Figure 4 shows the biogas production setup.

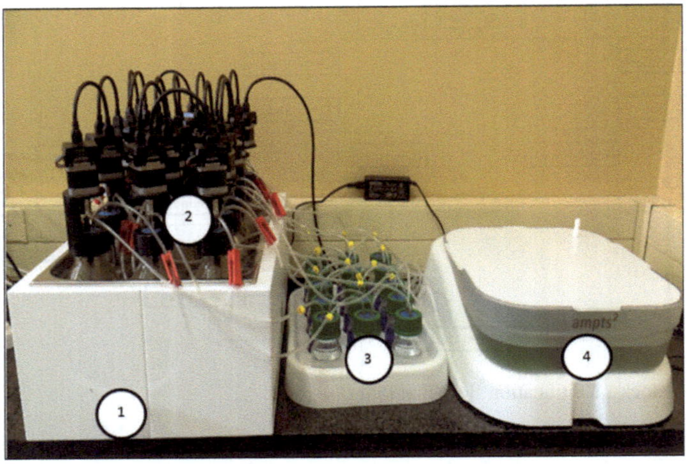

Fig. 4 Biogas production setup [biochemical methane potential (BMP) test, (*1*) thermostatic water bath, (*2*) glass bottle reactor, (*3*) CO_2-fixing unit and (*4*) gas volume measuring device]

2.3 Waste to Biogas Process Design

Using the results obtained from the feedstock analysis and literature, the appropriate size of the biogas digester was determined using standard procedure considering feedstock quality and quantity.

2.3.1 Multi-Criteria Decision Analysis

Multiple-criteria decision analysis (MCDA) technique was employed to select the most suitable biogas digester technology for OFMSW based on

- Cost of the digester
- Local availability of the digester
- Temperature regulation ability
- OFMSW suitability
- Ease of construction
- The presence of agitation accessory

The digesters investigated includes

- Complete mix—Continuous stirred tank reactor (CSTR)
- Up-flow anaerobic sludge blanket (UASB)
- Plug flow
- Covered lagoon
- Fixed film.

3 Results and Discussion

In this study, co-digestion of cow manure (CM) as inoculum and OFMSW were evaluated for the purpose of getting the biomethane potentials at optimum temperature (37 °C) and initial pH of 7.

3.1 Waste Quantification

Results showed that approximately 1.4 million ton per annum of domestic waste was generated in the City of Johannesburg, South Africa (Pikitup et al. 2015), of which from our investigation, the landfill comprised of 34% OFMSW portion made up of organic waste (Fig. 5). Of the total waste generated, 3, 1, 5, and 18% were the textile/fabric, special care waste, metals, and others general waste, respectively. Organic waste was the most abundant component of the MSW, accounting for 34%. Recyclables (plastics, glass, and paper/paperboards) was the second-largest component 19, 9 and 12% respectively.

Seasonal variation contributes to the difference in the organic waste. However general waste is influenced by the city population. In particular, seasonal variation affect the moisture content of the waste and hence the density of waste requiring disposal. Considering Fig. 5, OFMSW and compost were the main ingredients that were fed in the digesters for BMP.

Utilizing these organic wastes for energy production save disposal sites air space. In addition, there are MSW management benefits from AD which include reduction of cost of transportation and compression of waste to landfills sites. Even

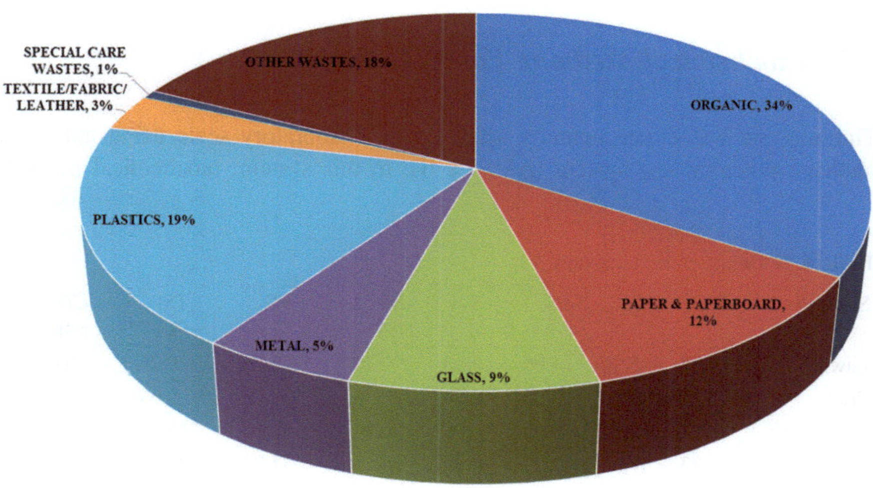

Fig. 5 Municipal solid waste quantification results, municipal landfill at City of Johannesburg

though there are countless benefits of energy production from MSW, the primary benefit of AD of MSW is twofold; to divert organic solid waste from municipal landfills and to mitigate climate change whereas energy production is only a secondary benefit which enhances the attractiveness of this technology. Hacker et al. (2010) reported that, 1 Mg of MSW is equivalent to 750 kWh power which can supply an average of 25 American household that uses an average of 30 kWh per day of electricity.

3.2 *Substrate Characterization*

Table 4 shows the substrate characterization of different organics biomass.

From damping landfill's substrates, it was observed that there was no significant difference in TS% between mixed waste and garden waste. TS% for mixed and garden waste was 27.33 and 29.26%, with moisture content of 72.67 and 70.74, respectively. The high TS of mixed MSW is due to the heterogeneous nature of the waste with elements of uncooked grains, some garden waste and other foreign bodies. VS (TS %) was relatively high, favouring anaerobic digestion, and ranged between 76.32 and 78.96%. The C/N ratio for both substrates was within the optimal range (10–30:1), indicating balanced nutrients (C/N) required by microorganisms during AD. Mixed waste had C/N ratio of 14.56 while garden had 10.1. From samples with high C/N ratio, co-digestion with substrate of low C/N ratio was recommended. From the ultimate and proximate analysis of the waste stream characterized, mono-digestion was possible as both sources were within acceptable range of parameters studied. However, for optimality and to reduce the need for high level control of process parameters, co-digestion of waste streams is recommended.

3.3 *Biochemical Methane Potential (BMP)*

The analysis was experimentally achieved in a laboratory scale using automatic methane potential test system (AMPTS II). In this system, biochemical methane

Table 4 Substrate characterization

Substrate	C	H	N	S	MC (%)	VS (%)	TS (%)	C/N ratio
Lawn	19.67	5.36	1.96	0.00	70.74	72.26	29.26	10.03
Fruits	5.17	2.90	0.95	0.00	90.30	82.78	9.69	5.44
Vegetables	5.97	2.52	0.99	0.00	86.24	87.65	13.75	6.03
Food waste	23.33	7.31	1.79	0.00	54.65	94.96	45.35	13.03
Mixed	13.25	6.25	0.91	0.00	72.67	80.69	27.33	14.56

potential (BMP) tests were performed to determine the anaerobic biodegradability and optimum methane potential of waste as well as the biodegradation rate in a laboratory scale. The ratios that were used for CM:OFMSW to perform methane analysis was 1:4, i.e., 30 g cow manure (CM) as inoculate and 200 g OFMSW with calcium hydroxide ($Ca(OH)_2$), sodium hydroxide (NaOH), and calcium carbonate ($CaCO_3$) used to regulate the pH. These ratios were chosen because of higher production of biogas when the fraction of organic fraction municipal solid waste (OFMSW) was higher than cow manure. This BMP assay was performed per day and presented as in Figs. 6 and 7.

It is observed in Fig. 7 that there was a sharp increase on day 1 to 2 and an equilibrium in the methane yield after day 2. After day 5, methane yields reduced. The yield in co-digestion was higher with $CaCO_3$ (Fig. 7a) than in mono-digestion with $Ca(OH)_2$ (Fig. 7b) as pH regulator. The results indicated a BMP average CH_4 concentration of 59.46%.

In Fig. 4.5 it is shown that CD and OFMSW under $CaCO_3$ (Fig. 7a) has higher biodegradability rate compared to ratios of CD and OFMSW under NaOH and $Ca(OH)_2$ (Fig. 7b) because calcium provides nutrients for microbial growth and minimize sharp deduction in pH. Methane production under $CaCO_3$ was approximately 4000 Nml/gVS compared with NaOH 150 Nml/gVS, co-digestion with grass 60 Nml/gVS and $Ca(OH)_2$ which was approximately 200 Nml/gVS. However, the ratio of 1:4 optimizes the efficiency for the production of methane. It has the highest methane yield which was also enhanced by the mixing for 60 s interval under retention time. The mixer speed was adjusted to 80 % to enhance homogeneity for microbial and substrates with the temperature kept constant within mesophilic range 37 °C and the possibility of overestimation was eliminated. The result indicated a BMP average CH_4 concentration of 59.49%.

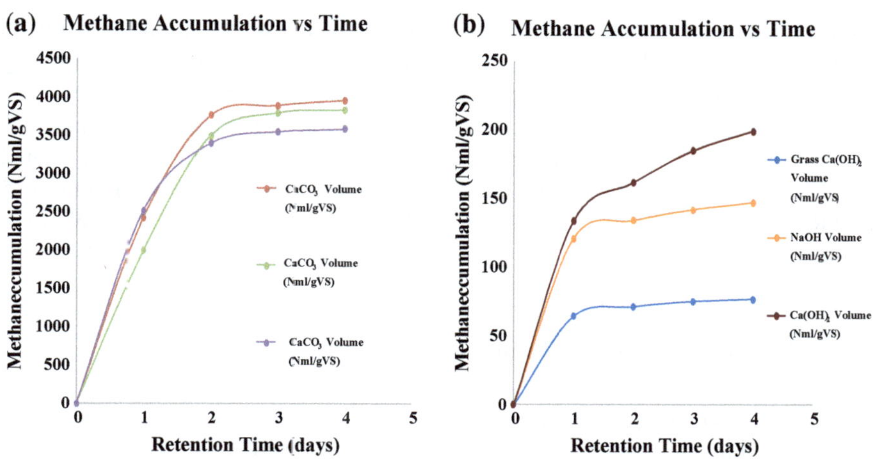

Fig. 6 Methane production from OFMSW with **a** $CaCO_3$ as pH regulator and **b** with NaOH, grass and $Ca(OH)_2$ as the pH regulator

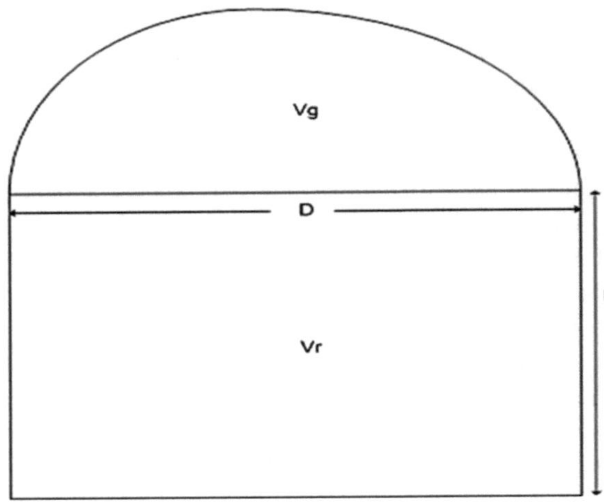

Fig. 7 Digester with single dome

3.4 Biodigester Design

Using the results obtained from the substrate analysis and literature, the appropriate size of the biogas digester was determined using standard procedure considering substrate quality and quantity.

Using MCDA techniques, a suitable biogas model was selected from a list of potential alternatives as showed in the subsequent sections. The developed list of biogas digesters alongside a summary of their attributes is presented in Table 5.

The project was fixed at OFMSW as a preselected type of feedstock. Therefore, the scalability of the plants and their suitability to handle OFMSW were taken to be the ruling factors for digester selection each having individual weighted factors of 0.2. Next in importance were the relative cost prices of the individual plants and their availabilities locally because both factors had a direct implication on the overall project cost. They weighed 0.17 and 0.18, respectively. Temperature regulation and ease of construction, operation and maintenance both weighed relatively lower at 0.1 because the technologies in consideration were relatively simple, easy to set up and therefore temperature as an operating factor can easily be regulated. The least important factor was the presence of agitation accessories weighing 0.05. CSTR scored highest with 0.785 and was selected for the design in OFMSW biogas production.

3.4.1 Design Using Waste Quantification and Characterization

Using data obtained by waste quantification and characterization, appropriate input parameters like the feedstock volumetric flowrates were ascertained that aided biodigester design using

Table 5 MCDA for biodigester selection

MCDA for biodigester selection															
Criteria	Cost		Local availability		Scalabilty		OFMSW suitability		Temperature regulation ability		Presence of agitation accessory		Ease of construction		
Weight		0.17		0.18		0.2		0.2		0.1		0.05		0.1	
Digester types	Score	Wt. score	Score	Wt. score	Score	Wt. score	Score	Wt. score	Score	Wt. score	Score	Wt. score	Score	Wt. score	Total score
1 Complete Mix-CSTR	0.65	0.11	0.80	0.14	0.85	0.17	0.80	0.16	0.80	0.08	0.90	0.05	0.75	0.08	0.78
2 UASB	0.50	0.09	0.75	0.14	0.65	0.13	0.30	0.06	0.75	0.08	0.80	0.04	0.75	0.08	0.60
3 Plug flow	0.70	0.12	0.60	0.11	1.00	0.20	0.40	0.08	0.60	0.06	0.30	0.03	0.75	0.08	0.67
4 Covered lagoon	0.80	0.14	0.80	0.14	0.40	0.08	0.50	0.10	0.50	0.05	0.30	0.02	0.80	0.08	0.61
5 Fixed film	0.65	0.11	0.70	0.13	0.40	0.08	0.60	0.12	0.70	0.07	0.75	0.04	0.75	0.08	0.62

Volume of digester

$$V_r = Q \times \text{HRT}, \tag{1}$$

where

V_r Volume of the digester
Q Volumetric flow rate
HRT Hydraulic retention time

Volume of gas holder is taken as half digester volume (V_g);

$$V_g = \frac{V_r}{2} \tag{2}$$

Total biodigester volume (V_d);

$$V_d = V_r + V_g \tag{3}$$

The digester cylindrical tank of volume (V_r);

$$V_r = \frac{\pi D^2 H}{4} \tag{4}$$

where

V_r Digester volume
D Diameter of the tank
H Height of the digester.

Assume height of the digester equal diameter of digester;

$$D \sim H \tag{5}$$

$$V_r = \frac{\pi D^3}{4} \tag{6}$$

Thus the diameter of the digester (D);

$$D = \sqrt[3]{\frac{4V_r}{\pi}} \tag{7}$$

Taking the gas holder/digester radial clearance to be 20 mm, gives a diameter (d) of the gas holder as;

$$d = (D - 0.04) = \left(\sqrt[3]{\frac{4V_r}{\pi}}\right) - 0.04 \tag{8}$$

Table 6 Organic waste in tonnage

Sites	Ton/day	Organic fraction	Ton of organic/day
Site A	817.79	0.34	277.88
Site B	23.71	0.14	3.43
Site C	49.9	0.93	46.4
Total	891.4		327.71

Table 7 Design 2 results from Waste Quantification (volume)

Q (m^3/day)	HRT (days)	V_r (m^3)	D (m)	H (m)	d (m)	h (m)	V_g m^3	V_d m^3
327.70	7.00	2293.90	14.30	14.30	14.26	7.19	1146.95	3440.85

Given a gas holder volume (V_g), the height (h) of the gas holder is therefore given by;

$$h = \frac{4V_g}{\pi d^2} = \frac{4V_g}{\pi} * \left(\left(\sqrt[3]{\frac{4V_r}{\pi}} \right) - 0.04 \right)^{-2} \tag{9}$$

A breakdown of the organic results is as shown in Table 6 from the damping sites and results from the design Table 7.

where

V_1 Volume of the digester top dome
f Height of dome
r Radius of the digester
V_2 Volume of the digester cylinder
H Height of digester
S_1 Surface area of the digester top dome
p Radius of the dome
S_2 Surface area of the digester cylinder body
D Diameter of the digester
S_3 Surface area of the digester bottom dome.

4 Conclusion

Co-digestion of substrate in 500 ml digester under mesophilic temperature had good productivity of methane of 59.46% with (54–62%) recommended in the literature. The rate of conversion increases as temperature increases and conversion of reactants increases as reaction time proceeds, until an equilibrium state was reached. The design of the biodigester in determining the correct dimensions and geometry of the biodigester parameters was comparable with the anaerobic digestion process.

Acknowledgements The authors are grateful to the City of Johannesburg, Pikitup, Robinson landfill, Johannesburg Market, Thabo Maahlatsi, Mlawule Mashego, Ms. Ireen Maile, and Renewable energy team members through the Process Energy and Environmental Technology Station (PEETS) at the University of Johannesburg for supporting this research.

References

[Online]. Available: http://www.e-inst.com/bimass-to-biogas. Accessed 15 Jan 2016.

Al Seadi, T., et al. (2008). *Biogas handbook.* University of Southern Denmark Esbjerg. ISBN: 978-87-992962-0-0.

Allan, H. (1998). *Grass productivity* (pp. 56–89). Washington DC: Island Press Conservation Classics Series.

Angelidaki, I., et al. (1996). *The biogas process. Lecture notes for energy from biomass,* (6362).

Babaee, A., & Shayegan, J. (2011). Effects of organic loading rates (OLR) on production of methane from anaerobic digestion of vegetable waste. In *Proceedings of the World Renewable Energy Congress*, Linköping, Sweden.

Bardiya, N., & Gaur, A. C. (1997). Effects of carbon and nitrogen ratio on rice straw biomethanation. *Journal of Rural Energy, 4*(1–4), 1–16.

Barron, F. H., & Barrett, B. E. (1996). The efficacy of SMARTER—Simple multi-attribute rating technique extended to ranking. *Acta Psychologica, 93*(1), 23–36.

Belton, V. (1986). A comparison of the analytic hierarchy process and a simple multi-attribute value function. *European Journal of Operational Research, 26*(1), 7–21.

Chai, J., Liu, J. N. K., & Ngai, E. W. T. (2013). Application of decision-making techniques in supplier selection: A systematic review of literature. *Expert Systems with Applications, 40*(10), 3872–3885.

Dangoggo, S., Aliyu, M., & Atiku, A. (1996). The effect of seeding with bacteria on biogas production rate. *Renewable Energy, 9*(1), 1045–1048.

Dodgson, J. S., et al. (2009). *Multi-criteria analysis: A manual.* London: Department for Communities and Local Government.

E.I.A. (2013). *International Energy Outlook, Annual Energy US, Energy Information Administration,* Washington, DC.

Kumar, S. (2012). *BIOGAS.* Rijeka: Intech Books.

Kuria, J., & Maringa, M. (2008). Developing simple procedures for selecting, sizing, scheduling of materials and costing of small bio-gas units. *International Journal for Service Learning in Engineering, Humanitarian Engineering and Social Entrepreneurship, 3*(1).

Latinwo, G. K., & Agarry, S. E. (2015). Modelling the kinetics of biogas production from mesophilic anaerobic co-digestion of cow dung with plantain peels. *International Journal of Renewable Energy Development (IJRED), 4*(1), 55–63.

Leake, D. B. (1996). *Case-Based Reasoning: Experiences, lessons and future directions.* Cambridge: MIT Press.

Lfu. (2007). *Biogashandbuch Bayern, Materialienband.* Augsburg. Online unter: http://www.lfu.bayern.de/abfall/biogashandbuch (letzter Zugriff am 06.12. 2012).

Matheri, A. N., Belaid, M., Tumisang, S., & Ngila, C. J. (2015a). The kinetic of biogas rate from cow dung and grass clippings. *7th IIENG international conference of latest trends in engineering and technology (ICLTET'2015),* Pretoria, South Africa, Nov 2015.

Matheri, A. N., Belaid, M., Tumisang, S., & Ngila, C. J. (2015b). Optimising biogas production from anaerobic co-digestion of chicken manure and organic fraction of municipal solid waste. *Renewable and Sustainable Energy Review Journal* (Submitted).

Matheri, A. N., Belaid, M., Tumisang, S., & Ngila, C. J. (2015c). Analysis of the biogas productivity from dry fermentation of organic fraction of municipal solid waste. *Renewable and Sustainable Energy Review Journal* (Submitted).

Matheri, A. N., Belaid, M., Tumisang, S., & Ngila., C. J. (2016). Role of impact of trace elements on anaerobic co-digestion in biogas production. *24th World Congress on Engineering (WCE 2016)*, London-UK, 29 June 2016.

McDougall, F. R., et al. (2008). *Integrated solid waste management: A life cycle inventory*. Hoboken: Wiley.

Nordberg, U. (2006). *Biogas–Nuläge och framtida potential* (pp. 1653–1248). Värmeforsk.

Pikitup, C.o.J., http://joburg.org.za. Accessed 2015.

Pohekar, S. D., & Ramachandran, M. (2004). Application of multi-criteria decision making to sustainable energy planning—a review. *Renewable and Sustainable Energy Reviews, 8*(4), 365–381.

Rapport, J., et al. (2008). *Current anaerobic digestion technologies used for treatment of municipal organic solid waste*. University of California, Davis, Contractor Report to the California Integrated Waste Management Board.

Rice, E. W., Bridgewater, L., & Association, A. P. H. (2012). *Standard methods for the examination of water and wastewater*. Washington, DC: American Public Health Association.

Saaty, T. L. (2004). Decision making—the analytic hierarchy and network processes (AHP/ANP). *Journal of Systems Science and Systems Engineering, 13*(1), 1–35.

Sambo, A. S., Garba, B., & Danshehu, B. G. (1995). Effect of some operating parameters on biogas production rate. *Renewable Energy, 6*(3), 343–344.

Schnurer, A., & Jarvis, A. (2010). Microbiological handbook for biogas plants. *Swedish Waste Management U, 2009*, 1–74.

Song, Y. C., Kwon, S. J., & Woo, J. H. (2004). Mesophilic and thermophilic temperature co-phase anaerobic digestion compared with single-stage mesophilic-and thermophilic digestion of sewage sludge. *Water Research, 38*(7), 1653–1662.

Sundararajan, R., Jayanthi, S., & Elango, R. (1997). Anaerobic digestion of organic fractions of municipal solid waste and domestic sewage of Coimbatore. *Indian Journal of Environmental Health, 39*(3), 193–196.

White, P., Dranke, M., & Hindle, P. (2012). *Integrated solid waste management: A lifecycle inventory*. New York: Springer.

Part II
Environment

Zymomonas mobilis—Towards Bacterial Biofuel

Kim G. Clarke, Thapelo Mokomele, Linda H. Callanan and Jeanne Groenewald

Abstract Bioconversion of lignocellulosic hydrolysates to ethanol is a promising solution to energy from renewable non-food sources. While utilisation of all sugars in these hydrolysates by *Zymomonas mobilis* has been facilitated through genetic modification of this organism, and molecular biology and fermenting capabilities of xylose-utilising *Z. mobilis* recombinants have been extensively documented, there is considerably less information on fundamental process optimisation and intensification studies to maximise these organisms' potential so that the advances in molecular biology can be fully realised. In this study, process optimisation using the *Z. mobilis* 8b recombinant has been conducted by optimising both the process conditions and the process strategy, and process intensification has been examined by coupling a membrane filtration unit to a continuous process. Optimum process conditions on glucose–xylose substrates were quantified as 33.5 °C and pH 6.5 using multiple response optimisation. Under these optimum operating conditions, *Z. mobilis* 8b exhibited both a superior ethanol yield (94%) and ethanol productivity (1.43 gL^{-1} h^{-1}) during batch culture on 100 gL^{-1} sugar (glucose = xylose) when compared with *Saccharomyces cerevisiae* strains. Continuous operation of the *Z. mobilis* 8b culture under the optimum conditions enhanced ethanol productivity fivefold to 6.74 gL^{-1} h^{-1}, the highest yet reported. Importantly, significant enhancement in ethanol productivity (threefold) was also achieved during continuous operation at dilution rates below 0.15 h^{-1}, whilst maintaining the ethanol concentration above the threshold for cost-effective distillation. Cell separation and recycling was facilitated via a membrane unit for process intensification.

Keywords Biofuel · *Zymomonas mobilis* · Optimum process conditions · Process kinetics

---

K.G. Clarke (✉) · T. Mokomele · L.H. Callanan
DST-NRF Centre of Excellence in Catalysis (c*change), Rondebosch, South Africa
e-mail: kclarke@sun.ac.za

K.G. Clarke · T. Mokomele · L.H. Callanan · J. Groenewald
Department of Process Engineering, Stellenbosch University, Stellenbosch, South Africa

W. Leal Filho and D. Surroop (eds.), *The Nexus: Energy, Environment and Climate Change*, Green Energy and Technology, https://doi.org/10.1007/978-3-319-63612-2_13

1 Introduction

Concerns of environmental security, economic stability and global warming have escalated the realisation of a need for a fuel source other than the finite supply of fossil fuels (Hahn-Hägerdal et al. 2007; Chander et al. 2011; Menon and Rao 2012; Van Rensberg et al. 2013). A most promising alternative fuel is ethanol from lignocellulosic hydrolysates, and considerable effort has been directed towards developing the commercial potential of the biological production of ethanol from hydrolysate sugars (Lynd et al. 2008; Lau and Dale 2009).

Generally, the biocatalyst of choice for conversion of lignocellulosic hydrolysates has been the yeast *Saccharomyces cerevisiae*. Yet *Z. mobilis* has been reported to exhibit excellent ethanol production kinetics with respect to ethanol productivity, ethanol tolerance and sugar uptake, thereby suggesting the significant promise of this bacterial sp. as a biocatalyst for renewable energy production (Karsch et al. 1983). However, wild-type strains of *Z. mobilis* are incapable of metabolising pentose sugars to ethanol and consequently are limited in their use as biocatalysts mediating conversion of these hydrolysates which contain both pentose and hexose sugars (mainly glucose and xylose) (Swings and De Lay 1977; Zhang et al. 1995; Gunasekaran and Chandra 1999).

More recently, xylose-fermenting *Z. mobilis* strains capable of utilising both five and six carbon sugars have been engineered and evaluated for their ethanol production capabilities (Zhang et al. 1995; Rogers et al. 1999; Rogers and Joachimsthal 2000). However, while the molecular biology and fermenting capabilities of xylose-utilising *Z. mobilis* recombinants have been extensively documented, there is considerably less information on fundamental process optimisation and intensification studies to maximise the these organisms' potential so that the advances in molecular biology can be fully realised.

It is in the context of process development of xylose-fermenting *Z. mobilis* that this paper provides value. Here, a combination of statistically optimised process conditions and a systematically optimised process strategy which support the highest kinetic parameters yet are reported, moreover with ethanol yields sufficient to support cost-effective downstream purification by distillation. Potential further kinetic enhancement through process intensification has been demonstrated using a membrane filtration unit coupled to a continuous system to facilitate cell retention.

2 Materials and Methods

2.1 Microorganism and Media

Z. mobilis ATCC 31821 (ZM4) and *Z. mobilis* ATCC-PTA-6976 (8b) were used in this study. The wild-type *Z. mobilis* ZM4 is able to utilise hexoses only. The recombinant *Z. mobilis* 8b contains xylose assimilation (xylose isomerase, xylukinase) and pentose metabolism (transketolase, transaldolase) genes from *Escherichia coli* and utilises both pentoses and hexoses (Zhang et al. 2007). *Z. mobilis* 8b was obtained from the ATCC with kind permission from Dr. Min Zhang, National Renewable Energy Laboratory, Colorado. These organisms were stored in cryovials containing 25 gL^{-1} glucose, 25 gL^{-1} xylose, 5 gL^{-1} yeast extract and 20% (v/v) glycerol at −70 °C.

Liquid media for the first stage inoculum consisted of 25 gL^{-1} glucose and 25 gL^{-1} xylose for *Z. mobilis* 8b (or 50 gL^{-1} glucose for *Z. mobilis* ZM4), 10 gL^{-1} yeast extract, 2 gL^{-1} KH_2PO_4, 1 gL^{-1} $(NH_4)_2SO_4$, and 0.5 gL^{-1} $(MgSO_4)\cdot 7H_2O$ (Rogers et al. 2000). A phosphate buffer (capacity 100 mM) was provided in the second stage inoculum and fermentation test culture media by replacing the 2 gL^{-1} KH_2PO_4 with 3.3 gL^{-1} KH_2PO_4 and 11 gL^{-1} K_2HPO_4.

2.2 *Experimental Procedure*

A two-stage inoculum protocol was used to maximise reproducibility. The first stage inoculum was incubated in a shaker-incubator (Labcon) for 14–16 h (30 °C, 200 rpm) and the second stage until an optical density (OD) of approximately 1.5 (6–10 h) under the same conditions. All transfers were conducted at 10% (v/v).

Temperature and pH_0 (where pH_0 was the initial pH value) optimisation experiments were performed in 1 L Erlenmeyer flasks in buffered media with a working volume of 300 mL and incubated in a shaker-incubator (Labcon) at 150 rpm under controlled temperature conditions (as indicated in the text).

Kinetic evaluation at optimum pH and temperature during batch and continuous operation was conducted in a custom-built instrumented 1.5 L bench-top bioreactor with 500 (continuous) and 750 mL (batch) working volumes. The pH in the bioreactor was controlled through the addition of 4 M KOH using an automated peristaltic pump (Watson Marlow) connected to a pH controller unit (Alpha pH 1901). A PID pH controller operating with a feedback control mechanism was used to maintain the culture pH at the desired set-point. (Strictly, the control pH would be marginally lower than the pH_0 so a pH control value of 0.25 less than the optimum pH_0 used as the set point in the bioreactor cultures.) The temperature was controlled by circulating water from an external temperature-controlled water bath in a glass jacket around the reactor.

Continuous culture with cell retention was facilitated by the separation and recycling of cells via a membrane recycle unit. A bespoke autoclavable borosilicate glass shell (GlassChem, South Africa) was designed to accommodate a tubular ceramic membrane comprising a permeable layer of mean pore size 0.2 µm (Atech Innovations GmbH, Germany). The pressures at the feed and retentate side of the membrane were indicated by autoclavable low-pressure range SS needle gauges (SpiraxSarco, South Africa). When not in use, the membrane was stored in 0.0035% (m/v) sodium hypochlorite.

2.3 Analytical Techniques

Cell growth for all the experimental work was determined as OD using a spectrophotometer (Varian UV-Vis Superscan 3) at 660 nm. OD was standardised against cell dry weight (CDW). CDW was determined from the mass of filtered cells in a known volume (0.22 μm filter paper, 60 °C for 24 h). For OD < 0.8, CDW/OD = 0.3289 (R^2 = 0.9957).

Residual glucose and xylose, and ethanol produced, were determined from filtered supernatants using high-performance liquid chromatography (HPLC) (Water Breeze System, Milford, MA) fitted with a Bio-Rad Aminex HPX-87H ion exchange column with 5 mM H_2SO_4 (338 K, 0.6 mL/min) as the mobile phase. Peak detection was conducted using a Waters 2141 detector (Milford, MA). Sugar and ethanol concentrations were quantified using standard curves with known concentrations, repeated periodically to authenticate the calibration accuracy. Samples for HPLC analysis were prepared by adding 35% (v/v) perchloric acid (PCA) to a suitably diluted supernatant, cooling to 4 °C, then precipitating the PCA with 7 N KOH overnight at 4 °C to remove salts, finally centrifuging (Eppendorf Minispin, 14,000 rpm for 5 min) and filtering (0.22 μm syringe filters, Anatech) to remove the precipitate.

2.4 Experimental Design

The optimisation of process temperature and pH was performed in shake flasks using response surface methodology. A 3-level 2-factor (3^2) full factorial design was used to test the input variables at factorial and central points (Lawson and Erjavec 2000). A regression model predicted each of the three response variables (percentage ethanol yield, ethanol volumetric productivity and xylose utilisation rate) as a function of the two input variables (temperature and pH_0). Statistica™ (StatSoft Inc. 2002) software was applied for the design of experiments and regression analysis. Reproducibility was evaluated by comparing the variance in response variables from the mean of triplicate experiments, and expressed as a percentage. Percentage reproducibility was 95–100% in all cases.

3 Results and Discussion

3.1 Benchmarking of *Z.* mobilis *8b Against* Z. mobilis *ZM4*

The growth kinetics of *Z. mobilis* ZM4 and *Z. mobilis* 8b (Fig. 1) were quantified initially on 50 gL^{-1} glucose in duplicate shake flask cultures at 34 °C and initial pH

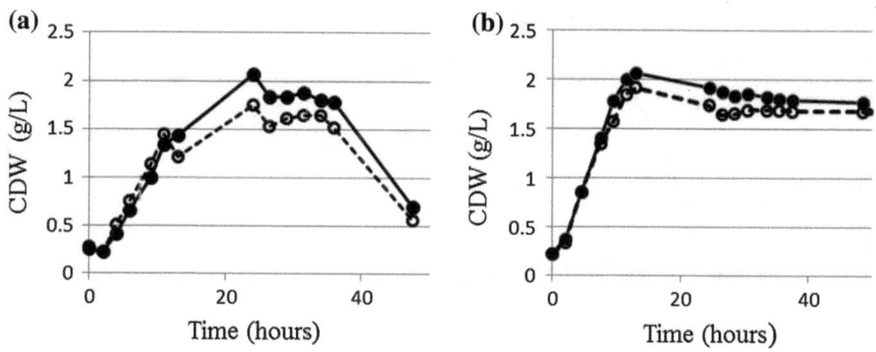

Fig. 1 Cell growth of *Z. mobilis* ZM4 (**a**) and *Z. mobilis* 8b (**b**) during shake flask culture at 34 °C and pH 6 on 50 gL^{-1} glucose (duplicate cultures showing reproducibility)

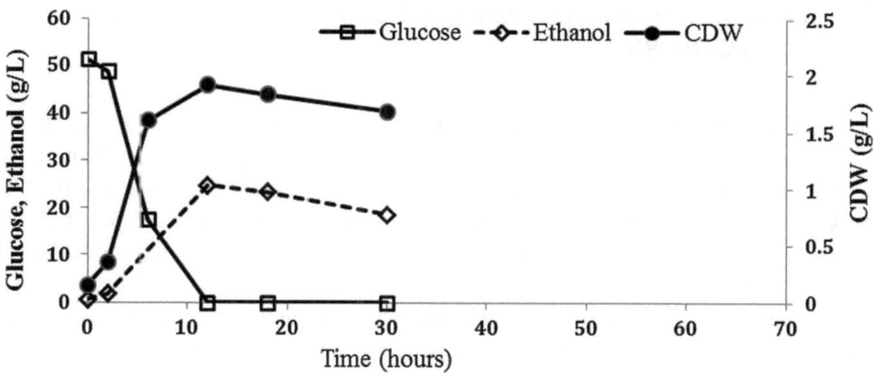

Fig. 2 Kinetics of *Z. mobilis* ZM4 during shake flask culture at 35 °C and pH 6 on 50 gL^{-1} glucose

6. Both strains exhibited a negligible lag phase, reaching a maximum cell concentration of approximately 2 gL^{-1} at 20 h. Consequently, similar maximum specific growth rates were obtained, namely 0.23 (ZM4) and 0.21 h^{-1} (8b). Co-cultures using both strains exhibited similar growth curves with a maximum specific growth rate of 0.21 h^{-1} (data not shown).

The sugar utilisation and ethanol production kinetics of these two strains were then individually quantified under similar conditions. *Z. mobilis* ZM4 was cultured on 50 gL^{-1} glucose (Fig. 2) and *Z. mobilis* 8b on a xylose-rich substrate (37 gL^{-1} xylose plus 13 gL^{-1} glucose), a ratio which is typical of the sugars in the hydrolysate from the hemicellulose fraction (Fig. 3). *Z. mobilis* ZM4 and *Z. mobilis* 8b were both efficient sugar utilisers with substrate exhaustion determined at about 15 h. In the case of *Z. mobilis* 8b, simultaneous glucose and xylose utilisation was observed with no evidence of diauxic growth. The final ethanol concentration for *Z. mobilis* 8b was similar to that obtained for *Z. mobilis* ZM4.

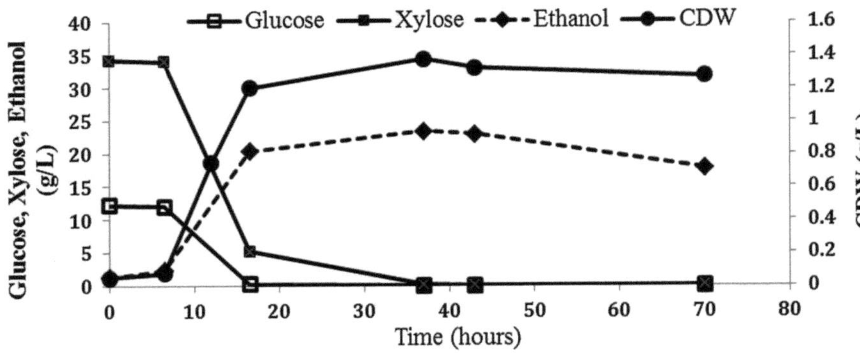

Fig. 3 Kinetics of *Z. mobilis* 8b during shake flask culture at 35 °C and pH 6 on a xylose-rich substrate (37 gL^{-1} xylose, 13 gL^{-1} glucose)

3.2 Determination of Optimum pH and Temperature for Z. mobilis 8b

To determine the pH and temperature optima, shake flask cultures of *Z. mobilis* 8b were conducted on 13 gL^{-1} glucose and 37 gL^{-1} xylose at different pH and temperature combinations between the ranges of 6–8 and 30–37 °C, respectively. The model predictions exhibited strong curvature in both temperature and pH (Fig. 4), confirming that optimum temperature and pH were critical to the maximisation of ethanol production.

The degree of significance of each model term within a 95% confidence interval was quantified by analysing the *p* value using ANOVA. Model terms with a significant contribution were quantified by a *p* value ≤ 0.05. Statistical analysis with respect to ethanol yield showed that the linear and quadratic effects of pH_0 and temperature were statistically significant within a 95% confidence interval with the temperature-pH_0 interaction statistically insignificant. The linear effect of pH_0 had the largest effect, implying that strict pH control is necessary for maximising the ethanol yield. The effect of temperature on ethanol yield was smaller, although still significant, implying that temperature control was also necessary to maximise the yield. Moreover, a large quadratic effect of pH_0 signified that the ethanol yield response function had a high degree of curvature due to a change in pH_0. The significance of the effects of pH_0, temperature and the high degree of curvature (evident by the sharp decrease in ethanol yield as the pH_0 was increased beyond 7) is apparent in the surface and contour plots of the model prediction of ethanol yield (Fig. 4).

A similar approach was taken in analysing the ethanol productivity and xylose utilisation rate response functions, both of which showed linear and quadratic effects of pH_0 and temperature with no temperature-pH_0 interaction. The ethanol volumetric productivity and xylose utilisation surface and contour plots (Fig. 4) exhibit strong curvature in the model prediction of temperature as well as that of

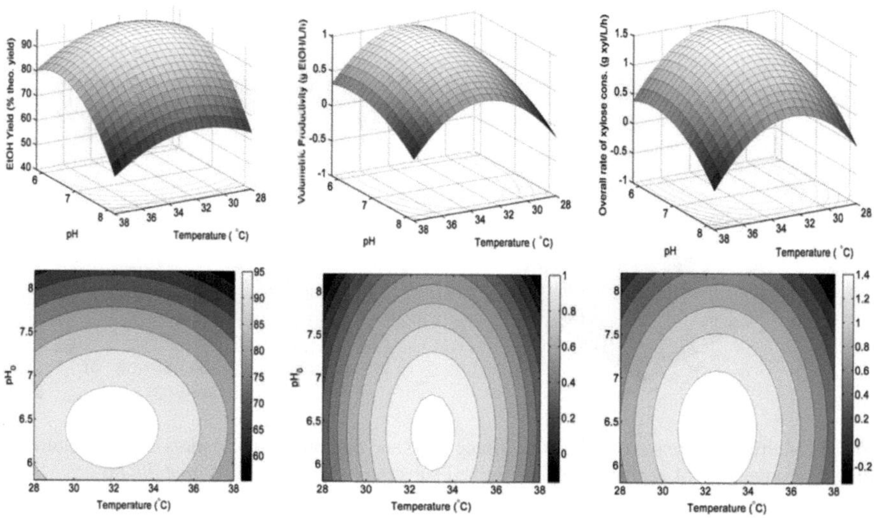

Fig. 4 The surface (*top*) and contour (*bottom*) plots of the response functions for ethanol % yield, ethanol volumetric productivity and xylose utilisation rate

pH_0. These results confirm temperature and pH control at optimum values are key to maximisation of fermentation potential.

A single temperature and pH_0 combination that resulted in the highest overall ethanol yield, ethanol volumetric productivity and xylose utilisation rate was obtained using a multiple response optimization approache (Grahovac et al. 2012). Since ethanol is a low-value high-volume product, the desirability was constrained such that the ethanol volumetric productivity was optimised first followed by the ethanol yield and then the xylose consumption rate. An additional constraint on the ethanol yield was imposed in that the product yield should have a minimum value of 90%.

The resultant surface and contour plots defining the optimisation of the temperature and pH_0 by the desirability function approach are presented in Fig. 5. The optimisation shows high overall desirability function values in the temperature region of 32–35 °C over a range of pH_0 from 6 to 7, defining the optimum conditions as 33.5 °C and pH_0 6.5, corresponding to the highest desirability function of 0.999. At these optimum conditions, the individual responses were an ethanol yield of 94%, ethanol productivity of 1.02 gL^{-1} h^{-1} and a xylose consumption rate of 1.486 gL^{-1} h^{-1}.

These results confirm that the temperature and pH for optimum process kinetics of recombinant *Z. mobilis* 8b on mixed glucose–xylose substrates are significantly higher than the 30 °C (Rogers et al. 1981) and pH of 5–5.5 (Swings and De Ley 1977) reported for bioconversion of glucose by the *Z. mobilis* wild-type and the 31.5 °C and pH 5.85 reported for the bioconversion of mixed sugar substrates during co-culture of the recombinants *Z. mobilis* ATCC 39676 (pZB4L) and *Z. mobilis*

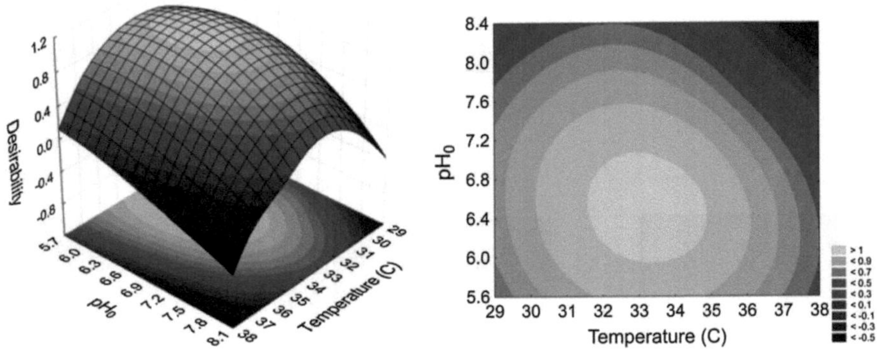

Fig. 5 Multiple response optimization surface (*top*) and contour (*bottom*) plots of temperature and pH_0 for conversion of a glucose–xylose mixed substrate to ethanol by *Z. mobilis* 8b

ATCC 39676 (pZB206) (Mohagheghi et al. 1998). Moreover, the higher optimum pH of the recombinant *Z. mobilis* 8b on a mixed glucose–xylose substrate potentially has additional advantages in that the higher optimum temperature would likely benefit the process performance through increased rates and the higher optimum pH could minimise the toxicity of some inhibitory compounds inherent in lignocellulosic hydrolysates (such as acetic acid). For example, at low pH, acetate is in its undissociated form (pKa = 4.75) and thus easily diffuses into the cell, lowering the intracellular pH and inhibiting cellular activity. At a higher pH, the extent of acetate on the intracellular pH and cell activity would be lessened.

3.3 *Z.* **mobilis** *8b Process Enhancement Through Optimised Process Conditions*

Once the optimum pH and temperature for *Z. mobilis* 8b had been established, a batch fermentation on 100 gL^{-1} sugars, with equal amounts of xylose and glucose, was conducted with *Z. mobilis* 8b in a fully instrumented bioreactor at the optimised pH and temperature conditions (Fig. 6). At the start of the fermentation, glucose and xylose were consumed simultaneously with glucose being consumed more rapidly. During the first 12 h, glucose was completely consumed, as was 20% of the xylose. Xylose was completely consumed by 37 h, by which time the ethanol concentration had reached its maximum of 48 gL^{-1}, representing an ethanol yield of 94% of theoretical and ethanol volumetric productivity of 1.43 $gL^{-1}\ h^{-1}$.

The batch culture kinetic parameters of *Z. mobilis* 8b were then compared to those of other recombinant strains of *Z. mobilis* and *S. cerevisiae* on glucose–xylose substrates with the same or similar ratios and concentrations (Table 1). To eliminate bias in the ethanol productivity (owing to inconsistencies in the actual sugar

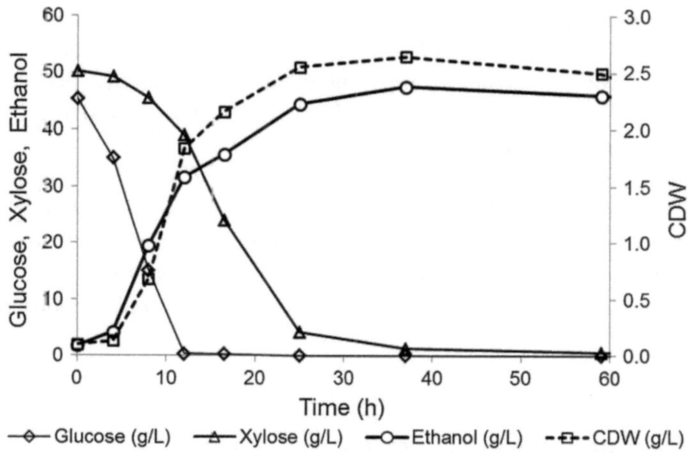

Fig. 6 *Z. mobilis* 8b batch culture on 50 gL^{-1} glucose and 50 gL^{-1} xylose controlled at optimised temperature and pH in an instrumented bioreactor

Table 1 Comparison of the kinetic parameters with literature during bioreactor culture with 45–62 gL^{-1} glucose and 50–56 gL^{-1} xylose

Strain	Cell yield (g cells g substrate^{-1})	Yield as a % theoretical (%)	Ethanol productivity (g ethanol L^{-1} h^{-1})	References
Z. mobilis ZM4 (pZB5)	0.03	94	1.60	Mohagheghi et al. (1998)
Z. mobilis 8b	0.03	94	1.43	This Work
S. cerevisiae (pLHN33)	0.08	78	1.22	Rogers et al. (1981)
Z. mobilis CP4 (pZB5)	0.02	90	0.97	Mohagheghi et al. (1998)
S. cerevisiae TMB3400[a]	0.07	67	0.52	Swings and De Ley (1977)
S. cerevisiae BH42[a]	0.04	71	0.30	Swings and De Ley (1977)
S. cerevisiae A4[a]	0.06	69	0.29	Swings and De Ley (1997)

[a]Industrial yeast strains

concentrations at initial conditions), the ethanol productivity per mass substrate (at initial conditions) was included in the analysis.

Z. mobilis 8b and *Z. mobilis* ZM4 (pZB5) demonstrated the highest ethanol yields (both 94%) which compared very favourably with the 69–78% exhibited by the yeast strains. The comparatively low cell yield (0.03) relative to those of the yeast strains (0.04–0.08) supports the preferred route of the substrate to ethanol rather than to cells in the *Z. mobilis* cultures. Further, the ethanol productivities of

Z. mobilis 8b and ZM4 (pZB5) ($\geq$ 1.43 gL^{-1} h^{-1}) were significantly higher than all other organisms investigated (0.29–1.22 gL^{-1} h^{-1}), even *S. cerevisiae* strains used in industrial processes. When comparing ethanol productivity per mass initial substrate, the *Z. mobilis* 8b and ZM4 (pZB5) strains (0.015–0.016 L^{-1} h^{-1}) were again superior to that of the other strains (0.004–0.011 L^{-1} h^{-1}).

With the maximum ethanol productivity and yield, *Z. mobilis* 8b and *Z. mobilis* ZM4 (pZB5) were not only the fastest but also the most selective ethanologic biocatalysts from this comparative study. While this is not necessarily an exhaustive comparison, it does serve to highlight the superior potential of *Z. mobilis* 8b over other pentose–hexose utilising biocatalysts when cultured under optimum process conditions.

3.4 Z. mobilis *8b Process Enhancement Through Optimised Process Strategy*

Since ethanol is a high-volume low-value product, high volumetric productivities are essential for the economic feasibility of the overall process, suggesting continuous culture as a preferential operational strategy. To this end, a *Z. mobilis* 8b continuous fermentation on 100 gL^{-1} substrate, with equal amounts of xylose and glucose, was conducted at the optimised pH and temperature conditions as determined in a fully instrumented bioreactor at several discrete dilution rates (D) under steady state conditions.

Continuous operation was initiated from batch operation by feeding sterile nutrient medium and withdrawing culture at an equal and constant rate to facilitate operation at a steady dilution rate. Steady state was established by examining CDW over a number (5–10) of volume changes at each of four discrete dilution rates until multiple stable CDW values were attained. Kinetic parameters relating to substrate, product and cell concentrations were then quantified at each dilution rate under steady state conditions (Fig. 7).

Characteristically, the CDW decreased with increasing dilution rates, but also decreased from D = 0.11 to 0.04 h^{-1}, suggesting significant diversion of the substrate for maintenance purposes at these low dilution rates. The residual xylose concentration increased from 1.6 to 12.6 gL^{-1} with an increase in the dilution rate over the range D = 0.04–0.2 h^{-1}, while residual glucose was negligible even at the highest dilution rate, strongly suggesting preferential glucose utilisation.

While ethanol concentration decreased slightly with increasing dilution rate, ethanol productivity increased such that it had reached 6.74 gL^{-1} h^{-1} at the highest dilution rate examined, already fivefold greater than that obtained in batch culture (1.43 gL^{-1} h^{-1}).

However, the selection of the optimum dilution rate is not only determined by that which relates to the maximum ethanol productivity, but also by the dilution rate at which, in addition, a minimum threshold ethanol concentration is achieved, since

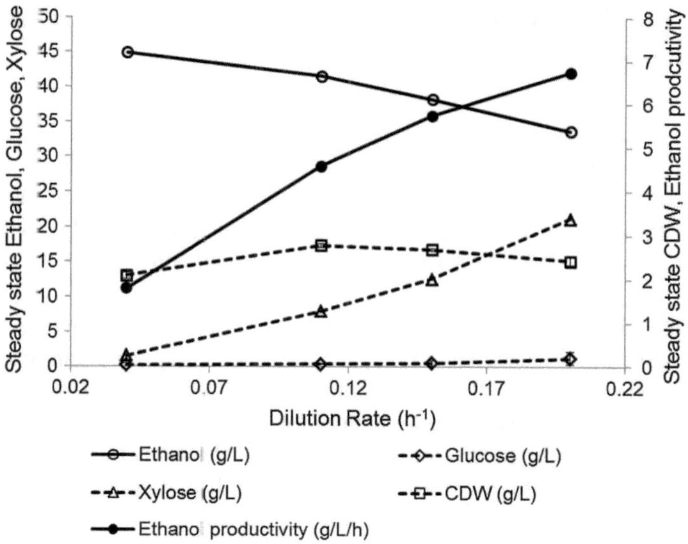

Fig. 7 *Z. mobilis* 8b continuous culture on 50 gL^{-1} glucose and 50 gL^{-1} xylose controlled at optimised temperature and pH in an instrumented bioreactor

it is the ethanol concentration which determines the energy input to distillation in the overall ethanol production process. A minimum ethanol concentration of 40 gL^{-1} has been reported as mandatory for minimising energy and economic requirements for ethanol recovery by distillation (Zacchi and Alelsson 1989). This suggests the optimum dilution rate lies between $D = 0.11$ and $0.15\ h^{-1}$, where the ethanol concentration ranged from 42 to 38 gL^{-1}. The corresponding ethanol productivities were 4.58–5.75 $gL^{-1}\ h^{-1}$, respectively, which are still threefold to fourfold greater than that obtained in batch culture.

Continuous culture kinetics of another recombinant *Z. mobilis* (strain 39676: pZB4L) on a glucose–xylose substrate (8 gL^{-1} glucose, 40 gL^{-1} xylose) has been previously reported at dilution rates from 0.04 to 0.1 h^{-1} (Lawford et al. 1998). Under conditions of temperature and pH controlled at 30 °C and 5.5, respectively, the maximum ethanol productivity attained was 2.1 $gL^{-1}\ h^{-1}$ at $D = 0.1\ h^{-1}$, less than half that obtained with *Z. mobilis* 8b (4.58 $gL^{-1}\ h^{-1}$) in this work at the same dilution rate. Culture of the recombinant *Z. mobilis* ZM4 (pZB5) on a glucose–xylose substrate (50 gL^{-1} glucose, 50 gL^{-1} xylose) incorporated cell recycle into the continuous system (Joachimsthal and Rogers 2000) where the maximum ethanol productivity (5 $gL^{-1}\ h^{-1}$) attained was comparable with that of *Z. mobilis* 8b obtained without recycle, but which could not be maintained for longer than 50 h due to loss of cell viability.

3.5 *Z. mobilis 8b Process Enhancement Through Process Intensification*

The continuous culture system was modified to incorporate a membrane filtration unit, comprising a ceramic membrane, to separate the cells from the stream exiting the bioreactor, thereby facilitating the return of the separated cells to the culture in the bioreactor (Fig. 8). The active layer of the ceramic membrane had a 0.2 μm cut off and a porosity of 0.3 and was situated on the lumen side of the membrane (Fig. 9).

Although glucose and xylose were utilised simultaneously, glucose was metabolised preferentially such that, at $D = 0.11–0.15\ h^{-1}$, approximately 8–13 gL^{-1} residual xylose remained in the continuous system without cell retention. Initial results suggest that the cell retention brought about by the coupling of the fermentation and membrane filtration units enhanced the capacity of the process for xylose utilisation. Kinetic studies of *Z. mobilis* 8b during continuous culture with recycling remains an ongoing focus of this research.

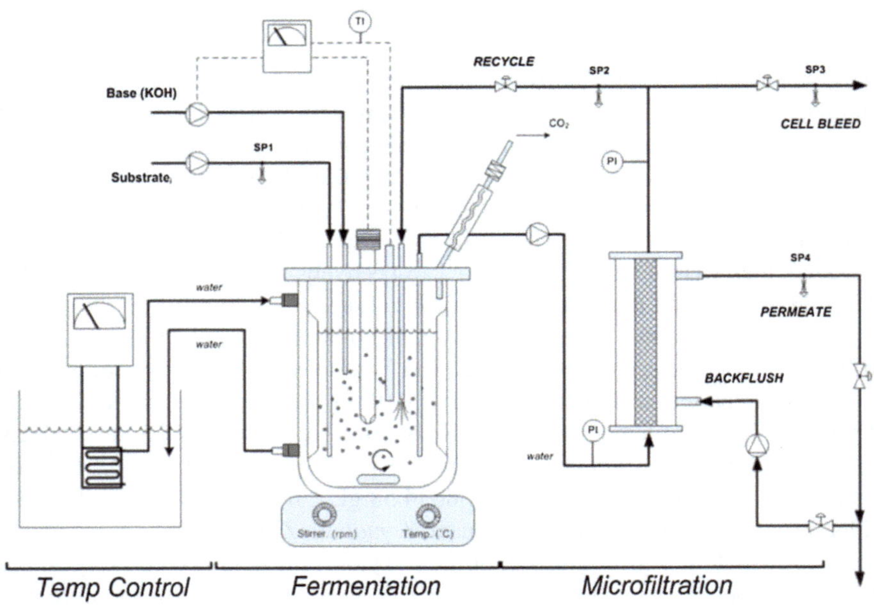

Fig. 8 Schematic layout of continuous culture with ultrafiltration separation and recycle

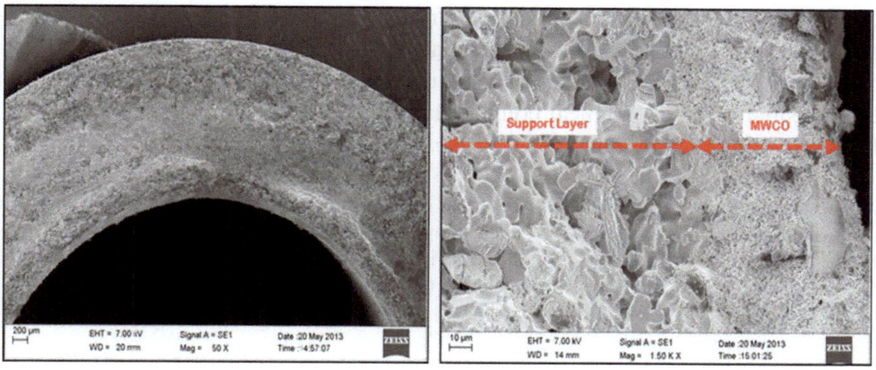

Fig. 9 SEM micrograph of the ultrafiltration membrane illustrating the cross section of the membrane with two distinct layers

4 Conclusion

Efficient ethanol production from the mixed hexose-pentose substrates of lignocellulosic hydrolysates is crucial to the success of biofuel production from renewable non-food sources. Recombinant *Z. mobilis* provides promising biocatalysts for bioconversion of these substrates for ethanol production. To realise the commercial potential of these biocatalysts, however, it is expedient to enhance the process performance through optimisation and intensification.

This work is the first quantification of the unique optimum process conditions of temperature and pH for bioconversion of a mixed glucose–xylose substrate by the recombinant *Z. mobilis* 8b. Statistical analyses further confirmed the temperature and pH control as key process parameters for maximisation of the bioconversion potential of xylose-containing hydrolysates to ethanol by this organism. Under optimised process conditions in batch culture, *Z. mobilis* 8b outperformed several competing xylose-fermenting biocatalysts. Continuous operation considerably enhanced ethanol productivity, attaining the maximum reported to date. Coupling cell separation and recycle with continuous operation facilitated cell retention and an increased capacity for xylose utilisation.

The high productivity attained during continuous culture operated under optimum process conditions, while still maintaining the ethanol concentration above the minimum required for cost-effective distillation, has major significance for the successful implementation of an industrial process. Integration of cell retention strategies to enhance xylose utilisation in the continuous configuration, would further improve the process economics by minimising substrate wastage. This strongly suggests continuous operation with cell retention as the way forward for the optimisation of process strategy for maximisation of ethanol production from mixed sugar hydrolysates by recombinant *Z. mobilis* strains.

Development of the continuous operational strategy with cell retention of *Z. mobilis* 8b from mixed glucose–xylose substrates under optimised process conditions remains a primary focus for further process intensification. It is envisaged that continued process enhancement will contribute to optimisation of bioconversion of the mixed xylose–glucose substrates and, ultimately, the unlocking of the potential of an economical commercial process for bioconversion of lignocellulosic hydrolysates by engineered *Z. mobilis*.

Acknowledgements The (DST-NRF) Centre of Excellence in Catalysis (c*change), South Africa (SA), and Stellenbosch University, SA, are acknowledged for funding this research. The Centre for Renewable and Sustainable Energy Studies, SA, is thanked for providing bursary funding for TM. Thanks also to Manda Rossouw for conducting the HPLC analyses.

References

Chander, R., Gupta, R., Pal, Y., Singh, A., & Zhang, Y. P. (2011). Bioethanol production from pentose sugars: Current status and future prospects. *Renewable and Sustainable Energy Reviews, 15,* 4950–4962.

Grahovac, J., Dodić, J., Jokić, A., Dodić, S., & Popov, S. (2012). Optimization of ethanol production from thick juice: A response surface methodology approach. *Fuel, 93,* 221–228.

Gunasekaran, P., & Chandra, R. K. (1999). Ethanol fermentation technology. *Current Science, 77,* 56–69.

Hahn-Hägerdal, B., Karhumaa, K., Fonseca, C., Spencer-Martins, I., & Gorwa-Grauslund, M. F. (2007). Towards industrial pentose-fermenting yeast strains. *Applied Microbiology and Biotechnology, 74,* 937–953.

Joachimsthal, E. L., & Rogers, P. L. (2000). Characterisation of a high-productivity recombinant strain of *Zymomonas mobilis* for ethanol production from glucose/xylose mixtures. *Applied Biochemistry and Biotechnology, 84–86,* 343–356.

Karsch, T., Stahl, U., & Esser, K. (1983). Ethanol production by Zymomonas and Saccharomyces, advantages and disadvantages known. *European Journal of Applied Microbiology, 18,* 387–391.

Lau, M. W., & Dale, B. E. (2009). Cellulosic ethanol production from AFEX-treated corn stover using Saccharomyces cerevisiae 424A (LNH-ST). *Proceedings of the National Academy of Sciences of the USA, 106,* 1368–1373.

Lawford, H. G., Rousseau, J. D., Mohagheghi, A., & McMillan, J. D. (1998). Continuous culture studies of xylose-fermenting *Zymomonas mobilis. Applied Biochemistry and Biotechnology, 70–72,* 353–367.

Lawson, J., & Erjavec, J. (2000). *Response surface methodology, modern statistics for engineering and quality improvement* (p. 465). Duxbury: Duxbury Press.

Lynd, L. R., Laser, M. S., Bransby, D., Dale, B. E., Davison, B., Hamilton, R., et al. (2008). How biotech can transform biofuels. *Nature Biotechnology, 26,* 169–172.

Menon, V., & Rao, M. (2012). Trends in bioconversion of lignocellulose: Biofuels, platform chemicals & biorefinery concept. *Progress in Energy and Combustion Science, 38,* 522–550.

Mohagheghi, A., Evans, K., Finkelstein, N., & Zhang, M. (1998). Cofermentation of glucose, xylose, and arabinose by mixed cultures of two genetically engineered *Zymomonas mobilis* strains. *Applied Biochemistry and Biotechnology, 70–72,* 285–299.

Rogers, P. L., & Joachimsthal, E. L. (2000). Characterization of a high-productivity recombinant strain of *Zymomonas mobilis* for ethanol production from glucose/xylose mixtures. *Applied Biochemistry and Biotechnology, 84,* 343–356.

Rogers, P. L., Joachimsthal, E. L., & Haggett, K. D. (1999). Evaluation of recombinant strains of *Zymomonas mobilis* for ethanol production from glucose/xylose media. *Applied Biochemistry and Biotechnology, 77,* 147–157.

Rogers, P. L., Kim, I. S., & Barrow, K. D. (2000). Kinetic and nuclear magnetic resonance studies of xylose metabolism by recombinant *Zymomonas mobilis* ZM4 (pZB5). *Applied and Environment Microbiology, 66,* 186–193.

Rogers, P. L., Lee, K. J., & Skotnicki, M. L. (1981). The effect of temperature on the kinetics of ethanol production by strains of *Zymomonas mobilis*. *Biotechnology Letters, 3,* 291–296.

Swings, J., & De Ley, J. (1977). The biology of Zymomonas. *Bacteriological Reviews, 41,* 1–46.

Van Rensburg, E., Gorgens, J. F., & Diedericks, D. (2013). Enhancing sugar recovery from sugarcane bagasse by kinetic analysis of a two-step dilute acid pretreatment process. *Biomass and Bioenergy, 7,* 149–160.

Zacchi, G., & Axelsson, A. (1989). Economic evaluation of preconcentration in production of ethanol from dilute sugar solutions. *Biotechnology and Bioengineering, 34,* 223–233.

Zhang, M., Chou, Y.-C., Howe, W., Eddy, C., Evans, K., & Mohagheghi, A. (2007). *Zymomonas pentose-sugar fermenting strains and uses thereof.* US patent 7223575B2.

Zhang, M., Eddy, C., Deanda, K., Finkelstein, M., & Picataggio, S. (1995). Metabolic engineering of a pentose metabolism pathway in ethanologenic *Zymomonas mobilis*. *Science, 267,* 240–243.

Author Biography

Prof Kim G. Clarke holds a Ph.D. and a B.Sc. Engineering (Chemical), both from the University of Cape Town, South Africa, and an MSc Engineering (Chemical and Biological) from the University of Birmingham, United Kingdom. She is a Registered Professional Engineer with the Engineering Council of South Africa, a Rated Scientist in the National Research Foundation, a senior member of the American Institute of Chemical Engineers and a chartered member of the Institution of Chemical Engineers. She leads a multidisciplinary research group directed towards the application of engineering principles in the development of biological processes, has authored a book on Bioprocess Engineering published by Elsevier as well as several scientific publications in leading international journals, and has presented her work at numerous conferences.

Co-digestion of Lawn Grass with Cow Dung and Pig Manure Under Anaerobic Condition

Suraya Singh, Anthony Njuguna Matheri, Mohamed Belaid and Edison Muzenda

Abstract The main objective of this study was to determine optimum biogas production by examining the effect of co-digestion of lawn grass with cow dung and pig manure. In order to conduct this experimental research, each of the substrates was co-digested with lawn grass at varying mixing ratios at stable, thermophilic (45 °C) temperatures, over a retention time of 16 days. Results indicated that co-digesting lawn grass with cow dung and/or pig manure may enhance biogas production and methane content. The results also imply that pig manure may be a more appropriate co-substrate to mix with lawn grass as it indicated a longer biogas potential for a relatively extended period of time, particularly when the lawn grass to pig manure mixing ratio was an even 30:30 g. Cow dung and lawn grass at a mixture 20:40 g had the highest biogas production, bio-methane potential. The C:N ratio, however, seems to be the main contributing factor in biogas yield hence the differences in yield for the different mixtures and substrates. Methane content of biogas was at its lowest levels for mono-digestion of lawn grass. However, mono-digesting lawn grass has a higher bio-methane potential and biogas accumulation over time when compared to a mixing ratio of pig manure:lawn grass of 40:20 g. The best possible bio-methane and biogas production were achieved when lawn grass was co-digested with both pig manure and cow manure at the ratio of, a 15:15:30 g which is attributed to the balanced nutrients and buffering in the digester.

Keywords Anaerobic · Biomenthane · Co-digestion · Kinetics · Thermophilic temperature

S. Singh · A.N. Matheri · M. Belaid (✉)
Department of Chemical Engineering, University of Johannesburg,
Johannesburg 2028, South Africa
e-mail: mbelaid@uj.ac.za

E. Muzenda
Department of Chemical, Material and Metallurgical Engineering, Botswana International University of Science and Technology, Private Bag 16, Palapye, Botswana

W. Leal Filho and D. Surroop (eds.), *The Nexus: Energy, Environment and Climate Change*, Green Energy and Technology, https://doi.org/10.1007/978-3-319-63612-2_14

Nomenclature

CH_3COOH	Acetic acid
NH_3	Ammonia
NH_4^+	Ammonium ion
AD	Anaerobic digestion
APHA	American Public Health Association
BMP	Biomethane potential
$CH_3CH_2CH_2COOH$	Butyric acid
CO_2	Carbon dioxide
CD	Cow dung
C:N ratio	Carbon nitrogen ratio
C_2H_5OH	Ethanol
GC	Gas chromatography
GHG	Greenhouse gas emissions
HRT	Hydraulic retention time
H_2S	Hydrogen sulphide
LG	Lawn grass
CH_4	Methane
NO_x	Nitrogen oxides
OLR	Organic loading rate
PM	Pig manure
PEETS	Process Energy Environmental Technology Station
CH_3CH_2COOH	Propionic acid
SRT	Solid retention time
SANEDI	South Africa National Energy Development Institute
SO_2	Sulphur dioxide
TIA	Technology Innovation Agency
VFA	Volatile fatty acids
VS	Volatile solids

1 Introduction

The rapid economic and population growth has left South Africa facing two major challenges which are in the form of energy supply and waste management. In order to overcome these challenges and support the governments (national, provincial and local) drive towards a green economy, innovative strategies are required to produce affordable, clean energy from renewable sources (S.A.B.I.A., South African Biogas Industry Association 2015; Matheri et al. 2015a). Biogas is seen as a renewable and sustainable source of energy. In South Africa, there seems to be an inclination towards anaerobic digestion for the production of biogas (S.A.B.I.A., South African Biogas Industry Association 2015).

Anaerobic digestion (AD) refers to the process of the natural biological decomposition of organic material in the absence of oxygen. The AD process results in the production of a biogas that consists essentially of methane (CH_4) and carbon dioxide (CO_2). Trace elements of hydrogen sulphide (H_2S), sulphur dioxide (SO_2), nitrogen oxides (NO_x) and volatile organic compounds may also be present. CH_4 is the important component as it is a highly flammable gas that can be utilized as fuel for cooking, lighting, water heating and, if the H_2S is removed, it can be used to run biogas-fueled generators to produce electricity (S.A.B.I.A 2015). Vehicular biogas also provides an important opportunity for reducing greenhouse gas (GHG) emissions (Goosen 2013). Hence, biogas may be used as a vehicle fuel or for the co-generation of electricity and heat thereby reducing GHG emissions. The biogas production process also encourages responsible waste disposal, production of nutrient rich by-products which can be used as organic fertilizer instead of chemicals and major environmental, safety and health benefits particularly when applied in a rural setting where the biogas will replace the usage of fuel wood and paraffin thereby reducing indoor air pollution, risk of fires, etc.

Co-digestion is the simultaneous digestion of two or more organic waste feedstock (Kangle et al. 2012). Co-digestion is used to increase methane production from low-yielding or difficult to digest materials (i.e. feed stocks) (Ryckebosch et al. 2011).

1.1 Biochemical Process of Anaerobic Digestion

Biogas production follows four fundamentals processes. These processes include hydrolysis, acidogenesis, acetogenesis and methanogenesis (Tiehm et al. 2001). Figure 1 shows a simplified generic anaerobic digestion process (Sreekrishnan et al. 2004).

The anaerobic system is as the result of complex interactions among different of bacteria. The major functional groups of bacteria according to their metabolic reactions are (Henze 2008): fermentative bacteria, hydrogen-producing acetogenic bacteria, hydrogen-consuming acetogenic bacteria, carbon dioxide reducing methanogens and aceticlastic methanogens.

1.1.1 Hydrolysis

This is an extracellular process (occurring on the exterior of the bacterial cell in the surrounding liquid) where the enzymes (such as cellulose, amylase, protease and lipase) of microorganisms breaks down organic matter or long chain molecules into smaller molecules (Forgács 2012). These enzymes are produced by different specialized bacteria that catalyze the decomposition process. Long chains of complex carbohydrates, proteins and lipids are decomposed or broken down into smaller molecules (Sabonnadière 2010). Examples of hydrolysis reactions (Fig. 2).

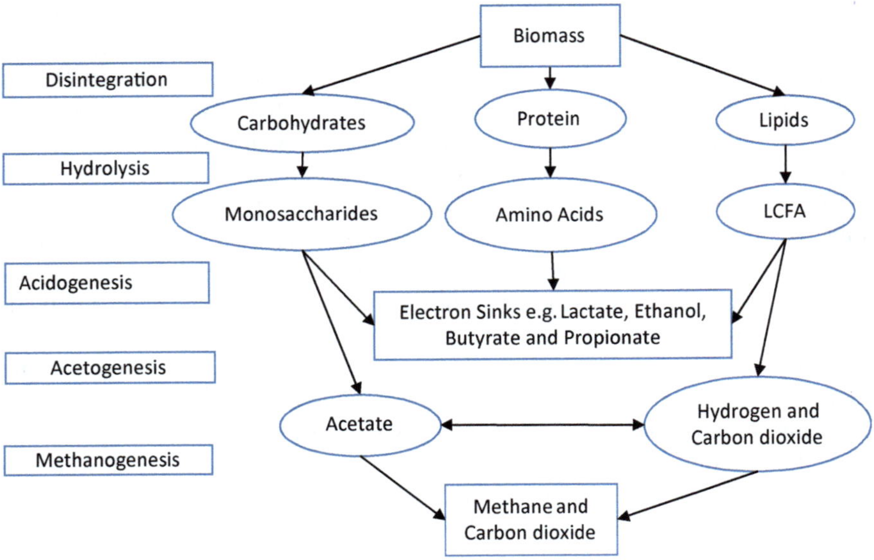

Fig. 1 Degradation steps of anaerobic digestion process (Angelidaki 1996)

Fig. 2 Hydrolyzed pathways for substrate in AD

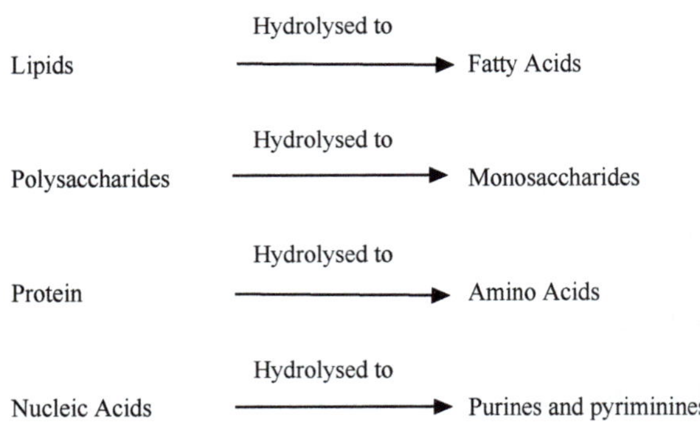

Anaerobic conditions are not always suitable for the hydrolysis of certain complex molecules. An example of such a case is lignin. Lignin, an important complex polymer that forms an integral part of the secondary cell walls of plants, cannot be decomposed under anaerobic conditions at all while proteins, simple sugars and starch, on the other hand, hydrolyse fairly easily under anaerobic conditions (Sabonnadière 2010).

Figure 3 shows the molecular makeup of complex polysaccharides cellulose, hemicellulose and lignin. Specialized bacteria are responsible for the relatively easy hydrolysis of cellulose and hemicellulose. However, when occurring in plant tissue, bacteria find it very difficult to hydrolyse the cellulose and hemicellulose because they are tightly packed within lignin and this may then be a plausible explanation as to the reason that during the biogas process, an estimated 40% of cellulose and

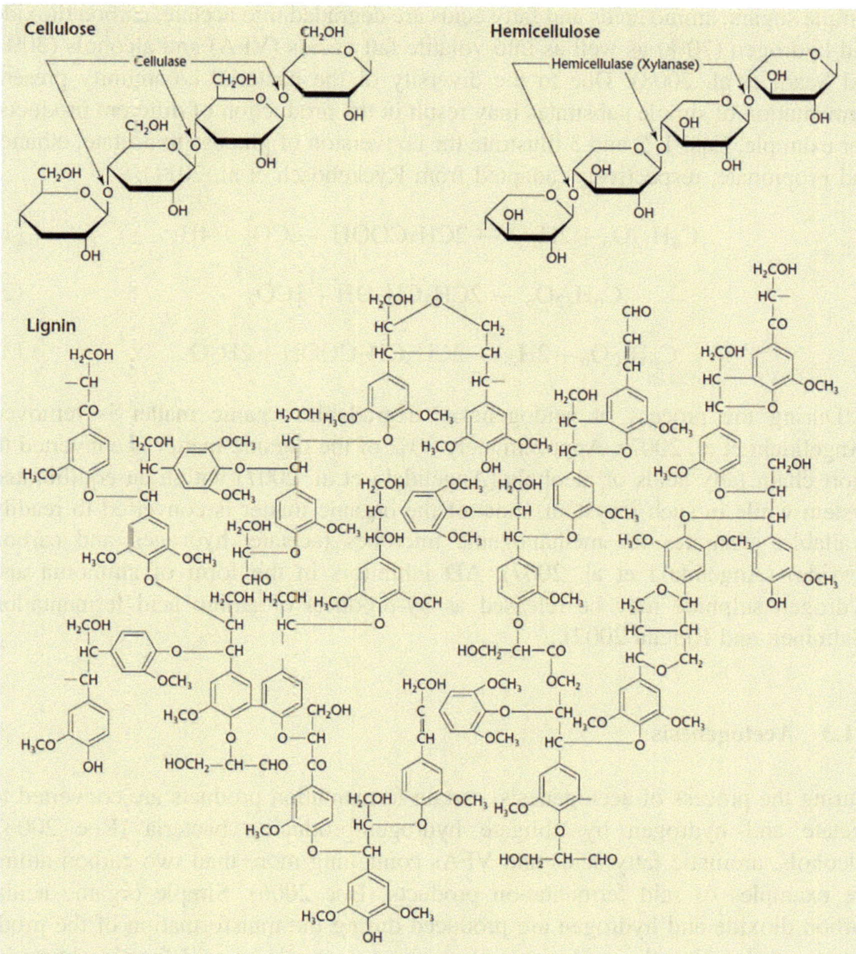

Fig. 3 Molecular make up of complex polysaccharides cellulose, hemicellulose and lignin (Biogas-Green Energy 2009)

hemicellulose in the pig slurry is decomposed (Biogas-Green Energy 2009). For animal manure and some other substrates with higher complex molecule concentrations, the breakdown of organic matter to methane and carbon dioxide is recurrently between 30 and 60% (Biogas-Green Energy 2009).

1.1.2 Acidogenesis

Due to the work carried out by acidogenic or fermentative bacteria, the products of hydrolysis are converted into methanogenic substrates (Al Seadi et al. 2008).

Simple sugars, amino acids and fatty acids are degraded into acetate, carbon dioxide and hydrogen (70%) as well as into volatile fatty acids (VFA) and alcohols (30%) (Al Seadi et al. 2008). Due to the diversity of the bacterial community present fermentation of simple substrates may result in the production of different products. For example, Eqs. 1, 2 and 3 illustrate the conversion of glucose to acetate, ethanol and propionate, respectively (adapted from Ryckebosch et al. 2011):

$$C_6H_{12}O_6 + 2H_2O \rightarrow 2CH_3COOH + 2CO_2 + 4H_2 \quad (1)$$

$$C_6H_{12}O_6 \rightarrow 2CH_3CH_2OH + 2CO_2 \quad (2)$$

$$C_6H_{12}O_6 + 2H_2 \rightarrow 2CH_3CH_2COOH + 2H_2O \quad (3)$$

During the process of acidogenesis, degradable organic matter is removed (Angelidaki et al. 2007). Approximately 30% of the organic matter is converted to short chain fatty acids or alcohols (Angelidaki et al. 2007) within an equilibrated system while in such a system, most of the organic matter is converted to readily available substrates for methanogenic microbes (acetate, hydrogen and carbon dioxide) (Angelidaki et al. 2007). AD inhibitors in the form of ammonia and hydrogen sulphide may be released as by-products of amino acid fermentation (Salminen and Rintala 2002).

1.1.3 Acetogenesis

During the process of acetogenesis, certain fermentation products are converted to acetate and hydrogen by obligate hydrogen-producing bacteria (Boe 2006). Alcohols, aromatic fatty acids and VFAs containing more than two carbon atoms are examples of said fermentation products (Boe 2006). Simple organic acids, carbon dioxide and hydrogen are produced during the transformation of the products created during the acidogenic phase, by acetogenic or acid forming bacteria. Acetic acid (CH_3COOH), propionic acid (CH_3CH_2COOH), butyric acid ($CH_3CH_2CH_2COOH$) and ethanol (C_2H_5OH) are the principal acids produced (Kangle et al. 2012), while the products of volatile fatty acids and alcohol are in the form of acetate, hydrogen gas and carbon dioxide produced by hydrogen-producing acetogenic bacteria. Homoacetogenic bacteria create acetate from carbon dioxide and hydrogen gas (Sterling et al. 2001), while hydrogen-producing acetogenic bacteria is responsible for the majority of the acetate produced (Angelidaki et al. 1996). An acetogenesis reaction is shown below:

$$C_6H_{12}O_6 \rightarrow 2C_2H_5OH + 2CO_2 \quad (4)$$

1.1.4 Methanogenesis

In order for a methanogenesis to be successful, an anaerobic digestion system requires that there is a variety of methane-forming bacteria (methanogens) present as there are no singular methane-forming bacterial species that can degrade all available substrates. During the process of methanogenesis, acetate is the source of an estimated 70% of the methane produced (Smith and Mah 1966). The reduction of carbon dioxide by hydrogen and other electron donors is responsible for the production of the remaining 30% of methane (Hashimoto et al. 1981). The division of methanogenesis into two main types is based on the substrates that are utilized by the methanogens (Bitton 2005). The two main types of methanogenesis are:

Hydrogenotrophic methanogenesis where hydrogen and carbon dioxide are converted into methane according to the following reaction (Kangle et al. 2012):

$$CO_2 + 4H_2 \rightarrow CH_4 + 2H_2O \tag{5}$$

Acetotrophic or aceticlastic methanogenesis where methane is formed from the conversion of acetate through the following reaction (Kangle et al. 2012):

$$CH_3COOH \rightarrow CH_4 + CO_2 \tag{6}$$

1.2 Parameters Affecting Anaerobic Digestion

The activity of biogas production depends on various parameters like temperature, available nutrients, pH value, hydraulic retention time, carbon/nitrogen (C/N ratio), volatile fatty acids, ammonia, agitation, inhibitory factors and concentration of substrate (Sreenivas et al. 2010; Sundararajan et al. 1997; Latinwo and Agarry 2015; Matheri et al. 2015b).

1.2.1 Temperature

The anaerobic process is so sensitive to temperature; change of acetic acid (acetate) to methane depends mostly on temperature but conversion to acetic acid will not affect much by slight temperature variations. Grimberg et al. (2015) reported that the environmental temperature has a major influence on the anaerobic microbial systems, which affects the metabolic rate, ionization equilibria, substrate solubility and fats. Higher temperature affects the activity of hydrogenotropic methanogens in the anaerobic process and enriches hydrogen-producing bacteria and spore forming bacteria (Grimberg et al. 2015). Mesophilic digestion temperature is considered to be most suitable for anaerobic digestion on the ranges of 35–37 °C In thermophilic digestion, 55 °C is considered to be ideal (Abbasi et al. 2011). Table 1 shows

Table 1 Thermal stages, process temperature and typical hydraulic retention times (Noshy 2013)

Thermal stages	Process temperature (°C)	HRT(days)
Psychrophilic	<20	From 70 to 80
Mesophilic	From 30 to 42	From 14 to 40
Thermophilic	From 43 to 55	From 14 to 20

different thermal stages, process temperatures and typical hydraulic retention times for the AD process.

1.2.2 Available Nutrients

Bacteria responsible for the production of methane have specific growth requirements. In order to optimize bacterial growth and methane production, various nutrients are required in addition to carbon, oxygen and hydrogen. Specific metals such as nickel, cobalt, molybdenum and iron are also required (Azbar et al. 2001). Additional trace metals, in media, such as molybdenum, selenium, manganese, boron and aluminium are suggested as additional components to enhance methanogenic activity (Azbar et al. 2001). Adequate amounts of the previously mentioned elements can be found within substrates such as agricultural waste and municipal sewage. It is important to note that a unique characterization of methanogenic bacteria is the unusual requirement for nickel within biological systems (Matheri et al. 2016).

1.2.3 pH

The pH value is the measure of alkalinity/acidity of a solution (Al Seadi et al. 2008). It affects the production of biogas because each group of the microorganisms has different optimum pH range. Methanogenic bacteria have an optimum pH between 6.5 and 7.5. They are extremely sensitive to pH. The fermentative microorganisms are less sensitive to pH since they have wide optimum pH range between 4.0 and 8.5. Low pH level favours the production of acids such as butyric acid, propionic acids and acetic mainly at pH of 4.0. At pH higher than 8.0, ammonia is mainly produced. The presence of volatile fatty acids (VFAs) tends to decrease the pH and can lower the methanogenic bacteria activity and hence the biogas production (Appels et al. 2008).

1.2.4 Volatile Fatty Acid

Volatile fatty acids (VFA) are intermediate compounds usually composed of a carbon chain made up of six carbon atoms and are produced during the acidogenesis process step of the AD process and whose concentration tends to indicate the stability or rather the instability of the AD process at hand (Al Seadi et al. 2008). Acetate, propionate, butyrate and lactate are examples of VFA produced during AD. In many cases, VFA accumulation inside the digester may result from the instability of the AD process and may contribute to the reduction of the pH value. The kind of biomass contained in the digester affects its (the digester/s) buffer capacity thus a decrease in the pH value may not automatically imply that there is in fact an accumulation of VFA (Al Seadi et al. 2008) For instance, animal manure has an excess amount of alkalinity in which case the detection of VFA accumulation indicated by a considerable reduction in the pH value may only take place if the VFA accumulation levels reaches and/or exceeds certain values, in which case the AD process may be significantly inhibited due to the elevated VFA concentration levels within the digester (Al Seadi et al. 2008). VFA concentration cannot be used as a stand-alone process monitoring parameter. This is due to the fact that it has been noted that the equivalent VFA concentration levels have an inhibitory effect in one digester while being at the optimum levels for another digester (Al Seadi et al. 2008).

1.2.5 Carbon/Nitrogen Ratio

The carbon/nitrogen (C/N) ratio represent the relationship between the amount of carbon and nitrogen present in the organic substrate. The optimal C/N ratio for anaerobic digestion is considered to be in the range of 15–30 (Annachhatre 2012). If the C/N ratio is too high, the nitrogen is consumed rapidly by the methanogens bacteria to meet their protein requirement and is no longer available to react on the left-over carbon content in the material. As a result, the biogas production is reduced (Annachhatre 2012). If the C/N ratio is too low, nitrogen is liberated and accumulates in the form of ammonia (Annachhatre 2012). This increases the pH of the digestates. When pH value rises higher than 8.5, it begins to exert a toxic effect on the methanogenic bacteria (Annachhatre 2012). To maintain the C/N level of the digester substrate at optimum levels, substrate of high C/N ratio can be co-digested with a substrate of low C/N ratio (Annachhatre 2012).

1.2.6 Retention (Residence) Time

Hydraulic retention time (HRT) is the time that the fluid element of the feed remains in the digester while the solid retention time (SRT) refers to the residence time of the bacteria (solids) in the reactor (Kangle et al. 2012). Variations in technologies, process temperatures and even the constituent make up of waste material plays an important role in determining the required retention time for the completion of the AD reactions (Kangle et al. 2012). In general, it is suggested that the retention time for substances undergoing fermentation within the mesophilic temperature range is between 15 and 40 days. For example, the approximate retention time for liquid cow manure is between 20 and 30 days while the approximate retention time for liquid pig manure and liquid chicken manure is between 15–25 and 20–40 days, respectively (Werner 1997). A lower retention time is suggested for fermentation in the thermophilic temperature range (refer to Table 1). The generation of methanogens takes a relatively long time. It is believed that the minimum SRT in order to avoid microbial washout is estimated at 12 days (Gerardi 2003). The production of higher biogas per volume will be seen at short retention times with less decomposition in organic matter. A decrease in digester volume may be obtained from shorter retention times but it is of paramount importance that a balance must also be made in order to achieve the required operating conditions (Kangle et al. 2012).

1.2.7 The Organic Loading Rate (OLR)

The organic loading rate (OLR) is the amount of volatile solids (VS) to be fed into the digester each day in a continuous process. As the OLR increases, the biogas yield increases to some extent but above the optimal organic loading rate, the volatile solids degradation and biogas yield decreases due to overloading (Rincon et al. 2007). The maximum possible OLR depends on the process temperature and its retention time.

1.2.8 Toxicity

Mineral ions, especially of trace elements, are among the materials that inhibit the growth of bacteria in a digester. Small amount of mineral (calcium, sodium, potassium, sulphur, magnesium and ammonium) stimulates the microorganisms growth, but higher concentrations have a toxic and inhibition effect (Abbasi et al. 2011). Heavy metals such as zinc, nickel, cobalt, copper, lead and chromium are essential for bacterial growth in very small quantities, but higher quantities have a toxic bacteria effect. Organic solvents and antibiotic also inhibit the bacteria. Recovery of digesters can only be achieved by flushing the content, cessation of feeding or diluting the contents to lower the concentration of inhibitory substances to below the toxic level (Abbasi et al. 2011).

1.2.9 Ammonia

Bacterial degradation of nitrogen-containing substances such as proteins results in the production of ammonia (NH_3) (Biogas-Green Energy 2009), an important endogenous inhibitor. Ammonia is an important source of nitrogen which is essential for the growth of bacteria. However, ammonia at high concentrations becomes toxic for the bacteria. Ammonium ion (NH_4^+) and a dissolved non-ionized form of ammonia (NH_3) are the two fundamental forms of ammonia that can exist in the system. Ammonia and ammonium are in equilibrium in an aqueous solution. Since ammonia is more toxic than ammonium, this equilibrium is of paramount

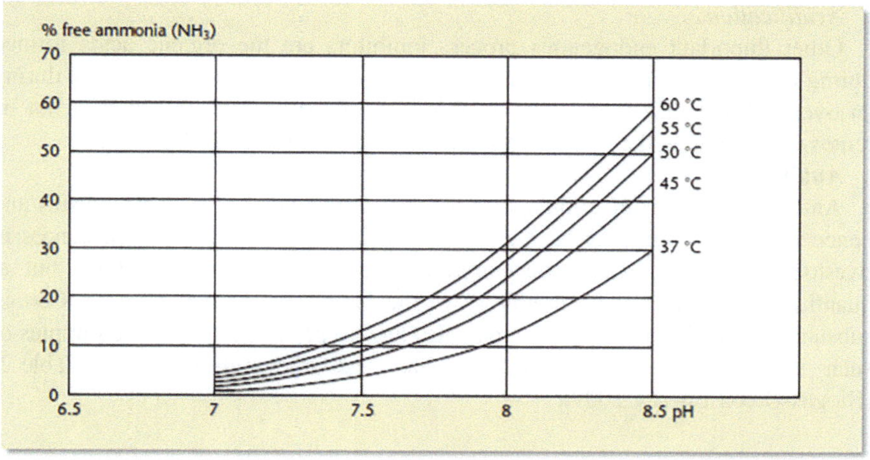

Fig. 4 Relative effect of pH and temperature on ammonia (Biogas-Green Energy. 2009)

importance and is determined by variations in certain conditions of the environment such as the acidity, temperature and pH (Biogas-Green Energy 2009):

$$NH_4^+ \leftrightarrow NH_3 + H^+ \quad (7)$$

Figure 4 shows the relative effect of pH and temperature on the balance between ammonium and toxic ammonia (NH_4^+/NH_3). From the graph, it can be seen that higher temperatures and pH values move the system's equilibrium to the right thus making the environment more toxic for the bacteria. The pH of the system has another significant effect on the level of ammonia inhibition, as the pH value determines the degree of ionization.

1.2.10 Agitation/Mixing

Mixing is required to maintain fluid homogeneity, hence process stability, temperature distribution, within a digester. The objectives of mixing are to combine the incoming substrate with the bacteria, to reduce the formation of scum, and to avoid pronounced temperature gradients within the digester. Very rapid mixing can disrupt the microbial balance while too slow stirring can cause short-circuiting and inadequate mixing (Abbasi et al. 2011).

1.2.11 Inhibitory Factor

Aside from ammonia and nitrogen inhibition, other inhibitory factors are in the form of:

Acidification:

Other important endogenous process inhibitors are the organic acids formed during the process. The development of organic acids is particularly possible during an overload and may lead to the acidification of the process should they not be removed as soon as they are formed (Biogas-Green Energy 2009).

Antibiotics

Antibiotics and disinfection agents act to disinfect and kill microorganisms and hence are evident inhibitors to the process. Due to the use of both substances in livestock production small concentrations may be found in the slurry but at quantities that may not be inhibitory to the biogas production process. Certain substances in high concentrations may also lead to process inhibition. Examples of such substances are heavy metals, salts and micronutrients (see Table 2) (Biogas-Green Energy 2009).

1.2.12 Solid Residue/Slurry

When the anaerobic degradation is nearly complete, the solid residue or digestate is removed and is normally cured aerobically and screened for items such as plastic pieces, glass, shards, etc., before being disposed on land as fertilizer (Abbasi et al. 2011).

Table 2 Examples of inhibitory factors (Biogas-Green Energy. 2009)

Chemical/formula	Inhibition level	Toxicity level
Ammonia, free, NH_3	50–100 mg N/l	100–200 mg N/l
Ammonia, total, $NH_4^+ + NH_3$	1000–6000 mg N/l	10,000 mg N/l (pH < 7.5)
Chloride, Cl^-	<8000 mg/l	10,000 mg/l
Cyanide, CN^-	2–20 mg/l	30 mg/l
Formaldehyde, H_2CO	100–400 mg/l	500–1000 mg/l
Phenol, C_5H_5OH	100–200 mg/l	
Chloroform, $CHCl_3$	>1 mg/l (single dose)	>50 mg/l (continuous feed)
Hydrogen, H_2	p(H; > ca. IO^4 atm.	
Copper, Cu^{+++}	10–250 mg/l	
Chrome, Cr^{+++}	50–100 mg/l	200–400 mg/l
Nickel, Ni^{++}	100–200 mg/l	300–1000 mg/l
Sodium, Na^+	3000–10,000 mg/l	
Calcium, Ca^{++}	8000 mg/l	
Magnesium, Mg^{++}	3000 mg/l	
Zink, Zn^+	350–1000 mg/l	
Sulphate, SO_4^{--}	500–4000 mg/l	
Sulphide (as sulphur)	200 mg/l	
Hydrogen sulphide, H_2S	250–1000 mg/l	

1.2.13 Substrate Solids Content and Agitation

Total solids content: Gradual retardation in methanogen movement within substrates may be the result of an increase in substrate solids content which may, in turn, have a negative effect on biogas generation. However, recent literature has reported occasions where high biogas yields were attained from landfill material that contained high solids content. Unfortunately, at this stage, the proposition of a generally valid guideline in relation to specific biogas production for a specific solids percentage is not in place (Werner 1997).

Agitation: Many substrates and various modes of fermentation require some sort of substrate agitation or mixing in order to maintain process stability within the digester (Werner 1997).

Mixing assists in preventing scum formation and in the avoidance of pronounced temperature gradients within the digester. Excessive mixing/rapid agitation is disruptive to the fermentation process hence slow stirring is preferred. The type of reactor and solids content in the digester are important factors to consider as they determine the kind of mixing equipment and the amount of mixing required.

The benefits of mixing include: (Adapted from Kangle et al. 2012):

- Eliminates or reduces scum build-up.
- Eliminates thermal stratification or localized pockets of depressed temperature.
- Maintains digester sludge's chemical and physical uniformity throughout the tank.
- Stimulates the rapid dispersion of metabolic wastes produced during substrate digestion that could otherwise inhibit methane production.
- Stimulates the rapid dispersion of any toxic material entering the tank (minimizing toxicity).
- Mixing also prevents deposition of grit.

1.2.14 Co-digestion

Studies show that co-digestion is a way of minimizing HRT and improving methane production (Ryckebosch et al. 2011). The other substrate should be manure which is dominated by high levels of organisms that have the ability to hydrolyse lingo-cellulose material. Co-digestion of biomass waste can produce more methane than manure itself, but the challenge in this process is to achieve completely break down of organic material in the stage of hydrolysis (Ryckebosch et al. 2011). The importance of co-digestion is to stabilize conditions or other parameters in digestion process such as C:N ratio as well as pH, macronutrients and micronutrients, inhibitors and dry material (Ryckebosch et al. 2011).

Other Advantages of Co-digestion

Improved Nutrient Balance

There are a number of positives linked to co-digestion which include an enhanced nutrient balance which in turn results in superior digester performance and ultimately greater biogas yields (Wu 2000). According to Desai et al. (1994) the maintenance of a proper C:N ratio in the reactor was possible by combining whey and poultry manure while Murto et al. (2004) reported that the co-digestion of solid slaughterhouse waste manure, and fruit and vegetable waste resulted in highly buffered system and produced gas yields of 0.8–1 $m^3\ kg^{-1}$ VS.

Optimization of Rheological Qualities

Some waste materials (such as a waste material with poor fluid dynamics, aggregating wastes, particulate wastes) that comprise of highly disturbing or inhibitory constituents can be utilized in a far more beneficial manner such as acting as co-substrates for co-digestion with well-performing sewage sludge or liquid manure (Braun 2002).

Effective Utilization of Digester Volumes in Sewage Plans

The utilization of free capacities is aided by the addition of co-substrates in existing digesters. Also, transportation costs are reduced due to the extensive distribution sewage treatment plants. There are potential economic benefits that are linked to the sharing of equipment. Therefore, plant economics can be vastly improved due to the application of co-substrates (Braun 2002).

Energy Crops as Co-substrate

One way of improving the biogas yield of agricultural digesters is to add energy crops or silage as co-substrates (Braun 2002). The natural glycolytic pathway can be inhibited by the ammonia content of animal manure. There are a number of potential benefits for co-digesting plant material and manures. A wide range of nutrients along with the provision of buffering capacity is supplied by the animal manure while adding high carbon content plant material goes a long way in balancing the carbon to nitrogen ratio of the feedstock which in turn decreases the risk of ammonia inhibition. Lehtomäki et al. (2007) and Somayaji and Khanna (1994) conducted a study with digesters fed with cow manure and varying proportions of wheat straw and conclude the highest specific methane yields were observed with 40% of wheat straw of total solids (TS) in the feedstock.

1.2.15 Substrate Pre-treatment

Pre-treatment is done to increase the efficiency of anaerobic digestion technology and increase the production of biogas (Wang et al. 2012). Pre-treatment can be classified as thermal, mechanical, biochemical pre-treatment. Pre-treatment is necessary since the nature of a substrate has an effect on the rate of biogas production (Wang et al. 2012).

2 Methodology

Cow dung, pig manure and lawn grass were collected from a farm in Gauteng province. Waste characterization was done to ascertain the composition. These included physical and chemical composition with regards to C/N ratio, volatile solids, total solids and elemental analysis for carbon, nitrogen, sulphur and hydrogen in accordance with the standard method (APHA 1995) (Rice et al. 2012). To determine biogas production rate, a batch digester was fed with the co-digested substrates and inoculum under pre-set conditions of 45 °C at pH of 7 as shown in Fig. 5. pH was neutralized by a solution of 8 g NaOH in 100 ml and H_2SO_4. The digester was flushed with nitrogen to expel the oxygen and make the process anaerobic. It was then immersed in the water bath and kept under constant temperature. The gas produced was measured using downwards displacement method on a daily basis until the end of retention time. Gas chromatography instrument was used to analyze the gas composition with flame-ionization detectors. The operation

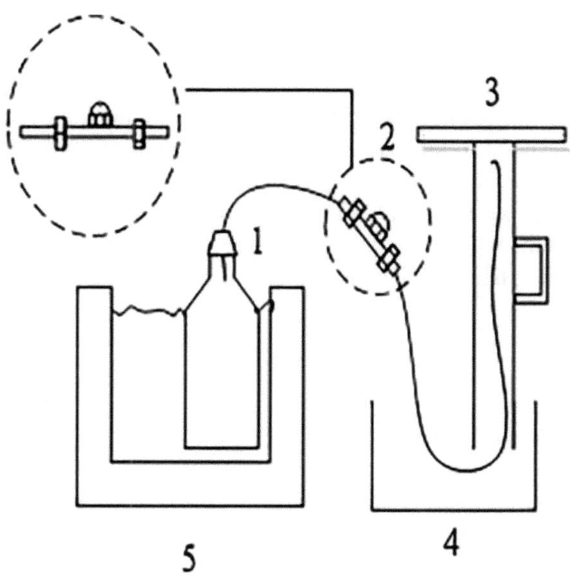

Fig. 5 Biogas production set-up (1 digester, 2 t-union, 3 measuring cylinder, 4 water bucket, 5 thermostatic water bath)

conditions were; helium was used as carrier gas (20 ml/min), oven temperature 70 °C, detector temperature 150 °C and injector port at 80 °C.

The main objective of this research was to estimate optimum biogas production by examining the effect of co-digestion of lawn grass with cow dung and pig manure under anaerobic condition.

3 Results and Discussion

3.1 Substrate Characterization

In this study, co-digestion of lawn grass (LG) with cow dung (CD), pig manure (PM) were evaluated for the purpose of getting the bio-methane potentials at optimum temperature (45 °C) and initial pH of 7. Table 3 shows the substrate characterization.

3.2 Influence of Co-digestion of Grass with Cow Dung

The accumulative biogas production, methane accumulative and methane content in biogas from anaerobic batch co-digestion process of cow dung (CD) and lawn grass (LG) at VS ratio of 0:60, 10:50, 20:40, 30:30 and 40:20 g (CD:LG) are shown in Fig. 6.

In Fig. 6a, the CD:LG ratio had a significant effect on cumulative biogas production hence biogas production increases with CD:LG. The highest cumulative biogas of approximately 405 mL was obtained in the digester with 20:40 g CD:LG ratio, followed by the digester with 10:50 g CD:LG ratio. These observations may be associated with the continuous growth of methanogens favoured by the prevalent environment and balanced C:N ratio. Although a ratio higher than 20:40 g CD:LG resulted in biogas decrement, at the ratio of 40:20 g CD:LG the process was more consistent, stable and prolonged. The cumulative biogas on the 14th day for digester with CD:LG ratio of 0:60, 10:50, 30:30 and 40:20 g were 2785, 3405, 2880 and 3175 mL, respectively. All the digesters with co-digested grass with cow dung had high biogas cumulative in comparison to the digester with mono digested

Table 3 Substrate characterization

Properties	Units	Cow dung	Pig manure	Lawn grass	Mixture
Moisture content	%	85.56	55.58	79.87	73.87
Total solids (TS)	%	14.44	44.42	20.13	26.13
Volatile solids (VS)	%	75.00	80.87	73.84	79.97
Ash content	%	3.61	2.55	1.58	1.57
C/N ratio	–	10.69	27.47	20.49	11.35

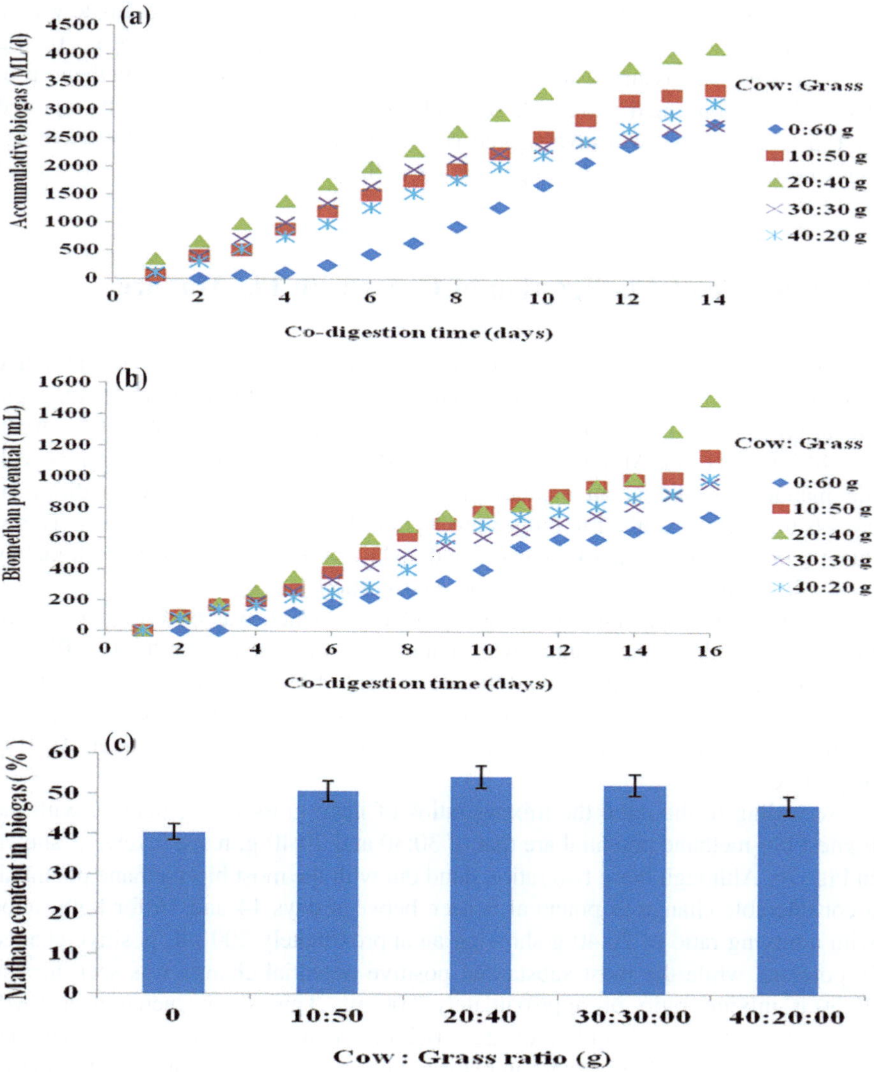

Fig. 6 Accumulative biogas production (**a**), methane cumulative (**b**) and methane content (**c**) obtained during mono-digestion of lawn grass and co-digestion of lawn grass with cow dung

grass. In addition, in this study, the biogas production was produced straight from day 1 of the experiment in all digesters, and increased with time.

The accumulative methane is shown in Fig. 6b, likewise the same trend like an accumulation of biogas production was observed and bio-methane increase with an increase in CD:LG ratio. Maximum cumulative bio-methane potential was obtained in the digester with the ratio of 20:40 g CD:LG. Figure 6c shows the methane content in biogas during anaerobic mono-digestion of grass and co-digestion of lawn grass with cow dung. An increase in methane content by approximately 10% was obtained when grass was co-digested with cow dung at the ratio of 20:40 g CD:LG. This was assumed to be due to the exponential growth of methane producing bacteria in the digester. These results are in agreement with results reported by Liu et al. (2008) who investigated rice straw as a substrate. The total average methane for the whole study ranges between 40 and 51%.

4 Influence of Co-digestion of Grass with Pig Manure

Figure 7 demonstrates the accumulative biogas production, accumulative bio-methane potential and methane content in biogas in the digester treating pig manure (PM) with lawn grass (LG) at the mixing ratio of 0:60, 10:50, 20:40, 30:30 and 40:20 g PM:LG. Although it was observed that co-digestion of lawn grass with pig manure resulted in an increase in biogas production, excessive addition of co-substrate (PM) resulted in biogas production decrement, refer to Fig. 7a. This is due to inhibition of microorganism and delay of hydrolysis, which ultimately results in delay of the whole anaerobic co-digestion.

The maximum accumulative biogas production which was 65% higher than of the digester with no pig manure was achieved in a digester with 30:30 g PM:LG ratio. Higher PM:LG ratio resulted in a decrement of biogas production. In Fig. 7a, b, it is observed that both the accumulative biogas production and bio-methane potential are at their highest when the mixing ratio of grass to pig is at 30:30 g, respectively.

According to the data, the mixing ratios of lawn grass to pig manure with the highest bio-methane potential are that of 30:30 and 20:40 g, respectively, as shown in Fig. 7b. Although these two ratios stand out with the most bio-methane potential, a considerable change in potential is seen between days 14 and 16 for both ratios with a mixing ratio of 20:40 g showing an approximately 200 mL positive change in potential while the most substantial positive potential change was seen for the 30:30 g mixing ratio of approximately 500 ml. This sharp increase in BMP (bio-methane potential) could indicate enrichment in methanogens in the system but the usual accompanying increase in pH is not observed. Considering that the system was at a constant thermophilic temperature of 45 °C, another potential reason for the sharp increase in bio-methane potential could be related to the retention time. According to Werner et al. (1997) the minimum retention time for AD to occur and reach its production peak is between 15 and 20 days, hence a sharp rise in BMP is seen as this experiment was stopped perhaps at the point of maximum BMP potential.

Figure 7c indicates the methane content across the mixing ratios. From the data gathered, the mixing ratio with the greatest methane content is that of 20:40 g (at 54.91%) while the lowest methane content was seen at a mixing ratio of 0:60 g (at 40.34%). Mata-Alvarez et al. (2000) suggested that co-digestion of animal manure with biomass will result in a higher methane yield than mono-digestion due to the

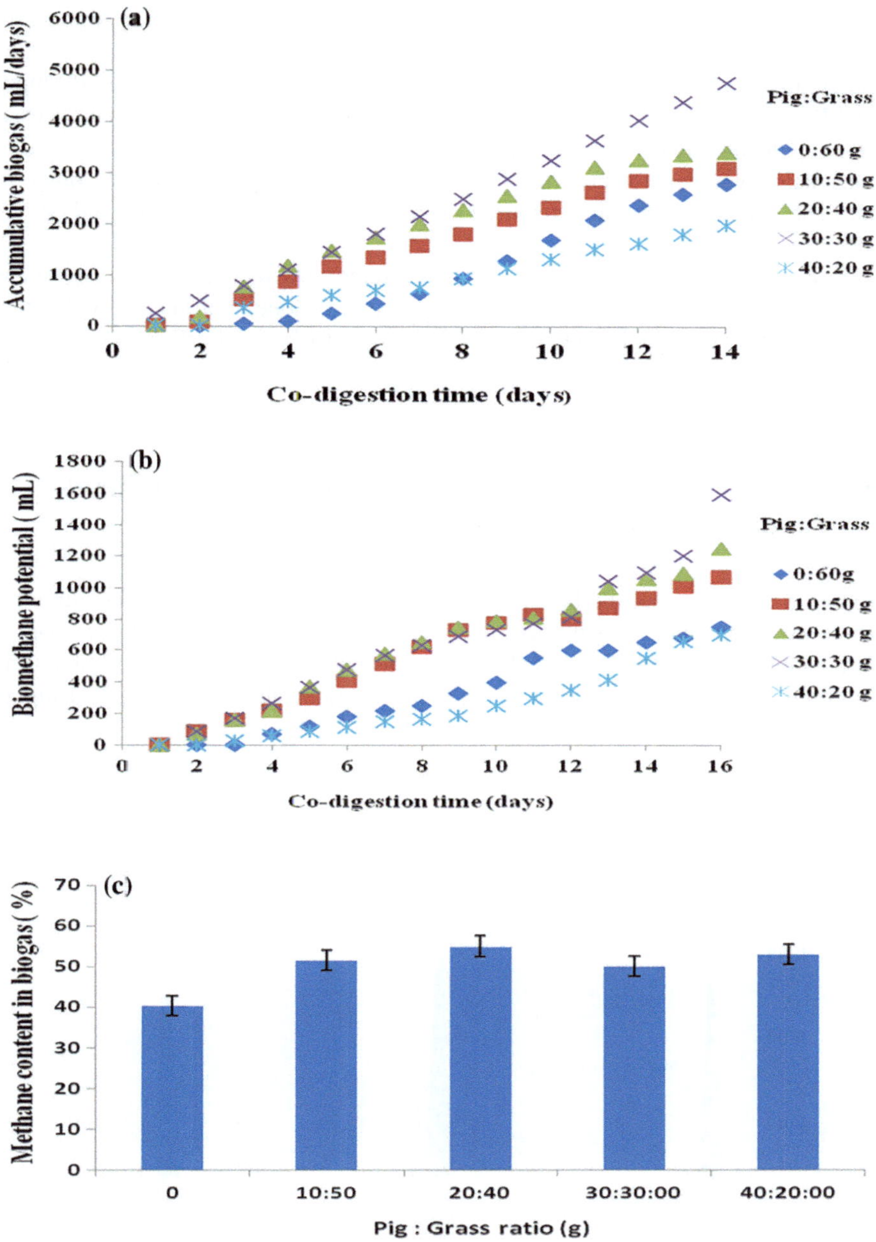

Fig. 7 Accumulative biogas production (**a**), methane cumulative (**b**) and methane content (**c**) obtained during mono-digestion of lawn grass and co-digestion of lawn grass with pig manure

synergistic effects of the co-substrates. The advantages of co-digestion of manures with co-substrates are: (1) due to the increase in the buffering capacity and the possibility in accumulation of volatile fatty acids (VFAs) during digestion (Campos and Palatsi 1999), pH value can be constantly suitable within the methanogenesis stage (Ten and Koster 1990); (2) it can avoid high concentrations of NH_3 in association with mono manure 38 digestion; (3) it can provide a wide range of nutrient contents (C:N ratios) required by the methanogens (Angelidaki and Ahring 1997).

It is noticeable, that animal manure has a rather low methane yield. This is the reason that most animal manure is not digested alone, but mixed with other co-substrates, with high methane yield, in order to boost the biogas production (Smith and Mah 1966).

5 Comparison of Biogas and Bio-Methane at the Optimal Co-digestion Mixing Ratios

Figure 8 indicates that the best possible bio-methane and biogas production were achieved when lawn grass was co-digested with both pig manure and cow manure at the ratio of, a 15:15:30 g (1:1:2). This is attributed to the balanced nutrients and buffering in the digester. Lawn grass to pig manure ratio of 30:30 g had higher

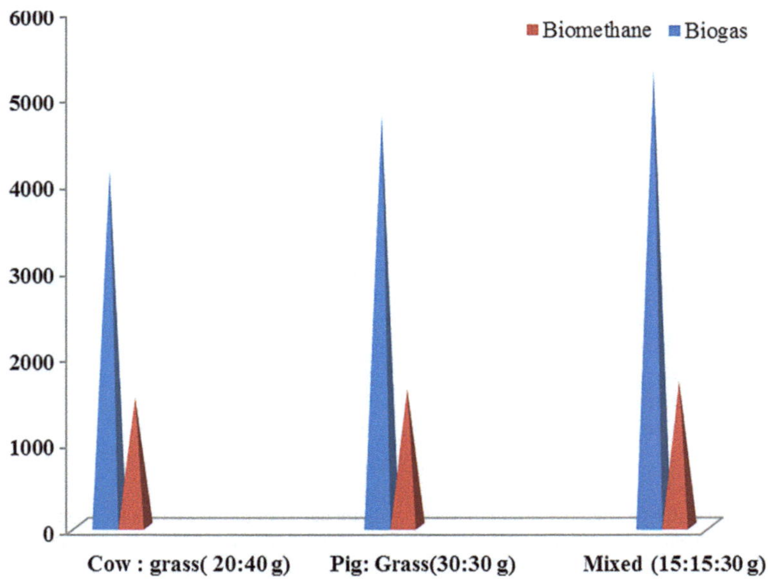

Fig. 8 Comparison of methane and bio-methane for maximum biogas and bio-methane production in co-digestion at CD:LG (20:40 g), LG:PM (30:30 g) and a mixture of CM:PM:LG (15:15:30)

biogas production in comparison to the digester with cow dung to lawn grass ratio of 20:40 g. This may be attributed to the fact that the C:N ratio of pig manure is above optimal as shown in Table 3, thus it was well blended and balanced when pig manure was mixed with lawn grass. Nevertheless, for both the substrates, a pattern of a 500 mL increase in bio-methane potential from days 14–16 for the best mixing ratios is clearly evident. Another interesting comparison is that the methane potential for both substrates is also similar, with a minimum of approximately 40% to a maximum of approximately 54–55%.

6 Conclusion

This study shows that co-digestion of lawn grass with cow dung or pig manure at thermophilic condition (45 °C) may enhance biogas production and methane content. Pig manure seems to be a more suitable co-substrate to mix with lawn grass as it showed a greater biogas potential for a longer period of time. This statement is especially true when considering a mixing ratio of 30:30 g pig manure to lawn grass. Cow dung and lawn grass at a VS mixture 20:40 g had the highest biogas production, bio-methane potential and methane content. Mono-digestion of substrates is not as effective as co-digestion between a substrate and a compatible co-substrate. This statement may hold true especially when the comparison relates to the mono-digestion of grass versus the co-digestion of cow dung and lawn grass at varying mixing ratios. In the results, it is evident that co-digesting cow dung and lawn grass yields higher bio-methane, biogas accumulation and a higher methane content than that of mono-digestion.

In the case of pig manure and lawn grass co-digestion, it is evident from the results that mono-digestion of lawn grass may have better biogas accumulation and bio-methane yields than is produced when co-digesting pig manure: lawn grass at a mixing ratio of 40:20 g. However, mono-digesting lawn grass had the lowest methane potential when compared to co-digestion.

Acknowledgements The author wish to express their appreciation to Process Energy Environmental and Technology Station (PEETS) funded by City of Johannesburg, South Africa National Energy Development Institute (SANEDI), Technology Innovation Agency (TIA) through Prof Charles Mbohwa, Mr. Thabo Maahlatsi, Mr. Mlawule Mashego, Chemical Engineering and Applied Chemistry Departments at the University of Johannesburg for allowing us to work in their laboratories. Ms. Noxolo Sibiya for assisting with some with analysis.

References

Abbasi, T., Tauseef, S. M., & Abbasi, S. A. (2011). *Biogas energy* (Vol. 2). Berlin: Springer.

Al Seadi, T., et al. (2008). *Biogas Handbook—University of Southern Denmark Esbjerg*. ISBN 978-87-992962-0-0.

Angelidaki, I., et al. (1996). *The biogas process. Lecture notes for: Energy from biomass (6362).*

Angelidaki, I., & Ahring, B. (1997). Anaerobic digestion in Denmark. Past, present and future. In *III Curs d'Enginyeria Ambiental*. Lleida (pp. 336–342).

Angelidaki, I., Sorensen, A., & Schmidt, J. (2007). *Notes for the course 12133: Environmental biotechnology*. Lyngby: Technical University of Denmark.

Annachhatre, A. P. (2012). *Dry anaerobic digestion of municipal solid waste and digestate management strategies*. Pathum Thani: Asian Institute of Technology.

Appels, L., et al. (2008). Principles and potential of the anaerobic digestion of waste-activated sludge. *Progress in Energy and Combustion Science, 34*(6), 755–781.

Azbar, N., Ursillo, P., & Speece, R. E. (2001). Effect of process configuration and substrate complexity on the performance of anaerobic processes. *Water Research, 35*(3), 817–829.

Bitton, G. (2005). *Overview of co-digestion study. Wastewater microbiology* (Vol. 3, pp. 1–13). Hoboken: Wiley.

Boe, K. (2006). *Online monitoring and control of the biogas process* (Ph.D. thesis). Technical university of Denmark.

Braun, R. (2002). *Potential of Co-digestion.* http://www.novaenergie.ch/iea-bioenergytask37/Dokumente/final.PDF. Access June 10, 2015.

Campos, E., Palatsi, J., & Flotats X. (1999). Co-digestion of pig slurry and organic wastes from food industry. In *Proceedings of the 2nd international symposium on anaerobic digestion of solid waste, Barcelona, Junio, Spain.*

Desai, M., Patel, V., & Madamwar, D. (1994). Effect of temperature and retention time on biomethanation of cheese whey-poultry waste-cattle dung. *Environmental Pollution, 83*(3), 311–315.

Forgács, G. (2012). *Biogas production from citrus wastes and chicken feather: Pretreatment and co-digestion*. Göteborg: Chalmers University of Technology.

Gerardi, M. H. (2003). *The microbiology of anaerobic digesters*. London: Wiley.

Goosen, R. (2013). Economic growth potential for vehicular biogas, Midrand: Industrial development corporation. In *Biogas national conference, industrial development, Midrand.*

Grimberg, S. J., et al. (2015). Anaerobic digestion of food waste through the operation of a mesophilic two-phase pilot scale digester—assessment of variable loadings on system performance. *Bioresource Technology, 178,* 226–229.

Hashimoto, A. G., Chen, Y. R., & Varel, V. H. (1981). *Theoretical aspects of methane production: State-of-the-art* [*Manures*].

Henze, M. (2008). *Biological wastewater treatment: Principles, modelling and design* (pp. 401–437). London: IWA Publication.

Peter, J. J. (2009). *Biogas-Green Energy*. http://scitech.au.dk/fileadmin/DJF/Kontakt/Besog_DJF/Oevelsesvejledning_og_baggrundsmateriale/Biogas_-_Green_Energy_2009_AU.pdf. Accessed June 9, 2015.

Kangle, K. M., et al. (2012). Recent trends in anaerobic codigestion: A review. *Universal Journal of Environmental Research and Technology, 2*(4), 210–219.

Latinwo, G. K., & Agarry, S. E. (2015). Modelling the kinetics of biogas production from mesophilic anaerobic co-digestion of cow dung with plantain peels. *International Journal of Renewable Energy Development (IJRED), 4*(1), 55–63.

Lehtomäki, A., Huttunen, S., & Rintala, J. A. (2007). Laboratory investigations on co-digestion of energy crops and crop residues with cow manure for methane production: Effect of crop to manure ratio. *Resources, Conservation and Recycling, 51*(3), 591–609.

Liu, Yuan, X. Z., Zeng, G., Li W., & Li J. (2008). Prediction of methane yield at optimum pH for anaerobic digestion of organic fraction of municipal solid waste. *Bioresource technology, 99* (4), 882–888.

Mata-A, J., Mace, S., & Llabres, P. (2000). Anaerobic digestion of organic solid wastes. An overview of research achievements and perspectives. *Bioresource Technology, 74*(1), 3–16.

Matheri, A. N., Belaid, M., Tumisang, S., Ngila, C. J. (2015a). Modelling the kinetics of biogas production from co-digestion of pig waste and grass clippings. In *24th World Congress on Engineering (WCE 2016) London-UK, 29 June 2015.*

Matheri, A. N., Belaid, M., Tumisang, S., & Ngila, C. J. (2015b). The kinetic of biogas rate from cow dung and grass clippings. In *7th IIENG international conference of latest trends in engineering and technology (ICLTET'2015) Pretoria, South Africa, November 2015*.

Matheri, A. N., Belaid, M., Tumisang, S., & Ngila, C. J. (2016). Role of impact of trace elements on anaerobic co-digestion in biogas production. In *24th World Congress on Engineering (WCE 2016) London-UK, 29 June 2016*.

Murto, M., Björnsson, L., & Mattiasson, B. (2004). Impact of food industrial waste on anaerobic co-digestion of sewage sludge and pig manure. *Journal of Environmental Management, 70*(2), 101–107.

Noshy, R. (2013). *Optimization of bioenergy solutions at different farm scales* (pp. 17–84).

Sreenivas, R., Retter, R. A. & Hobbs, P. J. (2010). Effect of biomass hydrolysis on biogas production. *Process Biochem, 28*(2), 119–123.

Rice, E. W., Bridgewater, L., & A.P.H.A Association. (2012). *Standard methods for the examination of water and wastewater*. Washington, DC: American Public Health Association.

Rincon, B., et al. (2007). The effect of organic loading rate on the anaerobic digestion of two phase, olive mill solid residue derived from fruits with low ripening. *Journal of Chemical Technology and Biotechnology, 82,* 2.

Ryckebosch, E., Drouillon, M., & Vervaeren, H. (2011). Techniques for transformation of biogas to biomethane. *Biomass and Bioenergy, 35*(5), 1633–1645.

S.A.B.I.A. (2015). *Biogas Info-SABIA Southern African Biogas Industry Association, Southern African Industry Association*. http://biogasassociation.co.za/biogas-info/. Accessed April 9, 2015.

S.A.B.I.A., South African Biogas Industry Association. (2015). biogasassociation.co.za. http://biogasassociation.co.za/website/wp-content/uploads/2015/01/The-State-of-Waste-to-Energy-Research-in-South-Africa_Part1. Accessed April 9, 15.

Sabonnadière, J. C. (2010). *Renewable energy technologies* (p. 146). London: Wiley.

Salminen, E. A., & Rintala, J. A. (2002). Semi-continuous anaerobic digestion of solid poultry slaughterhouse waste: Effect of hydraulic retention time and loading. *Water Research, 36*(13), 3175–3182

Smith, P. H., & Mah, R. A. (1966). Kinetics of acetate metabolism during sludge digestion. *Applied Microbiology, 14*(3), 368–371.

Somayaji, D., & Khanna, S. (1994). Biomethanation of rice and wheat straw. *World Journal of Microbiology & Biotechnology, 10*(5), 521–523.

Sreekrishnan, T. R., Kohli, S., & Rana, V. (2004). Enhancement of biogas production from solid substrates using different techniques—a review. *Bioresource Technology, 95*(1), 1–10.

Sterling, M. C., et al. (2001). Effects of ammonia nitrogen on H_2 and CH_4 production during anaerobic digestion of dairy cattle manure. *Bioresource Technology, 77*(1), 9–18.

Sundararajan, R., Jayanthi, S., & Elango, R. (1997). Anaerobic digestion of organic fractions of municipal solid waste and domestic sewage of Coimbatore. *Indian Journal of Environmental Health, 39*(3), 193–196.

Ten, B. E., & Koster, I. W. (1990). Enhancement of dry anaerobic batch digestion of the organic fraction of municipal solid waste by an aerobic pretreatment step. *Biological Wastes, 31*(3), 199–210.

Tiehm, A., et al. (2001). Ultrasonic waste activated sludge disintegration for improving anaerobic stabilization. *Water Research, 35*(8), 2003–2009.

Wang, X., et al. (2012). Optimizing feeding composition and carbon–nitrogen ratios for improved methane yield during anaerobic co-digestion of dairy, chicken manure and wheat straw. *Bioresource Technology, 120*, 78–83.

Werner, K., et al. (1997). http://biogas.ifas.ufl.edu/ad_development/documents/biogasdigestvol1.pdf. Accessed June 15, 2015.

Wu, W. (2000). Anaerobic co-digestion of biomass for methane production: Recent research achievements. *Optimization, !,* 1VS.

Application of Different Pre-treatment Techniques for Enhanced Biogas Production from Lawn Grass: A Review

N.T. Sibiya, H. Tesfagiogis and E. Muzenda

Abstract Biogas is considered as the best possible alternative source of energy. Biogas is clean, cheap, sustainable, and environmentally friendly. The biogas production potential from energy crops, grass, in particular, has been investigated widely. Although grass produces quality methane in comparison to other feedstock, low biogas yield has been a common challenge. This is because the grass is part of lignocellulosic biomass, consisting of cellulose and hemicelluloses, which are linked together by strong chemical bonds. These characteristics make it resistant to anaerobic digestion. Pre-treatment including chemical, mechanical, thermal, biological and/or a combination of them may be used to break the ether bonds, resulting in an easy accessibility of sugars present in a substrate for the enzyme. For these reasons, this chapter evaluates the various pre-treatment techniques to optimize the biodegradability of grass. All the pre-treatment techniques have been reported to increase biogas production rate. Among the pre-treatments options, chemical pre-treatment has been considered as the most preferred though its operational cost is still under investigation.

Keywords Biogas · Cellulosic · Grass · Hemicelluloses · Lignin Lignocelluloses

---

N.T. Sibiya (✉) · E. Muzenda
Department of Chemical Engineering, Faculty of Engineering and the Built Environment, University of Johannesburg, Doornfontein Campus, John Orr Building, 3160 Beit Street, Johannesburg, South Africa
e-mail: noxolo.sibiya@yahoo.com

H. Tesfagiogis
Agricultural Research Council—Institute for Industrial Crops, Private Bag X82075, Rustenburg 0300, South Africa

E. Muzenda
Department of Chemical, Materials and Metallurgical Engineering, College of Engineering and Technology, Botswana International University of Science and Technology, Private Mail Bag 16, Palapye, Botswana

W. Leal Filho and D. Surroop (eds.), *The Nexus: Energy, Environment and Climate Change*, Green Energy and Technology, https://doi.org/10.1007/978-3-319-63612-2_15

1 Introduction

The constantly increase in energy demand associated with population growth, industrial expansion, increase in commercial as well as transportation growth is a complex and important challenge that needs to be addressed with immediate effect (Singh and Parida 2015). In South Africa (SA) in particular, energy demand has been reported to increase by approximately 15% from 2005 to 2015 and is expected to double the current level by 2030 (Winkler et al. 2011). At the rate at which energy is being utilized, fossil fuels (petrochemicals, oil, and coal) which serve as a major source of energy worldwide will not be sufficient in the near future (Elias et al. 2010). Excessive fossil fuel consumption will not only lead to unprecedented pressure on the environment (Varma 2008) but also affect the human health adversely (Elias et al. 2010). Carbon dioxide (CO_2) emitted during combustion of fossil fuel leads to an increase in greenhouse gases (GHG) which in turn intensifies the hazards of global warming and air pollution (Varma 2008). The preceding issues raise an interest in research for an alternative source of energy, preferably renewable. Among renewable source of energy, biogas has been proven a promising prospect for the future use. Biogas is produced when bacteria convert organic matter to methane gas (55–70%), carbon dioxide (20–40%) and trace gases through the anaerobic process, referred as anaerobic digestion (Streekrishnan et al. 2004).

Although biogas can be produced from several organic matters, the utilization of energy crops for biogas and biochemical production is increasing rapidly, partly, due to their abundance and sustainability. Among energy crops used in anaerobic digestion, the grass is considered as more economically suitable feedstock for a process (Khor et al. 2015). Previous researchers reported that biodegradability of grass is limited due to the fact that grass is lignocellulosic biomass (Khor et al. 2015; Behera et al. 2014; Lagerkvist et al. 2012; Berglung et al. 2012); it is made up of structural carbohydrate cellulose and hemicelluloses which are connected together by complex of lignin (Behera et al. 2014; Nizami et al. 2009). In general, cellulose forms a frame which is surrounded by hemicellulose and lignin. The content of cellulose and hemicelluloses on grass depends on the species, variety, climate, soil fertility, fertilization and/or sometimes on the age of the substrate. Table 1 shows the composition of lignocelluloses in different types of grass. Normally, average grass composition ranges between 10 and 30, 15 and 50, 24 and 40% for Lignin, Hemicellulose, and Cellulose, respectively (Nizami et al. 2009). The set up of lignocelluloses biomass make it difficult for the enzymes to access sugar during biodegradation, thus the biodegradability of the lignocellulosic substrate is dependent on their complex structure (Zhao et al. 2012).

2 Pre-treatment

The biodegradation of lignocellulosic can be enhanced by pre-treatment (Yang et al. 2015). Generally, the pre-treatment options used before anaerobic digestion are mechanical (milling or microwave), chemical (alkali or acidic), biological (enzymes, bacteria, fungi), thermal (steam explosion, hydrothermal, liquid hot water), and/or combination pre-treatment. Figure 1 represents different pre-treatments for optimization of biogas production from the grass. Apparently, pre-treatment plays a huge role in destroying lignin wall protecting hemicelluloses and cellulose (Taherzadeh and Karimi 2008); decrease crystallinity of cellulose; and increase

Table 1 Composition of various grasses

Types of grass	Lignin (%)	Hemicellulose (%)	Cellulose (%)	References
Coastal bermuda grass	6.4	35.7	25.0	Sanchez (2009)
Switch grass	12.0	31.4	45.0	Sanchez (2009)
S32 rye grass (early leaf)	2.7	15.8	21.0	Sanchez (2009)
S32 rye grass (seed settling)	7.3	25.4	26.0	Sanchez (2009)
Orchard grass	4.7	40.0	32.0	Sanchez (2009)
Grass (average value for grasses)	10.0–30.0	15.0–50.0	25.0–40.0	Nizami et al. (2009), Xie et al. (2011), Sanchez (2009)
Grass esparto	17.0 > 19.0	27.0 > 32.0	33.0 > 38.0	Sanchez (2009)
Grass sabia	22.0	23.9	n/a2	Sanchez (2009)
Elephant/giant grass	23.9	24.0	22.0	Sanchez (2009)

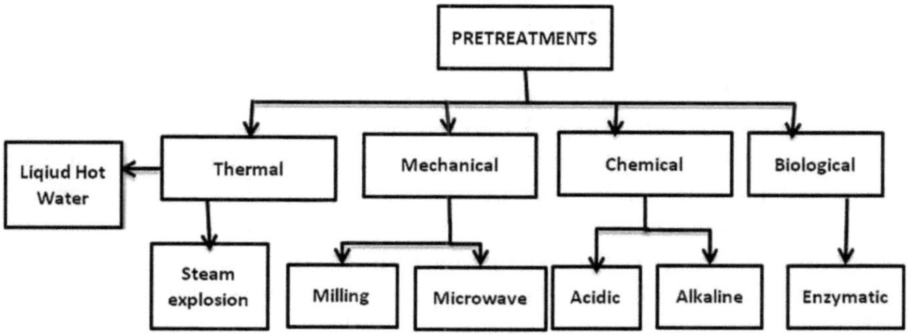

Fig. 1 Pre-treatment for improving degradability of grass

porosity and surface area of the substrate for the enzymes to access the substrate (Taherzadeh and Karimi 2008; Li et al. 2010). In addition, pre-treatment can result in high sugar recovery, increase in biodegradability of material, which ultimately results in an increase in methane and biogas production. According to Zheng and Rehman (2014), pre-treatment plays a huge role in the cost and the efficiency of enzymatic hydrolysis and fermentation; hence, it is the first major step in the process of anaerobic digestion. The following criteria should be taken into consideration in order to select an effective pre-treatment process: Evading the conversion of sugars and reducing the formation of inhibitors for further AD steps, conserving hemicelluloses fractions, reducing energy requirements and minimizing operational cost.

2.1 Mechanical Pre-treatment

Mechanical pre-treatment improves biogas yield through particle size reduction allowing enough specific surface area and availability of biomass for microorganisms (Bochmann et al. 2013). According to Kumarad et al. (2010), physical pre-treatment may also assist in the reduction of floating layer that may cause an equipment (reactor) blockage. The hydrolysis kinetics was raised about four times and 50–90% of glucose was increased when the particle size was reduced from 25.52 to 0.78 μm during hydrolysis of cellolusic biomass (Yeh et al. 2010). Tedesco et al. (2014) reported that when about 80% of the particles are sized below 1.6 mm^2, a biogas and methane yield of about 52 and 53% improvement could be achieved. Lindner et al. (2015) investigated the effect of mechanical pre-treatment after anaerobic digestion of maize using a ball mill to grind the substrate for four different pre-treatment time period (0, 2, 5, 10 min). Results showed an increase of 9% in methane production of two-stage maize silage digestate and of 17% of the two-stage hay/straw digestate after the mechanical pre-treatment was applied. Physical pre-treatment can be accomplished using: mills, blenders, screws, and crushers, etc. Though this technique is very attractive and easy to operate, its high-energy requirement during the grinding process makes it more expensive and unacceptable compared to other pre-treatments (Zhao et al. 2012). Kratky and Jirout (2011), studied biomass size reduction machines for enhancing biogas production. They observed that a knife mill used to reduce the particle size of wheat straw from 12.5 to 16 mm required 2.8–7.55 kWh/t. Similar results were reported by Tsapetos et al. (2015). In both studies, the mechanical pre-treatment test resulted in enhanced biogas and methane yields. The methane increase percentage ranged between (8–25%) in comparison to untreated ensiled meadow grass. It was suggested that the electrical conductivity may be used for BMP determination, hence during test, positive results were obtained for biodegradability of ensiled meadow grass but with relatively low R^2 of 0.68 (Tsapetos et al. 2015).

2.2 Chemical Pre-treatment

During chemical pretreatment, the substrate is soaked in a solution (alkali or acidic) at various temperatures. Although acid hydrolysis, ionic liquid, and oxidative pre-treatment are part of chemical pretreatment, chemical pre-treatment for biogas production improvement from grass has been alkaline (Bochmann and Montgomery 2013). Alkali pre-treatment increases the biodegradation of biomass by simply removing the acetate group that hinders hemicelluloses, causing it to be easily accessible to the enzymes (Bochmann and Montgomery 2013). It is also capable of solubilising a portion of lignin, which increases the accessibility of hemicelluloses and cellulose (Bochmann and Montgomery 2013). The efficiency of alkali pre-treatment is dependent on the level of lignocellulosic biomass and operation condition. The most effective and inexpensive alkali solution for chemical pre-treatment is lime (NaOH) and ammonia. The positive effect of NaOH pre-treatment on fiber reduction, total organic carbon and proteins solubilisation, and thereafter the anaerobic degradability of ensiled sorghum forage was reported (Sambusiti et al. 2012). That study showed a methane increase (from 8 to 19%) from when the substrate was compared to an untreated substrate. Moreover, the increase of sodium hydroxide dosage increased the soluble total organic carbon (TOC) (from 12 to 29%) and proteins (from 56 to 72%), reduction of hemicellulose was increased (from 37 to 70%), and lignin contents (from 26 to 70%). Sumblusite et al. (2013), also reported an increase in the first order kinetic constant when the alkaline dosage was increased by 40, 61, 64, 54, and 40% for sorghum. The effect of ammonia soaking pre-treatment on biochemical methane production of switch grass was also studied (Himmelsbach et al. 2010). The results showed that AAS-pre-treated switchgrass produced 65% more methane in comparison to the untreated switchgrass after 21 days of incubation.

Ionic liquid pre-treatment has a more positive effect and benefit in comparison to diluted acid pre-treatment process for switch grass. Li et al. (2010) studied the efficiency of dilute acid hydrolysis and ionic liquid pre-treatment of switchgrass in terms of biomass recalcitrance, delignification, and enzymatic saccharifications. It was reported that a rate of enzyme hydrolysis was optimized by 16.7 fold and a glucan yield of 96.0% when grass was pre-treated with the ionic liquid. Additionally, the pre-treatment of switchgrass with ionic liquid (dissolution and precipitation of cellulose by anti-solvent) resulted in reduced cellulose crystallinity, increased surface area, and decreased lignin content compared to dilute acid pre-treatment. Panneerselvam et al. (2013a, b) evaluated the effect of ozonolysis on *Miscanthus giganteus*, *Miscanthus sinensis* "Gracillimus", *Saccharum arundinaceum* and *Saccharum ravennae* under difference ozone concentration of (40, 50 and 58 mg/ml) in a two ozone flow configuration and reported that up to 59.9% of lignin was removed without cellulose degradation. It was also reported that product from ozonolysis, lignin degradation, in particular, caused the inhibition of enzymes. Hence, glucan conversion was lower when the substrate was pre-treated with Cellic® CTec2 at 0.06 g/g raw in comparison to untreated samples. It was further reported that when grasses were ozonated for maximum de-lignification and washed solid hydrolysed with 0.1 g/g Cellic® CTec2, optimum fermentable sugars were produced (My el Rendimiento Animal 2002).

2.3 Biological Pre-treatment

Biological pre-treatment is the addition of enzyme in a digester in order to break down the polymers in lignocelluloses biomass (Taherzadeh and Karimi 2008). This pre-treatment is considered as the more cost effective solution compared to other pre-treatments (Nizami et al. 2009). Cellulases, xylanases, pectinases, and amylases are normally used for this purpose (Borgstrom 2011). An increase in degradation of plant cell wall was observed when cellulose enzymes were used during the pre-treatment of the substrate (My el Rendimiento Animal 2002). A study conducted by Romano et al. (2009) where the effect of the addition of enzyme containing cellulose, hemicellulase and β glucosidas were evaluated on anaerobic digestion of Jose tall Wheat grass, showed a positive effect on the solubilization of wheat grass when used directly to treat the grass. During this test, the rate of biogas production was affected for the first few days, though biogas yield and methane yield were not improved. Additionally, Sonakya et al. (2001) reported an increase of 7–14% in methane production when wheat grains were pre-treated with Trizyme (cellulose, a-amylase, and protease) prior to anaerobic digestion. According to Nizani et al. (2009), inoculants containing hetero-fermentative bacteria facilitate the production of intermediates from ethanogens, thus its utilization may be beneficial for the efficiency of anaerobic digestion in comparison to homofermentative bacteria. In addition to cellulose, hemicellulase, glucosidas, and hetero-fermentative bacteria, utilization of fungi to enhance biogas production have been reported (Zhao et al. 2012). Fungi may be classified as brown fungi, white rot fungi, and soft fungi. According to Nizami et al. (2009), white rot Fungi has strong oxidative activity and low substrate specificity ligninolytic enzymes, which allow them to be able to degrade lignin to carbon dioxide and also degrade a variety of persistent environmental pollutants including chlorinated aromatic compounds, heterocyclic aromatic hydrocarbons and various dyes and synthetic high polymers. Brown rot fungi, on the other hand, may depolymerize cellulosic substrate while modifying lignin (Sanchez 2009).

2.4 Thermal Pre-treatment

Thermal pre-treatment can be achieved through the steam explosion and liquid hot water.

2.4.1 Steam Explosion

Zheng et al. (2014) reported that steam explosion has been used to treat various kinds of lignocellulosic biomass for optimization of methane production, including corn stalks, Miscanthus, hardwoods (Japanese cedar, willow, and birch), food processing wastes (citrus and potato pulp). The advantages of steam explosion include enhancement of biogas and methane yield and production rate, reduction of risks of floating layer in a bioreactor, when low-density substrates are treated, increase the convenience of substrate transportation and homogeneity may also be improved. The method may also be used for hemicelluloses hydrolysis (Mussatto and Teixeira 2010). This process is well-researched and tested in a lab- and pilot processes by several research groups and companies (Taherzadeh and Karimi 2008). The most interesting aspect of this method is that its energy cost is low while it meets all the requirements of the pre-treatment process. 20–30% of methane yield was achieved after treatment with a steam explosion. During this treatment the substrate in the reactor is fed under elevated pressure and temperature for about 5–10 min Zhao et al. (2012), then the pressure is suddenly reduced, which makes the material undergo an explosive decompression (Ramano et al. 2009). The equipment is normally operated at the temperature range of 160–260 °C and pressure range of 0.69–4.83 MPa.

Baure et al. (2014) studied different steam explosion condition on late harvested hay digestibility. Enzymatic hydrolysis under 220 °C and 5 min yielded maximum glucose. Maximum xylose was obtained at 175 °C in 10 min. Furthermore, an increase of 15.9% of methane was obtained in pre-treated hay compared to untreated hay. It was concluded that explosion pre-treatment is the suitable process prior to anaerobic digestion for the effective biodegradation of hay to methane. Shafiei et al. (2013) investigated the economic feasibility of steam explosion pre-treatment for improvement of biogas production from wheat straw and paper tube using Aspen plus for simulation for five different plant capacities. He reported that employing steam explosion pre-treatment resulted in 13% higher total capital investment while significantly improving the economy of the biogas plant and decreased the manufacturing cost of methane by 36%. Sui and Chen (2014) established a multi-stage heat transfer model of steam explosion process integrating technical features. The model showed the importance of pressure boost involved in total energy consumption.

2.4.2 Liquid Hot Water (LHW)

Liquid Hot Water pre-treatment is similar to the steam explosion. However, during this, process there is no sudden reduction of pressure, thus there is no explosion. According to Lui et al. (2003), this method is beneficial in a two-phase process (such as a SBLB-UASB digester), because higher concentrations of soluble carbohydrates such as xylan are achieved. In addition, during this method, the concentration of soluble hemicelluloses and lignin are reduced, hence high water input is added in the system (Nizami et al. 2009). Because of positive results from this treatment, it has been widely used to enhance methane production from lignicellulosic biomass such as sunflower stalks, sugarcane bagasse, Miscanthus (species of giganteus and sacchariflorus) and grass (Pennisetum hybrid), paper tube residuals, MSW, and microalgae (Mussatto and Teixeira 2010). Liquid hot water pre-treatment of soybean straw at 170–210 °C and 3–10 min resulted in almost 100% of cellulose removal. The increase in xylan dissolution from the raw material was proportional to an increase in pre-treatment temperature and time. The high amount of glucose yield (70.6%) and xylan (80%) removal was obtained at the at pre-treatment temperature of 210 °C in 10 min. LHW has been reported being more effective pre-treatment in comparison to NaOH soaking pre-treatment of soybean straw (Wan et al. 2011). Yu et al. (2013) investigated the liquid hot water pre-treatment of sugarcane bagasse and its comparison with chemical pre-treatment methods for the sugar recovery and structural changes. Though both treatments enhanced sugar recovery and enzymatic hydrolysis in comparison to an untreated sugar cane 90% of lignin was removed after NaOH pre-treatment. Thus, a novel pre-treatment method combined LHW and aqueous ammonia was developed to reduce energy input and enhance the sugar recovery.

2.5 Combined Pre-treatment

According to Zheg et al. (2014), a single pre-treatment is not 100% able to break the bonds between hemicelluloses and lignin, thus two or more methods may be combined to improve the performance. It has been also reported that combination of pre-treatment reduces energy consumption and enhance product yield (Yu et al. 2013). Zimbardi et al. (2007) investigated the synergistic effect of pre-impregnation by sulphuric acid and steam explosion. Nine conditions were used for the steam explosion treatment selecting the temperature of 180, 190, 200 °C and sulphuric acid loadings of 0, 1.5, 3 wt%. Sugar was recovered using water extraction and cellulose digestibility by enzymes. The results showed that there was an increase of 15.3% of sugar recovery when the process was combined. Basically, when corn was pre-treated at a temperature of 180 °C, only 1.5% of sugar was recovered. Then, after the addition of acid, 16.8% of sugar was recovered. In addition, the digestibility of cellulose was improved. Ferreira et al. (2014) investigated the influence of particle size, water impregnation, and thermal pre-treatment through biochemical methane potential tests (BMP). They obtained maximum methane (i.e., 27%) when straw was pre-treated at 200 °C for 5 min followed by steam explosion compared to non-treated straw. In addition, BMP was enhanced by 4–10% by the impregnation of the straw with water. Cutting (3–5 cm) also showed better results than milling (<1 mm). Effects of thermochemical pre-treatment of grass silage on methane production by anaerobic digestion were studied (Xie et al. 2011). At a temperature of 100 °C and the four NaOH loadings (1, 2.5, 5 and 7.5% by volatile solids), total COD was solubilized up to 45% and Lignin, hemicelluloses and cellulose of about 65.6, 36.1 and 21.2% were removed, respectively. Improvement in methane production by 10–39% was also achieved on the treated grass in comparison to the untreated grass silage. 5% of loading rate was reported as the optimal for grass silage pre-treated condition.

3 Conclusion

Various pre-treatment techniques that have been reviewed in this chapter have all shown to increase sugar recovery by simply breaking down the lignin structure and reducing the crystallinity of cellulose followed by easily accessibility of sugars by enzymes. Among the various pre-treatment, chemical pre-treatment is the most preferred though its operational cost is not clear. Under chemical pre-treatment, NaOH was found to be the best additive for increased substrate degradability. Ionic acid has a unique benefit in comparison to diluted acid. Ozone pre-treatment was found to remove lignin, without cellulose; however, the product from it inhibits enzymes which are responsible for methane production. Ammonia may also be used to increase methane yield and biogas yield. Mechanical pre-treatment is easy to operate although it requires a lot of energy and hence its limitation. Biological pre-treatment have been reported as the most cost effective treatment in comparison to other treatments although it does not always increase methane yield. The utilization of inoculum containing hetero-fermentative bacteria has been recommended as it facilitates the production of an intermediate form of ethanogens. White rot fungi is a strong oxidative agent compared to other fungi like brown rot fund and soft fungi. Steam explosion pre-treatment offers a lot of benefits including reasonable cost and it meets the entire process requirement. Multi-stage heat transfer model of steam explosion process has been proposed for the reduction of energy consumption. LHW is beneficial in a two-phase process (such as a SBLB-UASB digester) because higher concentrations of soluble carbohydrates such as xylan are achieved. Combining these pre-treatment processes can be used for the optimization of biogas and methane production.

Acknowledgements This work was supported by a research grant from South African Energy National Development Institute (SANEDI). The University of Johannesburg Global Excellence Strategy is acknowledged for providing Noxolo Sibiya with a bursary. The Botswana International University of Science and Technology is also acknowledged for financial and technical support.

References

Bauer, A., Lizasoain, J., Theuretzbacher, F., Agger, J. W., Rincón, M., Menardo, S., et al. (2014). Steam explosion pretreatment for enhancing biogas production of late harvested hay. *Bioresource Technology, 166*, 403–410.

Behera, S., Arora, R., Nandhagopal, N., & Kumar, S. (2014). Importance of chemical pretreatment for bioconversion of lignocellulosic biomass. *Renewable and Sustainable Energy Reviews, 36,* 91–106.

Berglund Odhner, P., Sárvári Horváth, I., Kabir, M. M., & Shabbauer, A. (2012). *Biogas from lignocellulosic biomass.*

Bochmann, G., & Montgomery, L. F. R. (2013). *Storage and pre-treatment of substrates for biogas production* (pp. 85–89). Sawston: The Biogas Handbook, Woodhead Publishing.

Borgström, Y. (2011). *Pretreatment technologies to increase the methane yields by anaerobic digestion in relation to cost efficiency of substrate transportation.*

Elias, J. (2010). *Study on renewable biogas energy production from cladodes of Opuntia ficus-indica* (Doctoral dissertation, aau).

Ferreira, L. C., Nilsen, P. J., Fdz-Polanco, F., & Pérez-Elvira, S. I. (2014). Biomethane potential of wheat straw: Influence of particle size, water impregnation and thermal hydrolysis. *Chemical Engineering Journal, 242,* 254–259.

Himmelsbach, J. N., Raman, D. R., Anex, R. P., Burns, R. T., & Faulhaber, C. R. (2010). Effect of ammonia soaking pretreatment and enzyme addition on biochemical methane potential of switchgrass. *Transactions of the ASABE, 53*(6), 1921.

Kamarad, L., Pohn, S., Harasek, M., Kirchmayr, R., Bochmann, G., & Braun, R. (2010). Hydrodynamic characteristics of the biogas plant digester using tracer tests and CFD methods. In *Proceedings of the International Water Association, 12th World Congress on Anaerobic Digestion, Guadalajara, Mexico.*

Khor, W. C., Rabaey, K., & Vervaeren, H. (2015). Low temperature calcium hydroxide treatment enhances anaerobic methane production from (extruded) biomass. *Bioresource Technology, 176,* 181–188.

Kratky, L., & Jirout, T. (2011). Biomass size reduction machines for enhancing biogas production. *Chemical Engineering and Technology, 34*(3), 391–399.

Lagerkvist, A., & Morgan-Sagastume, F. (2012). The effects of substrate pre-treatment on anaerobic digestion systems: A review. *Waste Management, 32*(9), 1634–1650.

Li, C., Knierim, B., Manisseri, C., Arora, R., Scheller, H. V., Auer, M., et al. (2010). Comparison of dilute acid and ionic liquid pretreatment of switchgrass: Biomass recalcitrance, delignification and enzymatic saccharification. *Bioresource Technology, 101*(13), 4900–4906.

Lindner, J., Zielonka, S., Oechsner, H., & Lemmer, A. (2015). Effects of mechanical treatment of digestate after anaerobic digestion on the degree of degradation. *Bioresource Technology, 178,* 194–200.

Liu, C., & Wyman, C. E. (2003). The effect of flow rate of compressed hot water on xylan, lignin, and total mass removal from corn stover. *Industrial and Engineering Chemistry Research, 42* (21), 5409–5416.

Mussatto, S. I., & Teixeira, J. A. (2010). *Lignocellulose as raw material in fermentation processes.*

My el Rendimiento Animal, M. (2002). *Effects of biological additives on silage composition of mott dwarf elephantgrass and animal performance.*

Nizami, A. S., Korres, N. E., & Murphy, J. D. (2009). Review of the integrated process for the production of grass biomethane. *Environmental Science and Technology, 43*(22), 8496–8508.

Panneerselvam, A., Sharma-Shivappa, R. R., Kolar, P., Clare, D. A., & Ranney, T. (2013a). Hydrolysis of ozone pretreated energy grasses for optimal fermentable sugar production. *Bioresource Technology, 148,* 97–104.

Panneerselvam, A., Sharma-Shivappa, R. R., Kolar, P., Ranney, T., & Peretti, S. (2013b). Potential of ozonolysis as a pretreatment for energy grasses. *Bioresource Technology, 148,* 242–248.

Romano, R. T., Zhang, R., Teter, S., & McGarvey, J. A. (2009). The effect of enzyme addition on anaerobic digestion of JoseTall Wheat Grass. *Bioresource Technology, 100*(20), 4564–4571.

Sambusiti, C., Ficara, E., Malpei, F., Steyer, J. P., & Carrere, H. (2012). Influence of alkaline pre-treatment conditions on structural features and methane production from ensiled sorghum forage. *Chemical Engineering Journal, 211,* 488–492.

Sambusiti, C., Ficara, E., Malpei, F., Steyer, J. P., & Carrère, H. (2013). Effect of sodium hydroxide pretreatment on physical, chemical characteristics and methane production of five varieties of sorghum. *Energy, 55,* 449–456.

Sánchez, C. (2009). Lignocellulosic residues: Biodegradation and bioconversion by fungi. *Biotechnology Advances, 27*(2), 185–194.

Shafiei, M., Kabir, M. M., Zilouei, H., Horváth, I. S., & Karimi, K. (2013). Techno-economical study of biogas production improved by steam explosion pretreatment. *Bioresource Technology, 148,* 53–60.

Singh, A. K., & Parida, S. K. (2015). A novel hybrid approach to allocate renewable energy sources in distribution system. *Sustainable Energy Technologies and Assessments, 10,* 1–11.

Sonakya, V., Raizada, N., & Kalia, V. C. (2001). Microbial and enzymatic improvement of anaerobic digestion of waste biomass. *Biotechnology Letters, 23*(18), 1463–1466.

Sreekrishnan, T. R., Kohli, S., & Rana, V. (2004). Enhancement of biogas production from solid substrates using different techniques—A review. *Bioresource Technology, 95*(1), 1–10.

Sui, W., & Chen, H. (2014). Multi-stage energy analysis of steam explosion process. *Chemical Engineering Science, 116,* 254–262.

Taherzadeh, M. J., & Karimi, K. (2008). Pretreatment of lignocellulosic wastes to improve ethanol and biogas production: A review. *International Journal of Molecular Sciences, 9*(9), 1621–1651.

Tedesco, S., Mac Lochlainn, D., & Olabi, A. G. (2014). Particle size reduction optimization of *Laminaria* spp. biomass for enhanced methane production. *Energy, 76,* 857–862.

Tsapekos, P., Kougias, P. G., & Angelidaki, I. (2015). Biogas production from ensiled meadow grass; effect of mechanical pretreatments and rapid determination of substrate biodegradability via physicochemical methods. *Bioresource Technology, 182,* 329–335.

Varma, R. (2008). *Characterization of anaerobic bioreactors for bioenergy generation using a novel tomography technique.* Ann Arbor: ProQuest.

Wan, C., Zhou, Y., & Li, Y. (2011). Liquid hot water and alkaline pretreatment of soybean straw for improving cellulose digestibility. *Bioresource Technology, 102*(10), 6254–6259.

Winkler, H., Hughes, A., Marquard, A., Haw, M., & Merven, B. (2011). South Africa's greenhouse gas emissions under business-as-usual: The technical basis of 'growth without constraints' in the long-term mitigation scenarios. *Energy Policy, 39*(10), 5818–5828.

Xie, S., Frost, J. P., Lawlor, P. G., Wu, G., & Zhan, X. (2011). Effects of thermo-chemical pre-treatment of grass silage on methane production by anaerobic digestion. *Bioresource Technology, 102*(19), 8748–8755.

Yang, L., Xu, F., Ge, X., & Yebo Li, Y. (2015). Challenges and strategies for solid-state anaerobic digestion of lignocellulosic biomass. *Renewable and Sustainable Energy Reviews, 44,* 82–834.

Yeh, A. I., Huang, Y. C., & Chen, S. H. (2010). Effect of particle size on the rate of enzymatic hydrolysis of cellulose. *Carbohydrate Polymers, 79*(1), 192–199.

Yu, Q., Zhuang, X., Lv, S., He, M., Zhang, Y., Yuan, Z., et al. (2013). Liquid hot water pretreatment of sugarcane bagasse and its comparison with chemical pretreatment methods for the sugar recovery and structural changes. *Bioresource Technology, 129,* 592–598.

Zhao, X. Q., Zi, L. H., Bai, F. W., Lin, H. L., Hao, X. M., Yue, G. J., & Ho, N. W. (2012). Bioethanol from lignocellulosic biomass. In *Biotechnology in China III: Biofuels and bioenergy* (pp. 25–51). Berlin: Springer.

Zheng, J., & Rehmann, L. (2014). Extrusion pretreatment of lignocellulosic biomass: A review. *International Journal of Molecular Sciences, 15*(10), 18967–18984.

Zheng, Y., Zhao, J., Xu, F., & Li, Y. (2014). Pretreatment of lignocellulosic biomass for enhanced biogas production. *Progress in Energy and Combustion Science, 42,* 35–53.

Zimbardi, F., Viola, E., Nanna, F., Larocca, E., Cardinale, M., & Barisano, D. (2007). Acid impregnation and steam explosion of corn stover in batch processes. *Industrial Crops and Products, 26*(2), 195–206.

Design and Operation of a Single-Stage Plug-Flow Digester Using Organic Wastes Generated from University of Mauritius Cafeteria

Geeta Somaroo, Santaram Venkannah, Anuksha Boojhawon, Sanjana Gunasee, Romeela Mohee and Hemraj Ramnarain

Abstract Anaerobic digestion of solid wastes is considered as an attractive technique for the conversion of organic wastes into energy. There is a need to upgrade towards the development of biodigesters that would be applicable in Mauritian context for bio-energy production. In this paper, a 3-zone plug-flow anaerobic digestion system was designed for the food wastes generated at around 20 kg/day from the cafeteria at the University of Mauritius. A rectangular septic tank of dimensions length 2000 mm, width 1450 mm and height 1400 mm was chosen and modified to meet the requirement of the designed digester. The anaerobic digester was fed with cow dung to act as inoculums. Parameters such as temperature, pH and volatile fatty acids (VFA) were monitored to assess the performance of the digester. The digester was operated under mesophilic conditions to produce biogas. A high specific average daily yield of biogas of 50.25 L/kg of food wastes added was produced. An optimum pH was maintained with the cow dung acting as a good buffer.

Keywords Biogas · Plug-flow digester · Energy · Organic wastes

G. Somaroo · S. Gunasee (✉)
Department of Chemical and Environmental Engineering, University of Mauritius, Reduit, Mauritius
e-mail: san_gunasee@hotmail.com

S. Venkannah
Department of Mechanical and Production Engineering, University of Mauritius, Reduit, Mauritius

A. Boojhawon
Ministry of Environment, Port-Louis, Mauritius

R. Mohee
University of Mauritius, Reduit, Mauritius

H. Ramnarain
Compagnie Mauricienne de Textile Ltee, La Tour Koenig, Mauritius

W. Leal Filho and D. Surroop (eds.), *The Nexus: Energy, Environment and Climate Change*, Green Energy and Technology, https://doi.org/10.1007/978-3-319-63612-2_16

1 Introduction

With the rapidly increasing global energy demand, many experts have come to the fore to predict that traditional fossil fuel will come to an end in the near future. Mauritius, being of volcanic origin, is highly dependent on imported petroleum such as gasoline, diesel, aviation fuel, kerosene, heavy fuel oil and liquefied petroleum gas to meet its energy demands in the domestic, industrial and transport sectors (Pillay et al. 2012; Mohee and Mudhoo n.d). With a rise in population growth and standard of living, the total energy requirement of the island increased from 1157.3 thousand tonnes of oil equivalent (ktoe) in 2002 to 1426.8 ktoe in 2011 leading to 33% ascend in the imported energy requirements (EEMO 2011). On the other hand, the increasing amount of solid wastes generation in Mauritius has led to more pressure on the current solid waste management strategy and a decrease in the lifespan of the only operating sanitary landfill in Mauritius.

Considering waste management and energy issues, solid wastes are being preferred as a substrate for energy recovery systems (Speight 2011). Several options such as incineration, sanitary landfill, gasification, pyrolysis and anaerobic digestion are adopted worldwide to convert solid wastes into energy. Food wastes are usually not suitable for energy conversion technology such as incineration, gasification and pyrolysis due to its high moisture content. Food waste management is of prime importance since it is known to cause odour, hygiene issues and adverse environmental effects (Mata-Alvarez et al. 2000). Anaerobic digestion is referred to as the most attractive technology for the treatment of food wastes given its comparatively less cost issues for the production of the same amount of energy as compared to the other technologies (Chaudary 2008). Anaerobic digestion can be biochemically defined as an engineered methanogenic decomposition of organic compounds into a biogas and microbial cells by a coordinated action of diverse microbial association in an oxygen free environment (Chaudary 2008; Corbit 1989; Appels et al. 2008). The whole process is usually divided into four stages namely enzymatic hydrolysis, fermentation, acetogenesis and methanogenesis (Zhang et al. 2014). The four types of microorganisms that are involved in the anaerobic process are hydrolytic, fermentative, acetogenic and methanogenic bacteria (Boone et al. 1993). The generation of the gas depends on several factors such as pH value, temperature, total solid content of the material, loading rate, seeding, diameter to depth ratio, carbon to nitrogen ratio, nutrients and retention time (Palatsi et al. 2009; Hartmann and Ahring 2006; Mashad et al. 2004; Wang et al. 2012). The biogas usually contains 65–70% of methane, 30–35% of carbon dioxide and trace amounts of nitrogen, hydrogen, hydrogen sulphide and water vapour, and has an energy potential of around 21 MJ/m^3 (Appels et al. 2008; Pourmovahed et al. 2011). The methane once purified and upgraded can be used as a transportation fuel (Yang et al. 2015). The digestate generated can be usually used as fertilizer since it contains nitrogen and sulphur as nutrients (Li et al. 2011).

Anaerobic digestion processes can be classified according to operating parameters and reactor design (Guendouz et al. 2008). In developing countries, anaerobic digesters are more often operated on low solids or wet system model to ease the waste feeding process (Lacour 2012). These digesters can further be broken down into three categories of feeding namely batch, semi-continuous and continuous feed that influence the design of the bio digesters (Héduit 1993). Batch systems are operated on dry modes, with Héduit (1993) and Vandevivere et al. (2000) reporting percentages of solids ranging from 15–30 to 30–40% respectively. Once the digestion process is completed, the reactor is emptied. The gas is allowed to be produced for a period of 40–50 days. The main disadvantage of the batch process is the lack of maintaining the correct microbial activity and thus affecting the production of the gas (Jain et al. 2015) Semi-continuous digestion is characterized by a random loading in terms of quantity and quality, a feeding rate of once or twice per day and a continuous discharge of effluent (Chaudary 2008; Héduit 1993). In a continuous system, feeding of wastes and withdrawal of effluent takes place non-stop leading to an uninterrupted digestion process with the benefits of a constant rate of biochemical reactions and biogas generation (Chaudary 2008). Two such common designs are the continuous stirred tank reactor (CSTR) model and the Plug-flow model. CSTR avoids the occurrence sediments deposits and scum formation, promotes intimate contact between the feed and active biomass and ensures a uniform prevalence of temperature and substrate concentration within the digester (Appels et al. 2008; Lacour 2012). The plug-flow model can be categorized into a vertical or horizontal arrangement with the latter model is the most commonly employed (Lacour 2012). Horizontal plug-flow model can be demonstrated by lower volatile fatty acids (VFA) concentrations in the effluent, a higher degree of digestate retention time, a more stable performance with higher biogas generation rate and lower investment costs (Rajendran et al. 2012; Van Lier et al. 1994; Anjan 1988).

Small scale digesters are mostly used in developing countries as a cooking fuel, heating water or for the generation of electricity (Ocwieja 2010). Fixed dome, floating drum and tubular are the most common types of digesters that can be found in developing countries (Kinyua et al. 2016). In order to properly design and operate a small scale digester to generate enough energy to meet the requirements of a specific community, several factors such as temperature inside the digester, organic load rate and retention times have to be carefully controlled (Kinyua et al. 2016).

Studies have been carried out in the local context in the field of biogas production from organic fraction of municipal solid wastes which could generate a power of 222 KWh/tonnes of biowaste digested (Surrop and Mohee 2012). However, stepping out of the research housing, the need to upgrade towards the development of reactors for anaerobic digestion of wastes is of utmost importance to gauge the significance of the pool of research based-results on a larger field of study and to assess the real-life application of bio-energy production. As such, the aim of this study was to determine the type of digester suitable for the targeted feedstock, as well as the design elements that would be required for the construction of such a digester and to assess the biogas production potential from the digester using organic wastes that are being generated at the Cafeteria of the University of Mauritius as an alternative to LPG use.

2 Materials and Methods

The cafeteria of the University of Mauritius generates around 20 kg of food wastes per day, which are eventually disposed of at the sanitary landfill at Mare Chicose. In view of finding alternatives to the utilization of LPG gas at the University, a project was initiated to investigate the potential of biogas production from these food wastes. As such, the design and construction of a low-cost digester was conducted.

2.1 *Feedstock and Preparation*

The targeted feedstocks for this study were the food wastes from the cafeteria of the University of Mauritius. The food wastes were being generated at a rate of 20 kg/day. The food wastes consisted mainly of vegetable and fruit wastes such as peelings. Cow dung, which was obtained from the University farm, acted as the inoculum in this study. The feedstocks used were characterized in terms of moisture content, total solid content (TS), volatile solids content (VS), and the fixed solids content (FS). These characteristics of the feedstocks enabled the determination of the amount of inoculum that will be required inside the digester.

2.2 *Design of Digester*

2.2.1 Model Selection

After the analysis of various types of digesters that are currently available, the horizontal plug-flow model (3-zone plug-flow process) was found to be better suited for the purpose of this study in terms of operation, cost and biogas production potential. Unlike continuous stirred tank reactors (CSTR), horizontal plug-flow digesters do not require agitation devices within the digester (Chaudary 2008). Also, plug-flow systems do not require heating when this type of digester is utilized in a tropical climate such as Mauritius due to its potential of retaining heat absorbed during winter conditions (Chara et al. 1999). The construction of plug-flow digesters is relatively cost effective as they operate on a low solid content of 11–14% (Appels et al. 2008; Vesilind 2003; Kumar and Singh 2006). Plug-flow digesters can either be operated in a continuous mode or in a semi-continuous mode (Rajendran et al. 2012). According to Hobson (1990), plug-flow reactors' performance is characterized by the higher organic matter removal efficiencies as compared to CSTR. In addition, plug-flow reactors can reduce the nitrogen and phosphorus concentrations to 35–70% (Lansing et al. 2008); control volatile fatty acids (VFA) concentrations against highly acidic degradation of food wastes Chanakya and Moletta (2004) and produce higher biogas yield (Hayes et al. 1980).

2.2.2 Material of Construction

The selection of the material of construction of the digester depends on the materials that are readily available in Mauritius, such as concrete, steel, polyethylene and Fibreglass. Steel was found to be the most recommended material for the construction of the digester. However, the cost of construction of a steel digester was found to be very high and thus steel did not satisfy the cost requirement of this study. Digesters made of concrete are normally constructed underground and thus a dedicated space was required for this type of installation and this was within the scope of this study. With time, digesters made of polyethylene may crack under different climatic conditions. As a result, Fibreglass was found to be three times lighter than a steel tank and demonstrated the structural strength of 3–6 times greater than that of a steel tank. Digesters made of fibreglass can easily be transported without being subjected to any damage due to its rigidity and durability. Hence, the plug-flow digester used in this study was made of fibreglass.

2.2.3 Design Parameters

In order to achieve the construction of a low-cost and effective digester, various types of available tanks made of Fibreglass were analyzed in order to meet the required design elements. After consultation with various companies, it was found that it would be more appropriate to modify an existing tank to meet the requirements of this study as the construction of a new mould is expensive and this would have doubled the cost of construction of the digester. Consequently, an existing rectangular septic tank model having dimensions as follows: length (L) = 2600 mm, width (W) = 1450 mm, height (H) = 1400 mm, which was obtained from Fibre Marine Ltd. was chosen and modified so as to suit the requirements for the design of the digester. Also, a rectangular surface was found to be ideal by Lacour (2012) to adapt to the horizontal flotation of the degrading biomass which is further enhanced by a length to width (L/W) ratio of 2–2.5 (Jagadish et al. 1998). The dimensions of the rectangular septic tank proposed by Fibre Marine Ltd. hence closely met the desired L/W ratio of 1.8. Based on time and budget restrictions, this type of tank was also found to be to the most appropriate for this pilot-scale study.

The volumetric flow rate of the input to the digester was based on the amount of food wastes generated from the University Cafeteria. A retention time of 30 days was assumed as suggested by Chanakya et al. (2004) for the calculation of the volume of the digestion chamber. In addition, Chanakya et al. (2004) emphasized that a longer retention time did not lead to additional benefits in the biogas production. A safety factor of 0.25 was utilized for the design.

For sizing of the gas space, Chanakya et al. (2004), specified, a biogas generation rate of 50–60 L/kg of feedstock, having a moisture content of about 85% for a

3-zone plug-flow model. A retention time of 2 h was thus assumed in this study. The storage space required was calculated using Eq. (1).

$$\text{Storage space required} = \text{Biogas produced from feedstock} \times \text{retention time} \quad (1)$$

A headspace was also calculated based on a submerged period of 5 days (Lacour 2012) which is a worst case scenario with the assumption that there has been no volume reduction of the feedstock within 5 days of anaerobic treatment. A submersion percentage of 30 of the wastes for a period of 25 days was assumed for the calculation of the volume of wastes floating. The total headspace required was thus calculated as the sum of the space for gas storage and for the floating biomass.

Biogas production was estimated at 1.25 m^3/day. A total storage capacity 3 m^3 of the digester was hence determined to take into account the fluctuations in biogas generation, as well as, the non-utilization of the biogas produced during the weekend and public holidays.

2.2.4 Operation and Monitoring

Once the digester was constructed, it was installed and operated at the University Farm. Initially, the digester was filled with cow dung slurry at a cow dung to water weight ratio of 0.44 to maintain an optimum solid content of 8–10% within the digester. The top of the digester was then sealed using temperature resistant red silicon to prevent air from entering and loss of biogas. The acclimatization of the cow dung inside the digester was allowed for a period of 35 days before the addition of the food wastes. Cafeteria food wastes were then added at an average rate of 7.96 kg per working day depending on the availability of the wastes generated from the cafeteria.

Samples were taken on alternate days from the start of the process to test parameters such as temperature, pH and VFA so as to assess the performance of the digester.

The daily biogas production was monitored by making use of the pump and the flow metre. For the purpose of monitoring biogas production, the biogas was stored in the biogas bags which were purged into the atmosphere by the use of the pump. The amount of biogas purged was recorded by the flow metre.

3 Results and Discussions

3.1 Initial Characteristics of Feedstock

The initial characteristics of the feedstocks are presented in Table 1. The moisture content was determined using ASTM E790-87. The initial moisture content of food

Table 1 Characteristics of feedstock

Feedstock	Moisture content (%)	Total solids (%)	Volatile solids (%)	Fixed solids (%)
Food wastes	92	8	7.2	0.8
Cow dung	80	20	17.3	2.7

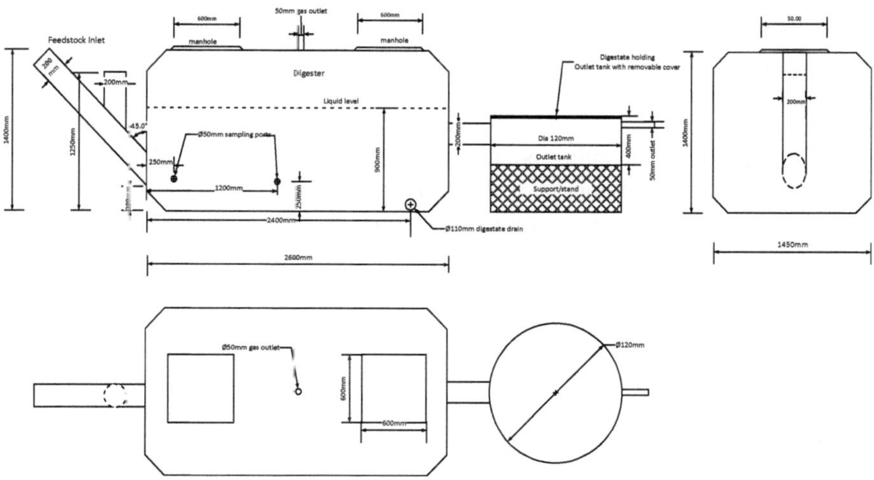

Fig. 1 Schematic representation of digester

wastes was found to be 92%. Other studies (Babaee and Shayegan 2011) also reported high values of moisture contents for food wastes. The moisture content of fresh cow dung generally varies between 80 and 90% by weight (Vaiieln 1969) and this is in accordance with the value obtained in this study. The initial value of the total solids content (TS) of the cow dung was determined to be 20%. In order to reduce the TS of the cow dung, water was added to bring its value to 8%. A TS of 8%, TS of the mixture promotes good flow characteristics inside the plug-flow digester. The TS of food wastes was found to be 8% (Babaee and Shayegan 2011). The volatile solids content (VS) provides a good indication of the fraction of solids that can be converted into biogas. From Table 1, it can be observed that the fixed solids (FS) content of food wastes and cow dung were 0.8 and 2.7%, respectively.

3.2 Digester Design

For a food wastes generation rate of 20 kg/day and a retention time of 30 days inside the digester, a working volume (or digestion chamber) of 3.36 m^3 and a headspace (or gas dome) of 1.70 m^3 were obtained. An inlet pipe diameter of 200 mm was introduced at an angle of 45° with the lower edge of the pipe 200 m above the digester base, as shown in Fig. 1. A 200 mm inlet pipe was chosen so as to prevent any clogging inside the inlet pipe when the wastes were introduced as-received inside the digester. An inclination of 45° was chosen (Chanakya and Moletta 2004) in order to ensure proper submersion of the buoyant wastes under the digester liquid so as to promote acid digestion, volatile solids destruction and dispersion of volatile fatty acids (VFA). A gas outlet of diameter 10 mm was fitted at the centre on the top of the digester.

Sampling taps were fixed at a height of 250 mm from the base at three different positions namely: 0.25, 1.20 and 2.40 m along the length of the digester. The first sampling point covers the acidification process of the submerged wastes, the second one encompassed the flotation section of the fermenting mass which started to manifest itself after 5 days of submersion (Lacour 2012; Chanakya and Moletta 2004) and the third sampling point was used to collect the stabilized sample as shown in Fig. 2. In addition, an outlet drainage pipe of 110 mm was introduced at the base of the digester for the removal of solids which had settled out. A circular outlet tank was connected to the digester with a pipe of 200 mm diameter. The digester dimensions are represented in Table 2. The gas outlet was connected to two biogas storage bags having capacities of 1500 and 2500 L, respectively. The biogas produced was measured using a flow metre and was purged to the retrofitted stove by the biogas pump. Figure 2 shows the process flow diagram of the whole system when in operation. A total cost of 131,304 MUR was required for the setting up and operation of the digester, as shown in Table 3.

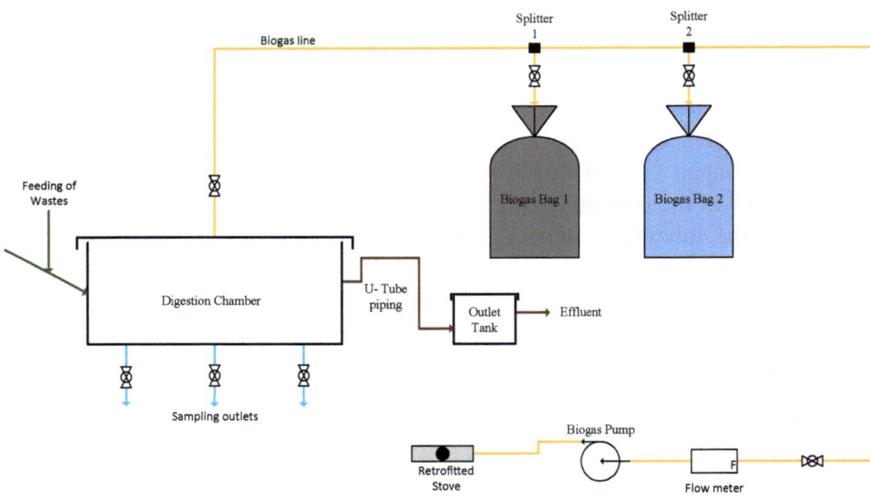

Fig. 2 Process flow diagram for the whole system

Table 2 Digester dimensions

Components	Length (m)	Width/dia (m)	Height (m)	Volume (m^3)
Digestion chamber	2.6	1.45	0.9	3.393
Gas dome	2.6	1.45	0.5	1.885
Outlet tank	–	1.20	0.40 m	0.451
Sampling ports	–	0.05	0.25 above ground	–
Digestate drain	–	0.11	–	–
Connecting pipe (main chamber—outlet tank)	–	0.20	–	–
Inlet of tank	–	0.20	0.20 above ground	–

Table 3 Digester costing

Equipment	Cost (MUR)
Biogas bags, biogas pump, pressure gauge, pressure relief valves, gas flow metre including air freight	28,044
4″ ball valves, 2″ ball valves and seal tapes	18,057
Air valves, 1/2″ (Rs. 350/piece for 10 pieces)	3500
Construction of digester tank	69,000
Stove	1713
Pipes	990
Other fittings	10,000

3.3 Process Parameters

3.3.1 Biogas Production

Figure 3 represents the daily biogas production. During the first week of operation, no biogas was produced due to leakage of the gas over the cover seal. Hence, biogas production started immediately after solving this leakage problem. The availability of wastes ranged from five to 13 kg/day. The digestion of food wastes occurred over a period of 59 days. The addition of food wastes led to an increase in biogas production as shown in Fig. 3. The volume of biogas produced ranged from 132.6 to 920 L/day. The biogas specific yield averaged to 50.25 L/kg of food wastes added and this was in line with Chanakya and Moletta (2004) findings. After 60 days of digestion, an average biogas yield of 1 m^3/kg of food wastes added was obtained. The biogas constituted of 59.3% CH_4 with a methane yield of 1 m^3/kg VS added. Literature values for methane yield from food wastes ranged from 0.396 to 0.472 m^3/kg VS added (Cho and Park 1995; Zhang et al. 2007, 2011). The

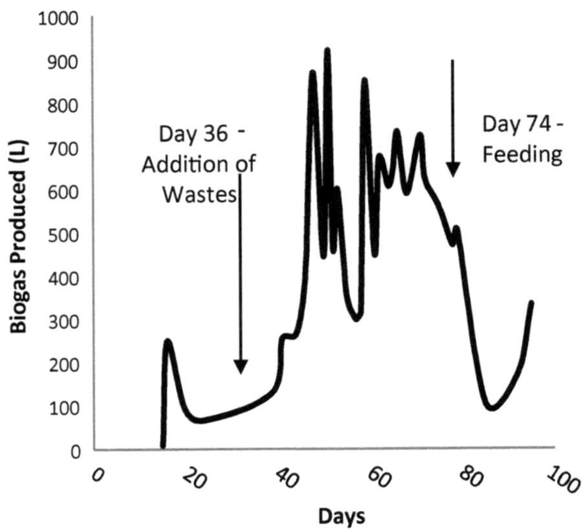

Fig. 3 Daily production of biogas

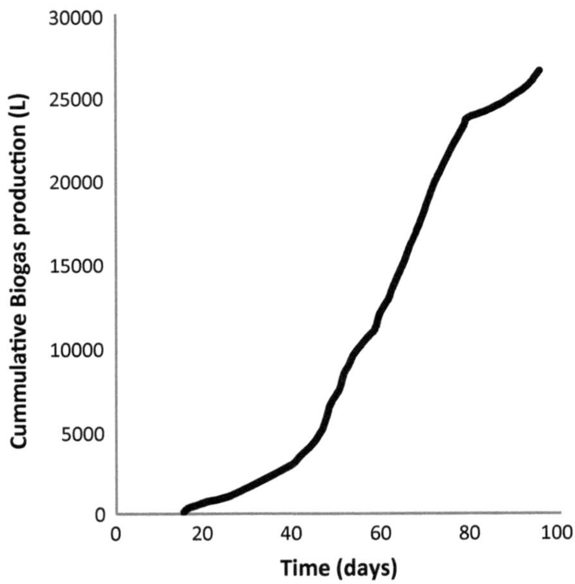

Fig. 4 Cumulative biogas production

results from this study largely exceed the reported yield which was obtained from other studies. Aragaw et al. (2013) also observed an increased in yield of biogas production from 22 to 27% when co-digesting cow manure and food wastes. The cumulative biogas production is represented in Fig. 4. I should be pointed out that, the 3-zone plug-flow system did not require the addition of cow dung during the course of its operation.

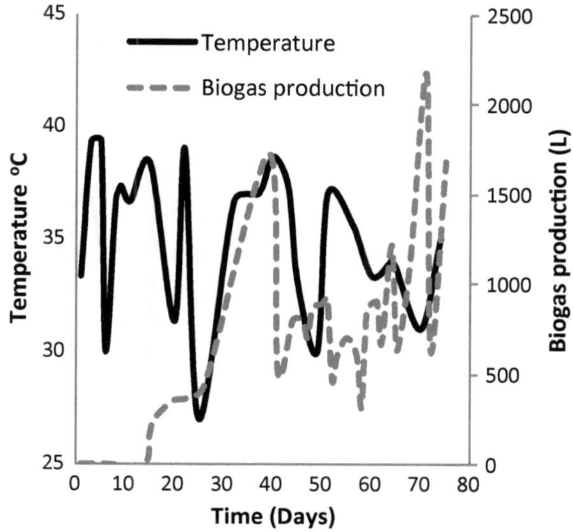

Fig. 5 Temperature profile

3.3.2 Temperature Evolution with Time

Figure 5 shows the temperature profile within the digester over time. The temperature was monitored at three different points, namely sampling points 1, 2 and 3. Temperature variations are known to exert significant effects on the microbial community, process kinetics, stability and methane yield (Riau et al. 2010; Dela-Rubia et al. 2002). The temperature inside the digester varied from a minimum of 27 °C to a maximum of 39.5 °C, with an average daily temperature of 35.0 °C which was the optimum operating temperature of the mesophilic range (20–42 °C) (Fernandez et al. 2008; Ward et al. 2008). The variation in temperature was due to the fact that the digester was left to ambient conditions in the sun and a change in the ambient temperature also brought about a deviation in the temperature within the digester. The lowest temperature (27 °C) was noted on day 25 and this eventually led to a decrease in biogas production. Hulsoff (1995) reported that for each 10 °C drop below 35 °C, a one and a half decrease in bacterial activity occurred. It should be noted that no external heating system was applied to the digester for maintaining the mesophilic condition prevailing inside the digester.

3.3.3 pH Evolution with Time

The pH variation during the anaerobic digestion process is represented in Fig. 6. It can be observed that the lowest pH recorded was on day 3 (6.66) and the highest pH value on day 74 (7.48). On day 4, the pH value was found to be on the lower limit of operability of anaerobic bacteria present inside the digester. In order to adjust the pH, Calcium Carbonate powder was added. A total 5 kg of powdered $CaCO_3$ was

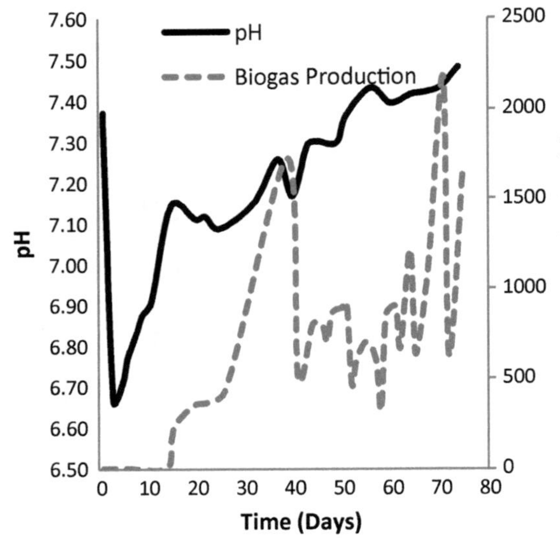

Fig. 6 pH variation

added in a small portion on alternate days for a duration of 1 week as from day 4. Then, the pH started to increase again constantly. Usually, a pH of 6.5–7.5 is recommended to achieve enhanced methanogenic activity and favourable gas production (Hulsoff 1995). The addition of food wastes as from day 35 to the digester did not change the ability of the substrate to maintain the pH level at optimum operating condition even if food wastes is very acidic in nature.

3.3.4 VFA

The volatile fatty acids (VFA) variation profile is shown in Fig. 7. The VFA level within the reactor was monitored at sampling points 1, 2 and 3, respectively. The formation of VFA was enhanced inside the digester. However, the accumulation of the acids hindered the biogas production. This was due to the action of the acidogenic consortia (Madsen et al. 2011). The sudden decrease in VFA level shows that the VFA produced by the acidogenic microorganisms was rapidly depleted by the methanogenic consortia. As from day 35, feeding the digester resulted in an increase in VFA. The highest amount of VFA obtained was 1326 mg/L and the lowest VFA level recorded was 246 mg/L. The amount of daily biogas production increased with VFA concentrations as methanogens have more material to convert to mainly methane and carbon dioxide. The rise in the concentration of VFA in the digester was not a limiting factor as the cow manure was a very good buffer to pH changes.

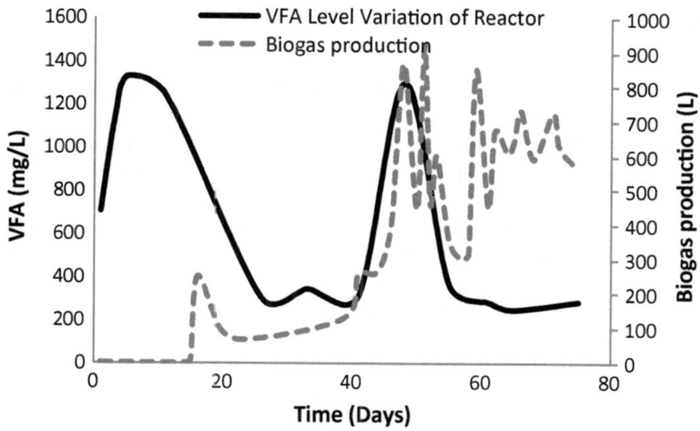

Fig. 7 VFA variations

4 Conclusions

As a means of upgrading lab-scale anaerobic digestion studies to field application, a 3-zone plug-flow anaerobic digestion system was designed to sustainably dispose of the cafeteria's food wastes. The performance of this digester was assessed in terms of certain parameters of importance for anaerobic digestion to occur. For a food wastes digestion period of 59 days, the average daily specific yield of biogas amounted to 51.03 L/kg day. A total of 23,967 L of biogas which is equivalent to a specific yield of 1.672 m^3/kg VS was collected from the digestion of 199 kg of food wastes. The high yield of biogas obtained was attributed to the efficacy of the 3-zone plug-flow system in maintaining an optimum pH level during the digestion process. The biogas produced can be used for cooking purposes. This 3-zone plug-flow anaerobic digestion system can easily be reproduced and maintained for the treatment of food wastes and production of biogas in the local community and in other African countries, as well as, in small island developing states. The system was designed to be aesthetic, can be located anywhere in the backyard, and can be moved easily at any time. In addition, the construction of this type of digester required low investment cost. There have not been any major environmental issues during the normal operation the system. Little odour production was detected. This project is a stepping-stone in the field of biogas production from organic wastes. The results of this project show that there is a big potential of producing biogas, at the same time, reducing the amount of organic wastes going to our sole sanitary landfill at Mare Chicose and most importantly could be a substitute to LPG consumption.

References

Anjan, K. K. (1988). Development and evaluation of a fixed dome plug flow anaerobic digester. *Biomass, 16,* 225–235.
Appels, L., Baeyens, J., Degreve, J., & Dewil, R. (2008). Principles and potential of the anaerobic digestion of waste-activated sludge. *Progress in Energy and Combustion Science, 34,* 755–781.
Aragaw, T., Andargie, M., & Gessesse, A. (2013). Co-digestion of cattle manure with organic kitchen waste to increase biogas production using rumen fluid as inoculums. *International Journal of Physical Science, 8*(11), 443–450.
Babaee, A., & Shayegan, J. (2011). Anaerobic digestion of vegetable waste. *Chemical Engineering Transactions, 24,* 1291–1296.
Boone, D. R., Chynoweth, D. P., Mah, R. A., Smith, P. H., & Wilkie, A. C. (1993). Ecology and microbiology of biogasification. *Biomass and Bioenergy, 5*(3–4), 191–202.
Chanakya, H. N., & Moletta, R. (2004). Performance and functioning of USW plug-flow reactors in a 3-zone fermentation model. In *Proceedings of the fourth international symposium on anaerobic digestion of solid wastes.* DTU, Copenhagen.
Chara, J., Pedraza, G., & Conde, N. (1999). The productive water decontamination system: A tool for protecting water resources in the tropics. *Livestock Research for Rural Development, 11*(1), 25–35.
Chaudary, B. K. (2008) *Dry continuous anaerobic digestion of municipal solid waste in thermophilic conditions*. Thesis (M.Eng.). Asian Institute of Technology.
Cho, J. K., & Park, S. C. (1995). Biochemical methane potential and solid waste in digestion of Korean food wastes. *Bioresource Technology, 52*(3), 245–253.
Corbit, R. A. (1989). *Standard handbook of environmental engineering*. New York, NY: McGraw-Hill.
Dela-Rubia, M. A., Perez, M., Romero, L. I., & Sales, D. (2002). Anaerobic mesophilic and thermophilic municipal sludge digestion. *Chemical and Biochemimical Engineering Quartely, 16,* 119–124.
Energy Efficiency Management Office (EEMO). (2011). Energy observatory report. http://eemo.govmu.org/English/report/Documents/Energyobservatory11.pdf. Accessed February 21, 2015.
Fernandez, J., Perez, M., & Romero, L. I. (2008). Effect of substrate concentration on dry mesophilic anaerobic digestion of organic fraction of municipal solid waste. *Bioresource Technology, 101,* 1628–1634.
Guendouz, J., Buffiere, P., Cacho, J., Carrere, M., & Delgenes, J. P. (2008). High-solids anaerobic digestion: Comparison of three pilot scales. *Water Science Technology, 58,* 1757–1763.
Hartmann, H., & Ahring, B. K. (2006). Strategies for the anaerobic digestion of the organic fraction of municipal solid waste: An overview. *Water Science Technology, 53*(8), 7–22.
Hayes, T. D., Jewell, W. J., Dell'orto, S., Fanfoni, K. J., Leuschnen, A. P., & Shemon, D. E. (1980). *Anaerobic digestion of cattle manure* (pp. 255–288). London: Applied Science Publishers.
Héduit, M. (1993). *La filière biogaz dans les pays en development* (p. 82). Quebec: IEPF and ACCT.
Hobson, P. N. (1990). The treatment of agricultural wastes. *Critical Reports on Applied Chemistry, 31,* 93–138.
Hulsoff, P. (1995). Waste characteristics and factors affecting reactor performance. In *International course on anaerobic wastewater treatment.* Wageningen Agriculture University, Netherlands.
Jagadish, K. S., Chanakya, H. N., Rajabapaiah, P., & Anand, V. (1998). Plug flow digesters for biogas generation from leaf biomass. *Biomass and Bioenergy, 14*(5–6), 415–423.
Jain, S., Jain, S., Wolf, I. T., Lee, J., & Ton, Y. W. (2015). A comprehensive review on operating parameters and different pretreatment methodologies for anaerobic digestion of municipal solid waste. *Renewable and Sustainable Energy Reviews, 52,* 142–154.

Kinyua, M. N., Rowse, L. E., & Ergas, S. J. (2016). Review of small-scale tubular anaerobic digesters treating livestock wastes in the developing world. *Renewable and Sustainable Energy Reviews, 58*, 896–910.

Kumar, R., & Singh, R. N. (2006). *Municipal water and wastewater treatment*. New Delhi: Teri Press.

Lacour, J. (2012). *Valorisation de résidus agricoles et autres déchets organiques par digestion anaerobic en Haïti* (Thesis Ph.D.). Université Quisqueya.

Lansing, S., Viquez, J., Martinez, H., Botero, R., & Martin, J. (2008). Optimising electricity generation and waste transformations in a low-cost, plug-flow anaerobic digestion system. *Ecological Engineering, 34*, 332–348.

Li, Y., Park, S. Y., & Zhu, J. (2011). Solid-state anaerobic digestion for methane production from organic waste. *Renewable and Sustainable Energy Reviews, 15*, 821–826.

Madsen, M., Holm-Nielsen, J. B., & Esbensen, K. H. (2011). Monitoring of anaerobic digestion processes: A review perspective. *Renewable and Sustainable Energy Reviews, 15*(6), 3141–3155.

Mashad, H. M., Zeeman, G., Loon, W. K. P., Bot, G. P. A., & Lettinga, G. (2004). Effect of temperature and temperature fluctuation on thermophilic anaerobic digestion of cattle manure. *Bioresource Technology, 45*, 191–201.

Mata-Alvarez, J., Mace, S., & Llabres, P. (2000). Anaerobic digestion of organic solid wastes. An overview of research achievements and perspectives. *Bioresouce Technology, 74*, 3–16.

Mohee, R., & Mudhoo, A. (n.d.). *Bio energy generation potential in Mauritius*. http://www.novator.se/boint/Mauritius.htm. Accessed March 15, 2015.

Ocwieja, S. M. (2010). *Life cycle thinking assessment applied to three biogas projects in central Uganda* (Thesis). Michigan Technological University, Houghton MI.

Palatsi, J., Laureni, M., Andrés, M. V., Flotats, X., Nielsen, H. B., & Angelidaki, I. (2009). Strategies for recovering inhibition caused by long chain fatty acids on anaerobic thermophilic biogas reactors. *Bioresource Technology, 100*, 4588–4596.

Pillay, C., Rajkumarsingh, B., & Goolaup, S. (2012). Energy harvesting from vehicle engine vibration for powering sensors. *Journal of Institution of Engineers Mauritius, 1*, 56–64.

Pourmovahed, A., Opperman, T., & Lemke, B. (2011). Performance and efficiency of a biogas CHP system utilizing a stirring engine. *International Conference of Renewable Energies and Power Quality*. doi:10.24084/repqj09.288.

Rajendran, K., Aslanzadeh, S., & Taherzadeh, M. J. (2012). Household biogas digesters—A review. *Energies, 5*, 135–156.

Riau, V., Dela-Rubia, M. A., & Perez, M. (2010). Temperature phased anaerobic digestion (TPAD) to obtain class A biosolids: A semi-continuous study. *Bioresource Technology, 101*, 2706–2712.

Speight, J. (2011). *The biofuels handbook*. London: Royal Society of Chemistry.

Surrop, D., & Mohee, R. (2012). Technical and economic assessment of power generation from biogas. In *International conference on environmental science and technology*, Singapore.

Vaiieln, I. (1969). An experimental study of the mortality factors of larval Musco autumnalis. *Ecological Monogrphs, 39*, 199–225.

Van Lier, J. B., Boersma, F., Debets, H., & Lettinga, G. (1994). High rate thermophilic anaerobic wastewater treatment in compartmentalized upflow reactors. *Water Science and Technology, 30*(12), 251–261.

Vandeviviere, P., Debaere, L., & Verstraete, W. (2000). Types of anaerobic digesters for solid wastes. http://www.adelaide.edu.au/biogas/anaerobic_digestion/pvdv.pdf. Accessed February 21, 2015.

Vesilind, P. A. (2003). *Waste water treatment plant design*. London: IWA Publication.

Wang, X., Yang, G., Feng, Y., Ren, G., & Han, X. (2012). Optimizing feeding composition and carbon–nitrogen ratios for improved methane yield during anaerobic codigestion of dairy, chicken manure and wheat straw. *Bioresource Technology, 120*, 78–83.

Ward, A. J., Hobbs, P. J., Holliman, P. J., & Jones, D. L. (2008). Optimization of the anaerobic digestion of agricultural resources. *Bioresource Technology, 99*, 7928–7940.

Yang, L., Xu, F., Ge, X., & Li, Y. (2015). Challenges and strategies for solid-state anaerobic digestion of lignocellulosic biomass. *Renewable and Sustainable Energy Reviews, 44,* 824–834.

Zhang, R., El-Mashad, H. M., Hartman, K., Wang, F., Liu, G., Choate, C., et al. (2007). Characterisation of food wastes as a feedstock for anaerobic digestion. *Bioresource Technology, 98,* 929–935.

Zhang, L., Lee, Y. W., & Jahng, D. (2011). Anaerobic co-digestion of food wastes and piggery wastewater, focussing on the role of trace elements. *Bioresource Technology, 101,* 5048–5059.

Zhang, W., Wei, Q., Wu, S., Qi, D., Li, W., Zuo, Z., et al. (2014). Batch anaerobic co-digestion of pig manure with dewatered sewage sludge under mesophilic conditions. *Applied Energy, 128,* 175–183.

Possible Absorbent Regeneration in Biogas Purification and Upgrading: A Review

O.I. Maile, H. Tesfagiorgis and E. Muzenda

Abstract South Africa has the highest emission levels of CO_2 equivalent to 315,957.24. CO_2 emissions are proportional to fuel consumption in a sense that 1% rise in fuel consumption results in 1% rise in CO_2 emissions. Many alternative energy sources to address this issue have been exploited, that is most preferably renewable. Biogas has been proved to reduce over 90% CO_2 emissions to the atmosphere. Its application is limited until its upgraded and chemical absorption is one of the techniques used. The following chemicals MEA, NH_4OH, CaO, Ca $(OH)_2$, KOH, NaOH, $FeCl_2$, $FeCl_3$, $FeSO_4$, Fe_2O_3, $Fe(OH)_3$ and ZnO are used for absorption. The alkalis can be regenerated but a penalty of high-temperature requirements has to be paid making their process not desirable. Considering the amines, they can be regenerated using $Ca(OH)_2$ or iron exchange resin with anionic resin. NaOH can be regenerated using electrodialysis and causticization is normally used to convert Na_2CO_3 to NaOH. Ammonia has proven to be a good absorbent because its purchase price is low and it can be regenerated with ease. Furthermore, it has an added advantage since its by-product can be applied directly as a fertilizer without any further treatment. However, MEA also stands to be a good absorbent because it has a potential of removing both CO_2 and H_2S.

Keywords Absorbent · Causicization · Decomposition · Electrodialysis Energy · Purification · Regeneration

O.I. Maile (✉) · E. Muzenda
Department of Chemical Engineering, Faculty of Engineering and the Built Environment, University of Johannesburg, Doornfontein Campus, John Orr Building, 3160 Beit Street, Johannesburg, South Africa
e-mail: ireenmaile@yahoo.com

H. Tesfagiorgis
Agricultural Research Council, Institute for Industrial Crops, Private Bag X82075, Rustenburg 0300, South Africa

E. Muzenda
Department of Chemical, Materials and Metallurgical Engineering, College of Engineering and Technology, Botswana International University of Science and Technology, Private Mail Bag 16, Palapye, Botswana

W. Leal Filho and D. Surroop (eds.), *The Nexus: Energy, Environment and Climate Change*, Green Energy and Technology, https://doi.org/10.1007/978-3-319-63612-2_17

1 Introduction

Energy is an important part of our daily lives; we use it for light, heat, as a fuel, electricity and other applications. The sources that the energy is derived from are depleting and cause climate change because they are non-renewable. Hence, we are experiencing energy shortages. As it is observed of recent load shedding stages implemented to try to save the country's energy. South Africa has the highest emission levels of CO_2 equivalent to 315,957.24 (United Nations Framework Convention on Climate Change 2005). This is solely because 90% of our energy is derived from coal (U.S. Energy Information Administration 2015).

The major contributors to GHG are the industrial and transport sector. The latter being our main concern. It is due to the increasing population and vehicle purchases. Besides, it was recorded by World Bank in 2010 that South Africa has 165 road motor vehicles per 1000 people (World Bank). Furthermore, the United States Environmental Protection Agency stated that CO_2 emissions are proportional to fuel consumption in a sense that 1% rise in fuel consumption results in 1% rise of CO_2 emissions (United States Environmental Protection Agency 2008). With that in mind, it can be seen that more vehicles on the road contribute to climate change.

Many alternative energy sources to address this issue have been exploited, that is most preferably renewable. For example; natural gas, biodiesel, and biogas. Biogas has been proved to reduce over 90% CO_2 emissions to the atmosphere and thus becomes our area of interest. Raw biogas consists of 50–70% CH_4 and 30–50% CO_2 as well as small amounts of other gases, which often limits its application. For instance, in order to use biogas as a fuel for vehicles, it needs to be upgraded to biomethane which can be equivalent to natural gas standard. As a result, it needs to be purified by applying purification techniques which as a part of this work chemical absorption was reviewed.

Purification by chemical absorption has been used effectively to purify biogas. However, for it to be commercialized regeneration of the spent absorbent is of utmost importance. In this work, regeneration of the chemicals is reviewed.

2 Background

Biogas technology has become a very attractive source of renewable energy which has a potential to ease the dependence on energy derived from fossil fuels as 1 m^3 of raw biogas has potential to provide for the cooking time of approximately 2 h, which is equivalent to 1.5 kWh electrical output. On the other hand, 2.5 m^3 of raw biogas is equal to 1 kg of LP gas. Chemical absorption has a potential to purify biogas to over 95% CH_4 vol which makes it even more effective and provide more energy output than when it is raw.

CO_2 absorption using chemicals was studied way back in 1943 and it was mainly to capture CO_2 from the atmosphere using sodium hydroxide. Potassium hydroxide is also used for the same purpose and is more effective but it is often not preferred because sodium hydroxide is cheaper. They both yield carbonate salt sodium carbonate and potassium carbonate, respectively. However, the challenge with the resultant carbonate is that it cannot be regenerated or decomposed with a change in heat or pressure. So chemical decomposition route is used with calcium hydroxide through a process called causticization (Zeman and Lackner 2004). This process was used since 1884 in the pulp and paper industry through a different name for the process called 'Kraft process'. The process occurs in a subsequence of steps which include; calcination whereby the calcium cation is transferral to the carbonate anion which results in a calcium carbonate precipitate, followed by drying and washing of the calcium carbonate precipitate and then thermal decomposition resulting in calcium oxide; the last step is slaking which is simply just hydrating the lime (Zeman and Lackner 2004).

Packed column reactors are usually used for regenerating monoethanolamine from a carbon dioxide-rich mixture. This occurs at temperatures higher than the boiling point. The use of monoethanolamine for carbon dioxide absorption has been employed from many years ago in industrial processes. It has also been used for natural gas sweetening (Kohl and Riesenfeld 1997). The type of columns that are mostly used for this processes is a standard packed column, trickle plate column and packed bed column, a typical column is shown in Fig. 2. Their operation may be limited by the physical processes, the solubility of carbon dioxide in the solution, diffusion or convection transport for the liquid–gas interface, liquid–gas interface desorption and diffusion into the gas stream (Curnow et al. 2005).

The factors that affect the rate of the regeneration process include the gas injection on the initial stage of regeneration, injection gas flow rate and the temperature of the incoming gas, lean solvent and the column. The chemistry involved in carbon dioxide absorption by MEA and desorption is shown in Eqs. (1)–(7). The process starts with water dissociation, hydrolysis of carbon dioxide, bicarbonate dissociation, hydrolysis of the carbamate and protonation of MEA (Zhang and Chen 2013).

$$2H_2O \leftrightarrow H_3O^+ + OH^- \tag{1}$$

$$CO_2 + 2H_2O \leftrightarrow HCO_3^- + H_3O^+ \tag{2}$$

$$HCO_3^- + H_2O \leftrightarrow CO_2 + H_3O^+ \tag{3}$$

$$MEACOO^- + H_2O \leftrightarrow MEA + HCO_3^- \tag{4}$$

$$MEAH^+ + H_2O \leftrightarrow MEA + H_3O^+ \tag{5}$$

$$MEA + CO_2 + H_2O \leftrightarrow MEACOO^- + H_3O^+ \tag{6}$$

$$CO_2 + OH^- \leftrightarrow HCO_3^- \tag{7}$$

The use of ammonia in the carbon dioxide capture has been gaining momentum especially in the coal-fired power plants which are the highest contributor to the CO_2 emissions (Zhang and Guo 2014). The absorption and regeneration process for CO_2 using ammonia is similar to that of MEA. However, the ammonia technology is becoming more competitive because of the advantages it carries. For example, it has higher carbon dioxide loading capacity, it requires lower regeneration heat, no corrosion of equipment involved, and no degradation of the absorbent (Darde et al. 2011; Zhang and Guo 2014).

The regeneration process can occur at temperatures ranges of 80–95 °C for ambient pressure and slightly higher for different pressures. The main four processes involved in the regeneration process are; hydrodynamics which involves the gas–liquid phases, flow and velocity characteristics and the packings surface; thermodynamics includes the pressure–volume–temperature relationship of the gas phase and the change in enthalpy of the system; the chemical reactions involve mainly the reactions that occur together with the equilibrium state reactions; and heat and mass transfer mainly is concerned with the mass and heat transfer in the process and the vapour and liquid phase equilibrium (Zhang and Guo 2014).

The process parameters that affect the regeneration process include the gas flow rate, solvent flow rate, temperature of both the inlet gas–liquid streams, and the temperature of the stripper. The reboiler and condenser duty also affects the process, hence there are often heat exchangers for preheating the solvent to the absorber and cooling the stripper inlet.

3 Methods of Regeneration

3.1 Ammonia Regeneration Processes

Ammonia is regenerated by thermal decomposition whereby the absorption by-products give off the CO_2 so that they can be sent to the absorber again. The mode of operation for the absorber–desorber unit as shown in Fig. 2 starts with the biogas from the anaerobic digester fed into the absorber at the bottom of the column. The solvent is fed at the top of the absorber, and it absorbs the CO_2 in the column from the gas stream. The biomethane exits at the top of the absorption column and the CO_2 rich solvent exits through the bottom of the column to be passed through a heat exchanger where it is heated before feeding into the regeneration unit. The absorber usually operates at ambient conditions. In the regenerator, the ammonium carbonate decomposes at a higher temperature and ambient pressure to give off the CO_2 which exits at the top of the column. The regenerated solvent exits at the bottom of the column where it is recirculated back to the absorber passing through a heat exchanger to lower the temperature before it is fed into the absorber (Yeh et al. 2005). The following equations illustrate what happens during digestion:

$$2NH_4HCO_3(aq) \leftrightarrow (NH_4)_2CO_3(aq) + CO_2(g) + H_2O \quad \Delta H_{rxn} = 6.4\,\text{kcal/mol} \tag{8}$$

$$NH_4HCO_3(aq) \leftrightarrow NH_3(aq) + H_2O + CO_2(g) \quad \Delta H_{rxn} = 24.1\ \text{kcal/mol} \tag{9}$$

$$(NH_4)_2CO_3(aq) \leftrightarrow 2NH_3(aq) + H_2O + CO_2(g) \quad \Delta H_{rxn} = 15.3\ \text{kcal/mol} \tag{10}$$

Yeh et al. (2004) carried out the absorption/regeneration cycle at 80 °F (27 °C) and 180 °F (82 °C), respectively, repeating to a total of three cycles. They observed a large drop in the loading capacity (g CO_2/g solution) between the first and second cycle, and that it seemed like equilibrium was reached after the second absorption run. The final measured specific absorption capacity was found to be 0.068, 0.053, and 0.040 g CO_2/g solution for the 14, 10.5 and 7% ammonia solutions, respectively. This is comparable to MEA technology that has a capacity of 0.036 g CO_2/g solution for a standard 20% MEA solution. Their results indicated that less than 100% regeneration of the initial aqueous ammonia solution is achieved and that the sensible heat of regeneration for the aqueous ammonia process may be as little as 45% that of the MEA process. Figures 1 and 2 represent ammonia absorption and regeneration processes which are normally used in CO_2 capture processes.

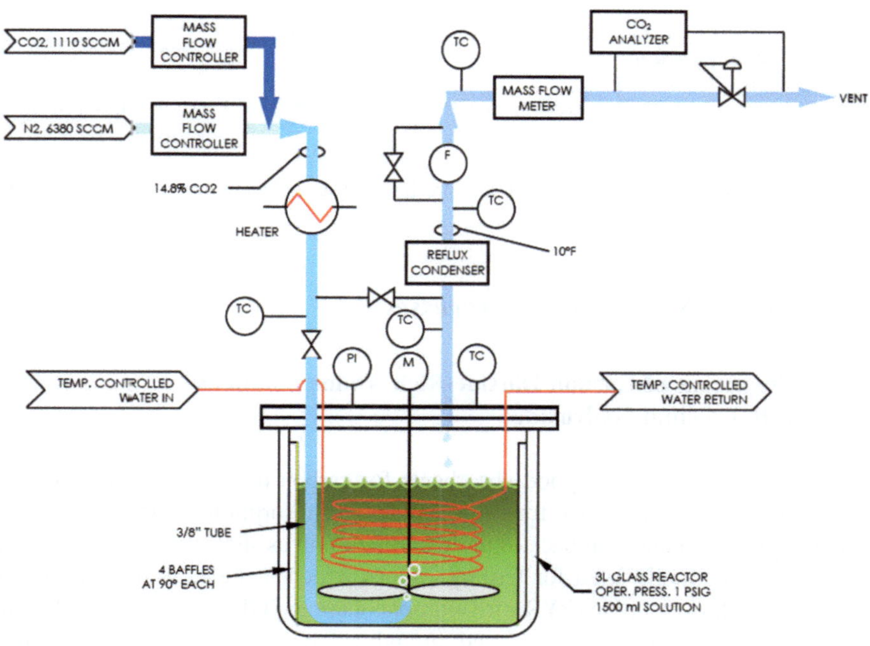

Fig. 1 Aqua ammonia semi-continuous reactor (Resnik et al. 2004; Yeh et al. 2004)

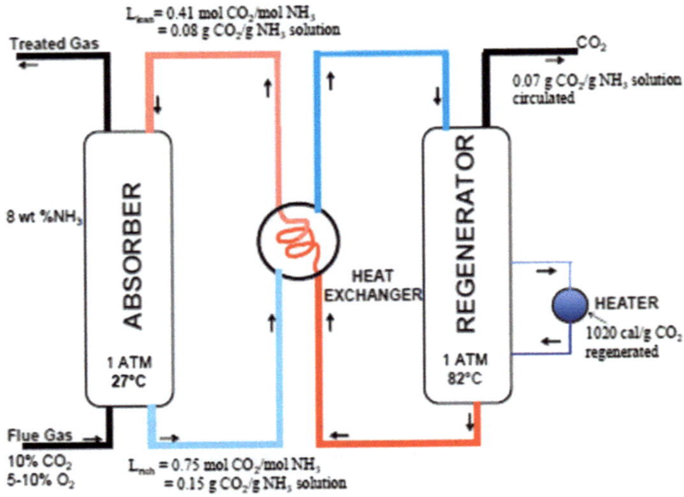

Fig. 2 Process flow diagram for aqua ammonia carbon dioxide capture process (Yeh et al. 2004)

Absorption	$Am + CO_2 + H_2O \rightarrow AmH^+ + HCO_3^-$
Regeneration	$AmH^+ + OH^- \rightarrow Am + H_2O$ $HCO_3^- + OH^- + Ca^{2+} \rightarrow CaCO_3\downarrow + H_2O$

Fig. 3 Reactions involved in absorption and precipitation regeneration processes (García-Abuín et al. 2014)

3.2 Amines Regeneration Processes

3.2.1 Treatment of Carbon Dioxide-Rich Amine Solutions with Calcium Hydroxide

Amines have proved to be good absorbents for carbon dioxide capture with high efficiency. Another advantage for using amines for absorption is the fact that they can be regenerated and sent back to the absorption unit as shown in Fig. 3. Calcium hydroxide is normally used for tertiary amines which include; methyldiethanolamine (MDEA), pyrrolidine (PYR), triisopropanolamine (TIPA) and triethanolamine (TEA). When carbon dioxide is passed through an aqueous solution of the amines, the main products include bicarbonate and carbamate. However, carbamate is often

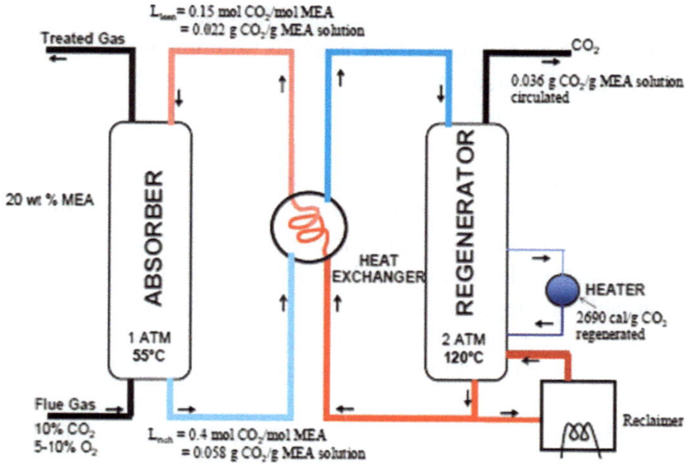

Fig. 4 Process flow diagram for typical MEA CO_2 capture process (Yeh et al. 2004)

produced as a result of low carbon dioxide loading rate (García-Abuín et al. 2014). The absorption/regeneration reactions are shown in Fig. 3.

Factors that affect the absorption behavior for the amines (Álvarez et al. 2000; Bougie and Iliuta 2012)

- The type of the amine used
- Chemical kinetics
- Physical properties of the aqueous solutions
- The flow behaviour in the reactor.

Aqueous methyldiethanolamine solutions have been observed to be a pseudo-first order regime (García-Abuín et al. 2014). An increase in the presence of bicarbonate is caused by the increase in the amount of carbon dioxide absorbed in the liquid phase. This results from the carbamate hydrolysis. Since most regeneration processes are designed targeting the bicarbonate, it is advisable to avoid carbamate formation for an effective regeneration process. It was found that triethanolamine and triisopropanolamine have high regeneration capacity approximately 100% when using calcium hydroxide (García-Abuín et al. 2014). The absorption of CO_2 using amines and the regeneration process of the absorbent are stipulated in Fig. 4 with a more detailed description of the process.

3.2.2 Ion Exchange Operation (Anionic Resin)

Garcia-Abuin et al. (2014) reported that a batch system for ion exchange operation does not give satisfactory results. Thus, cannot be used to replace conventional treatment methods. Therefore, a different configuration of the experiment was studied. This is a continuous regime. Though mode of operation is similar to the calcium hydroxide treatment, the gas and liquid flow rates are different and selected on a distinct basis. For example, PYR aqueous solution was used as a basis for selecting the liquid flow rate. Which forms two different kinds of products, i.e. carbamate and carbonate/bicarbonate (García-Abuín et al. 2012).

The regeneration efficiency of the regime decreases when the amine concentration is reduced in the solid phase by its adsorption. Bicarbonate ion saturation on the exchange resin causes the regeneration process to be of non-effect. When the resin is saturated, it leads to the disappearance of the reaction product in the liquid phase producing amine deprotonation. It was concluded that this kind of reaction product is suitable for regeneration since it was satisfactory in all cases and is comparable with results obtained from conventional processes. Hence, when applied it can help avoid the high cost of stripping processes (García-Abuín et al. 2012).

Another alternative method for amine regeneration is simply boiling the carbon dioxide solution for 5 min, i.e. for MEA (Kapdi et al. 2005).

3.3 Alkali Regeneration Processes

3.3.1 Electrodialysis to Regenerate Aqueous Solution of Sodium Hydroxide

This is an electro-membrane separation process in which an electric field is applied causing the ions to migrate from one solution to the other through permeable membranes. Normally used to demineralize, concentrate and concert salt-containing solutions. Electro-electrodialysis was applied as from ancient times but because of the environmental impact, it has an alternative process had to be studied which is bipolar membrane electrodialysis and they are both discussed below (Wei et al. 2012).

The principle of electrodialysis operates by the electrochemical potential difference. The typical electrodialysis component consists of cation and anion exchange membranes which lie between an anode and cathode which are positively and negatively charged, respectively.

A. Electro-Electrodialysis

It is used in the salt convention because it has an effective functional integration of the ion exchange membrane (Wei et al. 2012). Splitting of water occurs at the anode and cathode by oxidation and reduction as shown in Fig. 5. The negatively charged exchange membrane only allows the passage of cations while the positively charged exchange membrane allows the passage of anions as depicted in Fig. 5. Since the membranes are alternately charged, they often hinder the flow of the ions as the cathode attracts the cations and the anode attracts the anions. This leads to concentration of ions in some compartments of the membrane (Nezungai and Majozi 2016).

Fig. 5 Typical setup and indication of the EED operation (Wei et al. 2012)

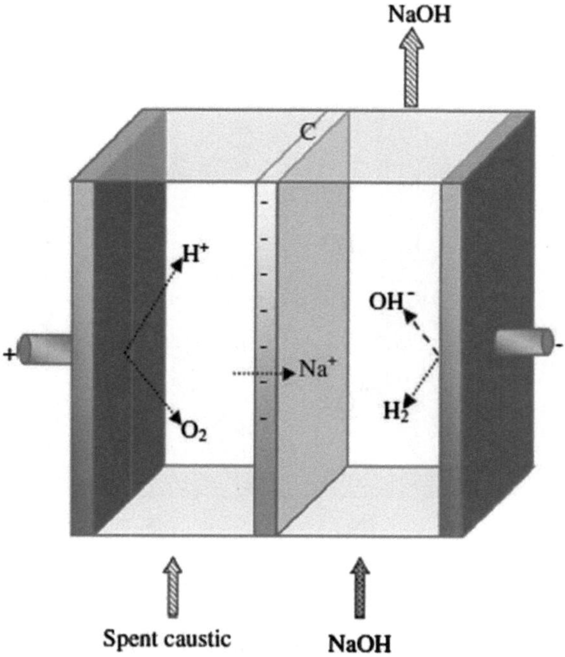

B. Bipolar Membrane Electrodialysis

It is used for the salt convention since it is effective in integrating water splitting in bipolar membranes. It has a lower voltage drop as compared to electro-electrodialysis which is attributed to the extra compartment in the stack, resulting in high electrical resistance (Wei et al. 2012). This can be seen in Fig. 6. The bipolar membrane electrodialysis has been used for the production of alkalis such as NaOH and acids like H_2SO_4 and has potential to reduce their cost. It is capable of simultaneously dissociate water into H+ and OH–. Furthermore, it can regenerate NaOH and result in enriched CO_2 during the separation occurring at the cation and anion exchange layer. Though this process can reduce the energy costs, the investment cost of bipolar membrane electrodialysis is high (Chen et al. 2013).

3.3.2 Caticization Conversion of Sodium Carbonate to Sodium Hydroxide

Causicization is a process which is used to regenerate sodium hydroxide from sodium carbonate in industrial carbon dioxide capture systems. The process can either be conventional or direct causticization using lime and titanium oxide, respectively. Both methods are discussed below. The chemical reactions that take place during the causticization have basically summarized a process where green

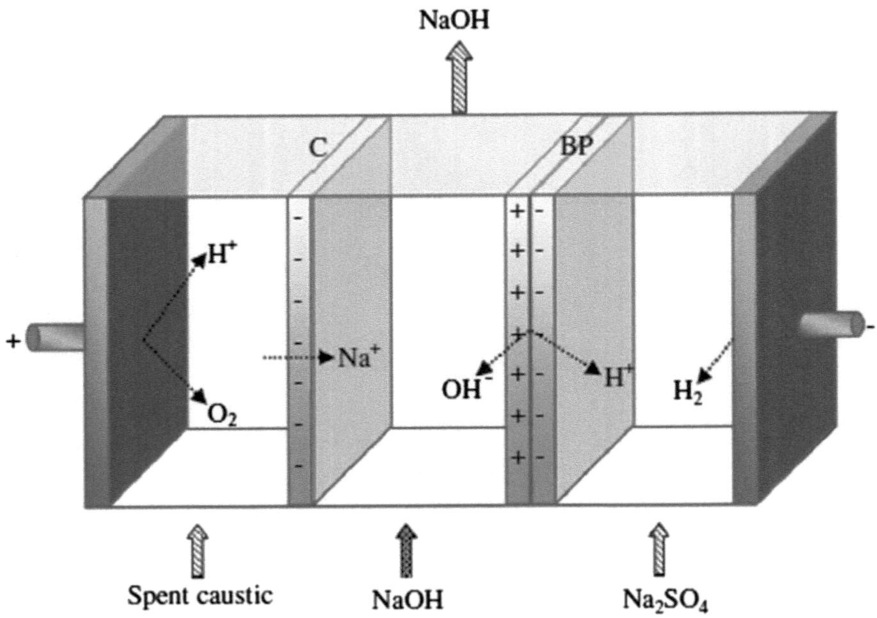

Fig. 6 Typical setup and indication of the BMED operation (Wei et al. 2012)

liquor (sodium carbonate) reacts with lime (calcium oxide) to form white liquor (sodium hydroxide) and lime mud (calcium carbonate):

$$Na_2CO_3 + H_2O + CaO \rightarrow 2NaOH + CaCO_3 \quad (11)$$

$$CaO + H_2O \rightarrow Ca(OH)_2 + \text{Heat} \quad (12)$$

$$Ca(OH)_2 + Na_2CO_3 \rightarrow 2NaOH + CaCO_3 \quad (13)$$

Equation (11) shows the overall reaction of the process and it is broken down in Eqs. (12) and (13) but Eq. (12) shows the reaction of CaO and water which is exothermic and gives off heat as it forms $Ca(OH)_2$. The reaction proceeds rapidly and it is an equilibrium reaction which is reversible (Sanchez 2007).

A. Causticization Using Lime

This process involves the causticization of sodium carbonate (Na_2CO_3) with lime producing sodium hydroxide (NaOH) and lime mud ($CaCO_3$). The process occurs in a series of liquid–solid reactions. 179 kJ/mol CO_2 energy is required to convert lime mud ($CaCO_3$) to lime (CaO). While that which is required for converting sodium carbonate (Na_2CO_3) to sodium hydroxide (NaOH) is 109 kJ/mol CO_2. Hence, the energy required for conventional causticization is far from the thermodynamic minimum (Mahmoudkhani and Keith 2009).

B. Causticization using titanium dioxide

This is a method of direct causticization. The reactions involved include the following:

$$7Na_2CO_{3(s)} + 5(Na_2O \cdot 3TiO_2)_{(s)} \leftrightarrow 3(4Na_2O \cdot 5TiO_2)_{(s)} + 7CO_{2(g)} \\ \Delta H_{850\,°C} = 90\,kJ/mol\,CO_2 \quad (14)$$

$$Na_2CO_{3(s)} \rightarrow Na_2CO_{3(l)} \\ \Delta H_{851\,°C} = 25\,kJ/mol\,CO_2 \quad (15)$$

$$7Na_2CO_{3(l)} + 5(Na_2O \cdot 3TiO_2)_{(s)} \leftrightarrow 3(4Na_2O \cdot 5TiO_2)_{(s)} + 7CO_{2(g)} \\ \Delta H_{850\,°C} = 65\,kJ/mol\,CO_2 \quad (16)$$

$$3(4Na_2O \cdot 5TiO_2)_{(s)} + 7H_2O \leftrightarrow 5(Na_2O \cdot 3TiO_2)_{(s)} + 14NaOH_{(aq)} \\ \Delta H_{100\,°C} = 15.2\,kJ/mol\,CO_2 \quad (17)$$

The final step in the process involves hydrolysis in a leaching unit operating at 100 °C to form sodium hydroxide and sodium tri-titanate. The advantage of this method is that it requires nearly half energy as compared to conventional causticization (Schomaker et al. 2000).

3.4 *Regeneration of Absorbents Used to Remove Hydrogen Sulphide*

The chemicals used in H_2S removal include NaOH, $FeCl_2$, $FeCl_3$, $FeSO_4$, Fe_2O_3, Fe $(OH)_3$ and ZnO. They react with H_2S according to the following reactions.

$$NaOH + H_2S \rightarrow NaHS + H_2O \quad \text{or} \quad 2NaOH + H_2S \rightarrow Na_2S + 2H_2O \quad (18)$$

$$FeCl_2 + H_2S \rightarrow FeS + 2HCl \quad (19)$$

$$2Fe(OH)_3 + 3H_2S \rightarrow Fe_2S_3 + 6H_2O \quad (20)$$

$$Fe_2O_3 + 3H_2S \rightarrow Fe_2S_3 + 3H_2O \quad (21)$$

H_2S can be removed from biogas as stipulated by the Eqs. (5)–(8), but it is important that regeneration of spent chemical be done since it can be commercially infeasible. However, sodium hydrogen sulphide (NaHS) or sodium sulphide (Na_2S) is non-regenerative and has to be disposed of, which might have an environmental impact. In addition to that, iron sulphide (FeS) is an insoluble salt that needs to be

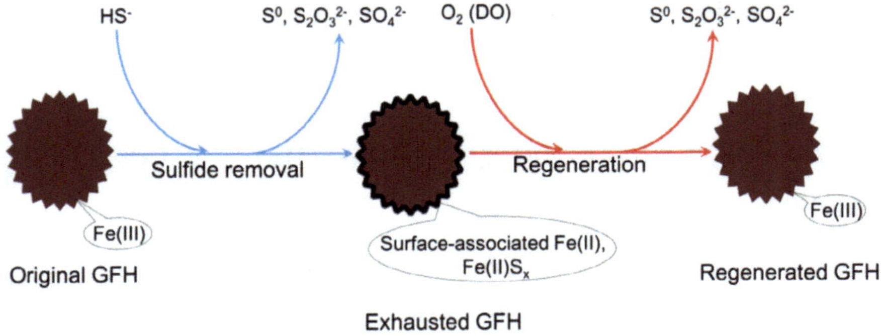

Fig. 7 Sulphide removal and regeneration process presentation (Bajracharya et al. 2009)

removed (Ryckebosch et al. 2011; Sun et al. 2014). As for the iron(II) sulphide Fe_2S_3 can be regenerated using oxygen as shown in Eq. (12).

$$2Fe_2S_3 + 3O_2 \rightarrow 2Fe_2O_3 + 6S \tag{22}$$

When ferric hydroxide is used for hydrogen sulphide removal it produces Fe_2O_3 as shown in Eq. (9) that can be regenerated as stipulated in Fig. 7.

4 Energy Requirements

Energy forms very important part of any process, because in order to determine the feasibility of a process we need to know the energy requirements and penalties involved. 1830 Btu per Ib CO_2 energy is used in ammonium solution regeneration compared to 4840 Btu per Ib CO_2 required for MEA (Resnik et al. 2004). This means there is approximately 62% energy savings for ammonium solution. In addition, this can also be due to the fact that MEA has a higher heat of reaction than ammonia. Furthermore, industrial grade ammonia solution costs one-sixth than MEA in the purchase price. Ammonia solution also has low absorbent regeneration energy requirements, i.e. up to 75% less. An added advantage for ammonia is the application of its by-product as scalable ammonium sulphate fertilizer (Resnik et al. 2004).

The alkalis are good absorbents and are often used to improve absorption in physical absorption processes. However, they have high-energy requirements for their regeneration. Calcium oxide is not preferred for because it requires a high temperature of 825 °C to be regenerated from $CaCO_3$ which needs to be heated to dissociate (Baciocchi 2006). The thermodynamic minimum requirement CaO regeneration is 179 kJ/mol CO_2 (Mahmoudkhani et al. 2009). Sodium hydroxide can be regenerated by causticization and electrodialysis processes. 109.4 kJ/mol thermodynamic minimum energy is required to regenerate sodium hydroxide from sodium carbonate (Zeman 2007; Stolaroff et al. 2008; Mahmoudkhani et al. 2009). 90 kJ/mol CO_2 enthalpy of reaction is required when using titanate as compared to when using lime. Therefore, titanate process has 50% savings on temperature requirements compared to the lime process (Mahmoudkhani et al. 2009).

Electrodialysis is also used for regeneration and bipolar membrane electrodialysis which yields 0.11 mol/L NaOH concentration after 2 h of operation. The process cost was estimated to be $ 0.97/kg NaOH recovered using laboratory scale setup (Wei et al. 2012).

5 Conclusions

Regeneration is a very important part of biogas purification and upgrading especially for chemical absorption purification techniques. The methods used for regeneration are; thermal decomposition for ammonia regeneration, treating carbon dioxide-rich amine solutions with calcium hydroxide, ion exchange operation using anionic resin, electrodialysis to regenerate aqueous solution of sodium hydroxide (EED and BMED), causticization conversion of sodium carbonate to sodium hydroxide (causticization using titanium dioxide, causticization using lime) and using NaOH, $FeCl_2$, $FeCl_3$, $FeSO_4$, Fe_2O_3, $Fe(OH)_3$ and ZnO for H_2S removal. Ammonia has proven to be a good absorbent because its purchase price is low and it can be regenerated with ease. Furthermore, it has an added advantage since its by-product can be applied directly as a fertilizer without any further treatment. However, MEA also stands to be a good absorbent because it has a potential of removing both CO_2 and H_2S.

Acknowledgements The authors acknowledge South African National Energy Development Institute (SANEDI), Global Excellence Scholarship (GES) and the City of Johannesburg (CoJ), the department of Chemical Engineering at the University of Johannesburg and the Process Energy and Environmental Technology Station (PEETS) for supporting this research.

References

Álvarez, E., Sanjurjo, B., Cancela, A., & Navaza, J. M. (2000). Mass transfer and influence of physical properties of solutions in a bubble column. *Chemical Engineering Research and Design, 78,* 889–893.

Baciocchi, R., Storti, G., & Mazzotti, M. (2006). Process design and energy requirement for the capture of carbon dioxide from air. *Chemical Engineering and Processing, 45,* 1047–1058.

Bajracharya, R., Dhungana, A., Thapaliya, N., & Hamal, G. (2009). Purification and compression of biogas. *Journal of the Institute of Engineering, 7,* 1–9.

Bougie, F., & Iliuta, M. C. (2012). Sterically hindered amine-based absorbents for the removal of CO_2 from gas streams. *Journal of Chemical and Engineering Data, 57,* 635–669.

Chen, M., Zhang, F., Zhang, Y., & Zeng, R. J. (2013). Alkali production from bipolar membrane electrodialysis powered by microbial fuel cell and application for biogas upgrading. *Applied Energy, 103,* 428–434.

Curnow, O. J., Krumdieck, S. P., & Jenkins, E. M. (2005). Regeneration of carbon dioxide saturated monoethanolamine-glycol aqueous solutions at atmospheric pressure in a packed bubble reactor. *Industrial and Engineering Chemistry Research, 44,* 1085–1089.

Darde, V., van Well, W. J. M., Fosboel, P. L., Stenby, E. H., & Thomsen, K. (2011). Experimental measurement and modeling of the rate of absorption of carbon dioxide by aqueous ammonia. *International Journal of Greenhouse Gas Control, 5,* 1149–1162.

García-Abuín, A., Gómez-Díaz, D., & Navaza, J. M. (2014). New processes for amine regeneration. *Fuel, 135,* 191–197.

García-Abuín, A., Gómez-Díaz, D., Navaza, J. M., & Rumbo, A. (2012). Carbon dioxide capture by absorption with pyrrolidine: Loading and speciation. *International Conference on Chemical and Biochemical Engineering, T8,* 13–14.

Kapdi, S. S., Vijay, V. K., Rajesh, S. K., & Prasad, R. (2005). Biogas scrubbing, compression and storage: Perspective and prospectus in Indian context. *Renewable Energy, 30,* 1195–1202.

Kohl, A. L., & Riesenfeld, F. C. (1997). *Gas purification.* Houston, TX: Gulf Publishing.

Mahmoudkhani, M., Heidel, K. R., Ferreira, J. C., Keith, D. W., & Cherry, R. S. (2009). Low energy packed tower and caustic recovery for direct capture of CO_2 from air. *Energy Procedia, 1,* 1535–1542.

Mahmoudkhani, M., & Keith, D. W. (2009). Low-energy sodium hydroxide recovery for CO2 capture from atmospheric air—Thermodynamic analysis. *International Journal of Greenhouse Gas Control, 3,* 376–384.

Nezungai, C. D., & Majozi, T. (2016). Optimum synthesis of an electrodialysis framework with a background process—I: A novel electrodialysis model. *Chemical Engineering Science, 147,* 1–9.

Resnik, K. P., Yeh, J. T., & Pennline, H. W. (2004). Aqua ammonia process for simultaneous removal of CO_2, SO_2 and NO_x. *International Journal Environmental Technology and Management, 4,* 89–104.

Ryckebosch, E., Drouillon, M., & Vervaeren, H. (2011). Techniques for transformation of biogas to biomethane. *Biomass and Bioenergy, 35,* 1633–1645.

Sanchez, D. R. (2007). *Recaustisizing—principles and practice* (pp. 1–30). Burlington: Vector Process Equipment Inc.

Schomaker, A. H. H. M., Boerboom, A. A. M., Visser, A., & Pfeifer, A. E. (2000). *Anaerobic digestion of agro industrial wastes: Information networks e technical summary on gas treatment* (pp. 1–31). Nijmegen: Nederland.

Stolaroff, J., Keith, D., & Lowry, G. (2008). Carbon dioxide capture from atmospheric air using sodium hydroxide spray. *Environmental Science and Technology, 42,* 2728–2735.

Sun, J., Zhou, S. C., & Kikkert, G. A. (2014). Removal of aqueous hydrogen sulfide by granular ferric hydroxide—Kinetics, capacity and reuse. *Chemosphere, 117,* 324–329.

United Nations Framework Convention on Climate Change. (2005). *Inventories of anthropogenic emissions by sources and removals by sinks of greenhouse gases* (p. 23). Montreal: United Nations.

United States Environmental Protection Agency. (2008). Average annual emissions and fuel consumption for gasoline-fuelled passenger cars and light trucks, office of transportation and air quality. EPA420-F-08–024: 1–6.

U.S. Energy Information Administration. (2015). Independent statistics and analysis for South Africa. Retrieved March 3, 2015 from http://www.eia.gov/countries/cab.cfm?fips=sf.

Wei, Y., Li, C., Wang, Y., Zhang, X., Li, Q., & Xu, T. (2012). Regenerating sodium hydroxide from the spent caustic by bipolar membrane electrodialysis (BMED). *Separation and Purification Technology, 86,* 49–54.

World Bank. Urban development. Retrieved March 3, 2015 from http://data.worldbank.org/indicator.

Yeh, J. T., Pennline, H. W., Resnik, K. P., & Rygle, K. (2004). Absorption and regeneration studies for CO_2 capture by aqueous ammonia. In *Third annual conference on carbon capture and sequestration*, Alexandria.

Yeh, J. T., Resnik, K. P., & Pennline, H. W. (2004b). Regenerable aqua ammonia process for CO_2 sequestration. *Preprints of Papers-American Chemical Society, Division of Fuel Chemistry, 49*(1), 247.

Yeh, J. T., Resnik, K. P., Rygle, K., & Pennline, H. W. (2005). Semi-batch absorption and regeneration studies for CO_2 capture by aqueous ammonia. *Fuel Processing Technology, 86,* 1533–1546.

Zeman, F. (2007). Energy and material balance of CO_2 capture from ambient air. *Environmental Science and Technology, 41,* 7558–7563.

Zeman, F. S., & Lackner, K. S. (2004). Capturing carbon dioxide directly from the atmosphere. *World Resource Review, 16*(2), 157–172.

Zhang, Y., & Chen, C. (2013). Modeling CO_2 absorption and desorption by aqueous monoethanolamine solution with Aspen rate-based model. *Energy Procedia, 37,* 1584–1596.

Zhang, M., & Guo, Y. (2014). A comprehensive model for regeneration process of CO_2 capture using aqueous ammonia solution. *International Journal of Greenhouse Gas Control, 29,* 22–34.

Wastewater Biorefineries: Integrating Water Treatment and Value Recovery

Robert Pott, Madelyn Johnstone-Robertson, Bernelle Verster, Shilpa Rumjeet, Lefa Nkadimeng, Tayana Raper, Sharon Rademeyer and Susan T.L. Harrison

Abstract The wastewater biorefinery (WWBR) bridges the gap between the concepts of the biorefinery (BR) and wastewater (WW) treatment. A WWBR aims to generate a product, or products, of sufficient value to make the process economically viable and enhance resource productivity, while simultaneously remediating wastewater to an acceptable quality. It is centred on the conversion of the organic carbon, nitrogen, phosphorous and associated trace nutrients in the wastewater stream to value added products, while concomitantly providing clean or 'fit for use' water as a product. The WWBR is increasingly recognised for its potential contribution to the bio-based economy, as well as its potential to augment the industrial sector. The concept of a WWBR contributes to valuing wastewater treatment as an integrated component of a wider system rather than a unit process for 'end-of-pipe treatment', as is generally the state of current operation, a significant paradigm shift which has the potential to increase plant profitability and reduce environmentally deleterious effluents. It provides a link between the users of water and those responsible for its management, leading to the recovery of resources in closed loop cycles and thus contributing to progressing towards the concept of a circular economy, where valuable nutrients and components are recovered and reused. This chapter examines the pedagogy around the design and implementation of WWBRs with a focus on the classification of wastewaters for WWBR design, selection of potential products, flow sheet development and the criteria for reactor selection.

Keywords Energy · Environment · Sustainable development · Wastewater biorefineries · Industrial ecology · Value from waste

---

R. Pott · M. Johnstone-Robertson · B. Verster · S. Rumjeet · L. Nkadimeng · T. Raper · S. Rademeyer · S.T.L. Harrison
Department of Chemical Engineering, Centre for Bioprocess Engineering Research (CeBER), University of Cape Town, Cape Town, South Africa

R. Pott (✉)
Department of Process Engineering, University of Stellenbosch, Stellenbosch, South Africa
e-mail: rpott@sun.ac.za

W. Leal Filho and D. Surroop (eds.), *The Nexus: Energy, Environment and Climate Change*, Green Energy and Technology, https://doi.org/10.1007/978-3-319-63612-2_18

1 Introduction

The provision of water at a quality fit for purpose and the accompanying treatment of wastewater to remove associated pollutants is of high priority for many countries owing to impending water scarcity in many parts of the world, growing water demand with population growth and quality of life, and growing wastewater generation with an increase in human activity. Typically, wastewater treatment is considered a requirement for industrial or municipal effluent release and so has an associated expense. It is focussed on the remediation of water to a requisite legislated effluent quality rather than necessarily to a quality suitable for the direct application for particular use further within an industrial ecology. On approaching wastewater from a different perspective, it can be seen as a potential feedstock for the production of a range of products, including water. This product range stems from utilising the organic carbon, nitrogen (N) and phosphorous (P) components of the wastewater streams. Such perspectives are aligned with water sensitive urban design (Donofrio et al. 2009; Verstraete and Vlaeminck 2011; Fisher-Jeffes et al. 2014) and with the principles of industrial ecology (Graedel and Allenby 2009).

The intersection between traditional bioprocessing and wastewater treatment, to utilise waste streams as a valuable substrate with concomitant water treatment, rather than as a hazard simply to be purified sufficiently, can be termed a 'wastewater biorefinery' (WWBR). The implementation of WWBRs moves industrial activity towards closing resource cycles, by capturing those components of wastewaters which have value, and re-inserting them into economic circulation through elemental cycling while at the same time remediating wastes and recovering clean water. The WWBR concept is gathering interest and is increasingly recognised for its potential contribution to the bioeconomy, as well as to the move towards an industrial ecology and cleaner production (Harrison et al. 2015).

The WWBR is designed and optimised with the aim of ensuring both the effective treatment of the water to the necessary standard (yielding 'clean' and 'fit for purpose' water as a product), and the conversion of the components removed from the wastewater to products of economic value. The purity of this water can variably depend upon what further use it will be put to, within the industrial ecology.

In this paper, the concept of the WWBR is examined. Further, the value and grouping of waste streams in terms of applicability to WWBRs is investigated. Waste streams that could potentially be used as feedstock for the WWBR are highlighted. Materials that can be produced from these streams and the technical requirements for such, including biological systems used and appropriate reactor configurations around specific waste streams, are considered.

2 The Wastewater Biorefinery Concept

A WWBR operates at the intersection of bioprocessing, which utilises biological agents to produce products of value, and wastewater treatment, which is focused on the degradation of contaminants in wastewaters to suitable effluent concentrations (Verster et al. 2013). Traditional examples of bioprocesses, geared towards replicating and replacing existing petro-chemical-based systems, include bioethanol production, biodiesel production, polylactic acid (PLA) or starch-based polymers replacing plastics (Verster et al. 2013; Stuart and El-Halwagi 2012). Typically, these processes require relatively expensive 'clean or pure substrates' as feedstocks, such as glucose for ethanol production and other cultivations producing desired materials (i.e. biopolymers) or vegetable oils for biodiesel production. These feedstocks may be constructed from pure chemicals (e.g. glucose, sucrose) or agricultural extracts (glucose syrup, starch and molasses). The requirement for high-grade feedstocks puts pressure on these processes within three spheres: (1) economically, as the feedstocks often account for a majority of the operating cost of the process for a commodity product, (2) environmentally, owing to the burden created through energy intensive agriculture [through requirement for combined nitrogen fertilisers for example (Harding et al. 2007, 2009)], and (3) through socio-economic implications due to the competition between food production and the production of agricultural feedstocks for bio-based chemicals or biofuel (Harding et al. 2009; Lobley and Winter 2009).

In addition to these 'traditional' bioprocesses, biological systems can also make use of 'dirty' streams; that is, streams of variable composition, variable flow rates, multiple and changing components that are non-sterile (Henze et al. 2013; Verstraete and Vlaeminck 2011). For the most part, these biological systems have been developed to deal with waste streams in order to produce sufficiently clean (as defined by the legislature) water as the product, with sparse regard for generating any other products. The notable exception to this is the biogas-producing anaerobic digestion (AD) which converts waste organic materials into methane for use as an energy source while reducing the COD of the waste stream (Lettinga et al. 1995).

Since one of the underlying principles of a WWBR is that it should be a revenue producing part of the plant, rather than simply a waste treatment expense, the WWBR needs to generate products of sufficient value and scale to be economically viable, as well as having unit processes to produce clean or 'fit for purpose' water as a product. This concept contributes to seeing wastewater treatment as an integrated system rather than as the traditional 'end-of-pipe' unit process, and potentially provides a link between the users of water and those responsible for its management where resources are recovered in closed loop cycles, thus contributing to progress towards a circular economy (IEA Bioenergy 2012; Carey et al. 2016). The concept of closing the nutrient loop is particularly pertinent to non-renewable mineral resources, such as phosphorous. Without a move towards a closed nutrient loop, a looming shortage of readily available phosphorous is a serious risk for the future (Carey et al. 2016).

The WWBR as a concept is still very much in development. Most approaches are still focussed on the 'first' and 'second' generation biorefinery stages, where the wastewater and sanitation products are directed towards biogas, compost or biofertiliser, or a combination of these. Consideration of other products has been minimal. However, significant benefit can be generated by considering processing steps which focus on both producing clean water, fit for purpose, as well as a suite of valuable products. In the WWBR, the unit processes typically found in a functional wastewater treatment work using biological nutrient removal are adapted to facilitate product recovery. Typically, multiple unit processes are present to enhance overall process performance and resilience, but only some are adapted for commercial production, depending on the characteristics of the incoming waste stream(s), surrounding market needs and similar factors.

3 Application of Wastewater Biorefineries

3.1 Types of Wastewater as Feed for WWBR

In order to design appropriate reactors and flow sheets, wastewaters need to be categorised sufficiently. Wastewaters may be categorised according to three factors: (1) volumetric flow rate, (2) concentration of the major components, and (3) complexity—the number and variability of components that can be expected in the system. The interplay of these three factors and the specific composition of each wastewater stream dictate the shape, scale and potential products that a proposed WWBR would aim to produce.

Many wastewaters (particularly municipal wastewater) have large flows, in the order of mega litres per day. These streams are also often dilute, with the most common components present in the order of milligrammes per litre. Potentially the most problematic characteristic of wastewater is its level of complexity. Some waters, like municipal wastewater, tend to be highly variable with changing concentrations and, in some instances, compositions (Henze et al. 2008). These waters are considered by municipal managers as 'receptacles', meaning that the compounds that make their way into the water are not controlled or predictable and are mixed from multiple sources (Coetzee 2012). The different groupings of wastewaters each have their specific challenges and opportunities, which dictate the applicability and shape of any potential WWBR based on that wastewater.

The types of wastewater treatment plants typically found, and their capacity can be broken down as follows according to their handling volumes, as shown in Table 1. This grouping speaks to the first factor (1) above, analysis of which gives the first insight into the applicability of a particular waste stream for use in WWBRs. By necessity, the WWBR developed around a *micro* scale flow will be significantly different from that build to accommodate a *large* flow. There is some examination in the literature which gives estimates and heuristics around what sorts

Table 1 Wastewater volume classifications Adapted from Van den Berg (2009)

Type of plant	Capacity (ml/day)
Micro	<0.5
Small	0.2–2
Medium	2–10
Large	10–25
Macro	>25

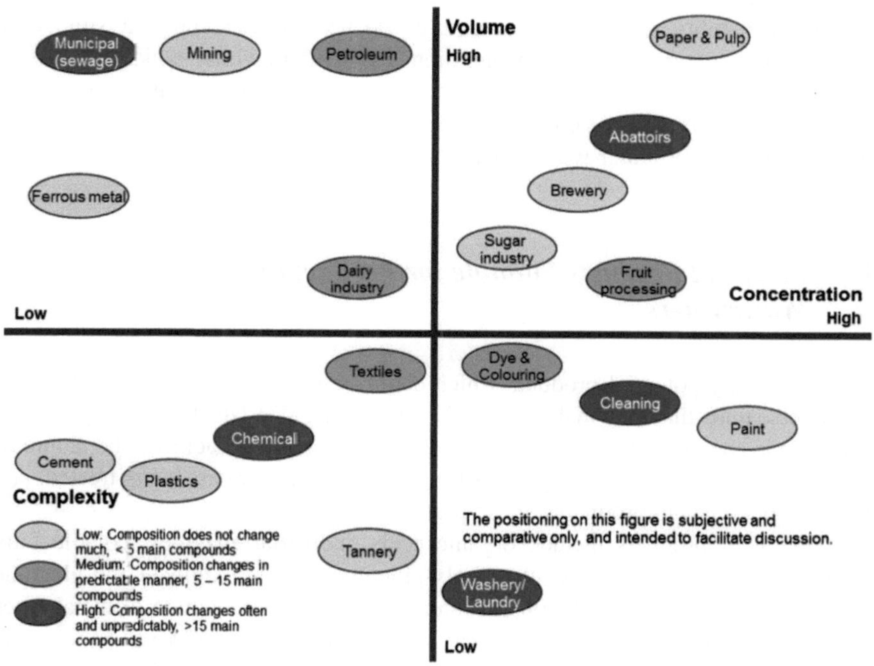

Fig. 1 Matrix illustrating grouping of wastewater in terms of volume, concentration and complexity

of wastewater generating processes will fall within each volume classification, for examples while municipal wastewater can be any order of magnitude, depending on the source settlement, sugar refineries tend to fall within a tighter band of flows (Harrison et al. 2015).

To examine points (2) and (3), we need wastewater-specific data. However, the challenge of presenting available data in a comparable and consistent fashion is non-trivial due to the scarcity and inconsistency of data sets. In addition, approaches to the validation of the data as well as its analysis on a national, provincial and local basis are essential to assess the potential to derive value from the wastewaters where the regional and local distribution of the resource is critical in determining practicality of its beneficiation.

Performing a conceptual plot of categorization variables [i.e. (1) volume, (2) concentration and (3) complexity], in Fig. 1, an initial grouping of a number of wastewaters of importance can be performed. This grouping allows for analysis of wastewaters with similar potential, which may be amenable to the development of similarly constituted WWBRs.

This three tiered analysis of each wastewater stream, depicted in Fig. 1 above is a necessary starting point for WWBR design and consideration. It is helpful to consider the clusters of potentially comparable wastewaters (for example those from the sugar industry, abattoirs, fruit processing and breweries), in preliminary assessments, since commonalities arise between these clusters. However, it is clear that the variability in complexity requires deeper analysis, beyond simply an understanding of component concentration and total flowrate, which have been primary considerations to date. The key result from an examination of complexity and composition leads directly to the next section: the appropriate selection of products from a particular wastewater.

3.2 Potential Products Suitable for Production from Specific Wastewaters

There are many potential products which could be produced within the integrated and diverse units that may make up a fully functioning WWBR. Key among these is clean water, the primary objective of a wastewater treatment facility. The nature of the other products produced depends strongly on the wastewaters being treated, and the biological agents producing these products.

Potential bioproducts include organic acids, industrial enzymes, volatile fatty acids, pigments, polymers including poly-glutamic acid and alginate and gluconic acid (Pandey et al. 2010). The practical approach for the production of metabolites and enzymes can be related to different applications (paper deinking, paper recycling, agricultural residue utilisation, pesticides biodegradation, fodders, olive and seed oils residues, pruning, fuels, paper pulp production, and so forth) and each of them require a different set of biotechnological conditions (Pandey et al. 2010).

The WWBR is established to maximise biomass productivity by ensuring that, not only is the wastewater treated to the necessary standard (yielding the outgoing water product), but that component removed from this wastewater are converted to products of value. Due to the dilute nature of wastewaters, highly energy intensive production processes, such as distillation recovery of fermentation products, are not likely to be appropriate. Because of the non-sterile nature of the wastewater environment, it is best to select microorganisms for product formation that also fulfil a role in the microbial ecology, thereby providing a selective pressure for their dominance. It is also beneficial to select culture conditions and product to contribute a selective advantage to the microbial community of interest. Hence, WWBRs are not suitable for all bioproducts.

Bioproducts from WWBRs can be grouped into three main categories. The first is produced by the breaking down of complex molecules into simple 'building block' chemicals, for example volatile fatty acids or ethanol, which can then be used for as the basis for more complex chemical synthesis. The second category includes function-based products that are composed of complex macromolecules, requiring minimal modification or purification. Examples of these include bioflocculants, biosurfactants or soil conditioners. Finally, category three includes bioenergy products, such as biogas produced in anaerobic digesters, which can be burnt for heat or electricity generation.

From an operational standpoint, products that fill an ecological niche in the reactor are advantageous, since the microorganisms producing these compounds are given a competitive advantage by their production. Further, the product needs to be easily and cheaply recoverable from the waste stream; either owing to it being produced into a different phase (gas or solid), or being recoverable through a cost efficient process (Verster et al. 2013). Hence, WWBRs are most suitable for products that fulfil both a defined role in the microbial ecology (allowing for natural selection towards our desired species) and are easily recoverable. In addition, products that are used in the functioning of the treatment works are favoured, as materials required for plant operation produced from its own waste resources allow a stable demand for the product and reduce operational costs.

Due to the fact that a considerable amount of fossil fuel derived energy is needed in a Biological Wastewater Treatment Works to supply aeration for the aerobic processes and to transport the large volumes of water and biomass from one unit to the next, the greenhouse gas contribution just to supply clean water is becoming environmentally unsustainable (Sheik et al. 2014). Stafford et al. (2013) and Burton et al. (2009) identified the technologies for recovery of energy from wastewater with applicability and potential in South Africa. The chemical energy contained in these influent waters (domestic or industrial) is at least tenfold higher than the amount of energy required to treat them (Heidrich et al. 2011; Shizas et al. 2004; McCarty et al. 2011), hence energy (or energy carrying products such as biogas) is also an important potential product to be considered in the development of a fully integrated WWBR. Indeed this inherent energy can be captured in a number of ways, from biogas production through anaerobic digestion (McCarty et al. 2011; Rabelo et al. 2011); to electricity production the microbial fuel cells (MFCs) (Borole and Hamilton 2010). Research in the area of energy reclaim from wastewaters is advancing rapidly, and novel methodologies are being developed and tested worldwide. It is likely that a successful WWBR will utilise appropriate energy recapture technologies as part of their industrial ecology.

4 Approach to Flow Sheet Development for Biorefineries

4.1 The Process Flow Sheet

Studies to date have recognised the need for multiple unit operations in BRs in general, and WWBRs specifically (Kamm et al. 2007). Multiple units allow both the harvesting and beneficiation of different components of the wastewater. Further, they allow the maximising of both the bioconversion of solutes to product and the quality of product water, through specialisation of unit operations (i.e. not everything can be accomplished successfully in one unit).

Due to the diverse nature of the potential feed wastewaters, as examined in Sect. 3.1 above, each biorefinery case is expected to have a unique process flow sheet, guided by heuristics that contribute to first order feasibility analysis and understanding of the potential of each WWBR design. Wastewater treatment generally consists of settling, primary treatment, secondary treatment and possibly polishing steps (Henze et al. 2008). A WWBR is expected to comply with these steps to produce clean or 'fit for purpose' water as a product. Producing co-products requires additional analysis in terms of product yield and efficiency, within the specialised units that can be added, or the modifications of current WWT unit models.

Figure 2 is an illustration of a potential WWBR flow sheet using municipal wastewater as its raw material, while Table 2 describes more fully the unit operations shown in Fig. 2. The unit processes typically found in a functional wastewater treatment works using biological nutrient removal have been adapted to facilitate product recovery. Typically, multiple unit processes are present to enhance overall process performance and resilience, but only some are adapted for commercial production, depending on the characteristics of the incoming waste stream(s), surrounding market needs and similar factors.

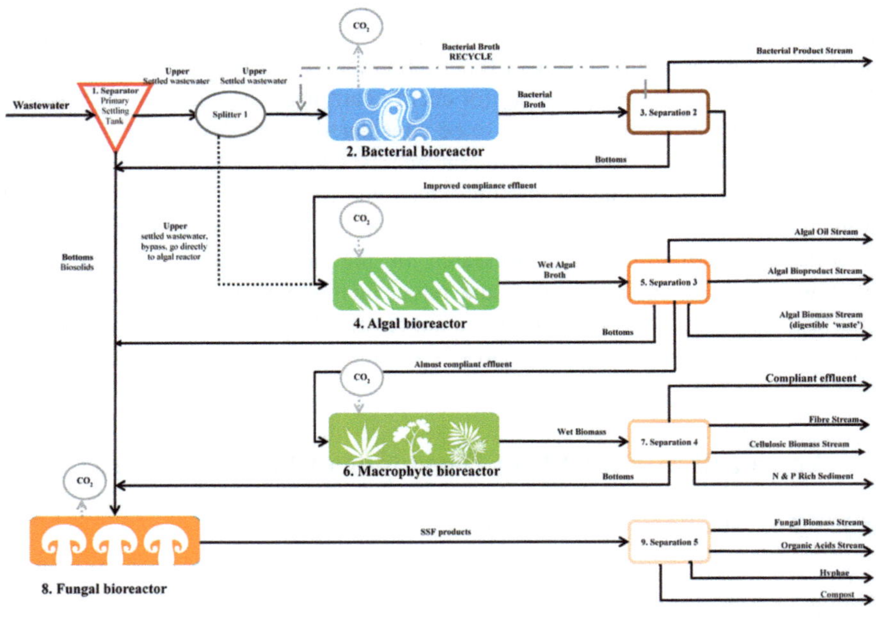

Fig. 2 A simplified flowsheet of a potential wastewater biorefinery

Table 2 List of unit operations for generalised wastewater biorefinery corresponding to Fig. 2

Unit number	Type	Unit description
1.	Separator	Separator 1: Primary Settling Tank (PST) settling raw wastewater
A	Splitter	Splitter 1: Settled, raw wastewater to bacterial and algal reactors
2.	Reactor	Bacterial bioreactor
3.	Separation Process	Separation 2: bacterial broth to biomass/product fraction, liquid to algal bioreactor, and solids to biosolids reactor
4.	Reactor	Algal bioreactor
5.	Separation Process	Separation 3: algal broth to biomass/product fraction, liquid to macrophyte bioreactor, and solids to biosolids reactor
6.	Reactor	Macrophyte bioreactor
7.	Separation Process	Separation 4: Macrophyte biomass harvested and fractionated into fibre and cellulosic biomass, solids to biosolids reactor
8.	Reactor	(Bio)solid substrate fermentation (SSF) reactor/fungal reactor
9.	Separation Process	Separation 5: Fungal biomass harvested and fractionated into products, solids used as compost

There is a multitude of potential configurations possible for WWBRs, which is both a hindrance to their implementation due to the difficulty of heuristic selection as well as a boon since complex configuration potentials make the concept extraordinarily flexible (Fava 2012). Life-cycle assessment may be suitable in detangling which system configuration is optimal. Beyond environmental implications, capital and operating costs must need to be the critical considerations in WWBR process design. Since the processes and unit operations available for use in a WWBR represent a wide range of technological maturity, scale and historical implementation, direct cost comparisons are inherently complex. Further, the intricate integration between units, and the sensitivities to feedstocks, operating conditions and most importantly market pricing for products all complicate a direct comparison of profitability.

Nonetheless, an exemplary illustration of a WWBRs implementation, as described in Fig. 2 and Table 2, can be illuminating in this context. In this case, an example wastewater is fed into a series of reactors (bacterial, algal, macrophytic and fungal), to produce water of suitable purity in addition to products from each reactor.

4.2 *Criteria for Reactor Selection*

Owing to the typically dilute nature of wastewater streams, their variability, the impracticalities of sterilisation and the need to handle large volumes, the selection of appropriate reactors needs to meet multiple process criteria. A significant challenge in developing a WWBR around a particular waste stream is the selection of suitable reactors, which match appropriate operating conditions, product collection and separation, cost and water purification requirements. Current wastewater treatment bioreactors are well designed to achieve nutrient removal through biomass retention, but typically not designed towards product accumulation or recovery. The requirement for efficient product formation and recovery has limited the application of bioprocesses using waste streams in existing plants.

Conventionally, bioreactor optimisation aims to reduce the reactor volume in order to reduce the energy invested per unit product (Harding 2009), and aims to achieve a higher biomass concentration, which results in less downstream processing (DSP) cost per unit product (Richardson 2011; Ramaswamy et al. 2013) as well as a reduced environmental burden (Harding et al. 2009). For bioproducts from wastewater to be feasible, reactors need to produce a product in the face of large volumes and a complex medium. The resultant suspension should not affect the environment adversely, and ease of downstream processing is important. Consequent requirements of the WWBR reactor are detailed in Table 3 and discussed further in the following sections. While DSP often contributes a large fraction to the total processing costs, the processing units are already very efficient and well developed (Ramaswamy et al. 2013).

One significant challenge of bioprocesses is the dilute nature of the medium, with both substrates and products present at a very low concentration, typically less than 5% of the total dissolved solids present. When using waste streams like municipal wastewater, which can be a thousand fold more dilute, this aspect is even more challenging. The apparent biocatalyst concentration must be increased to enhance process intensity over the current approach of huge dilute vats of water. This has major implications for mass and energy transfer needs. Aeration and heat transfer in dilute media is inefficient and energy intensive. Using biomass retention, these requirements can be better managed.

Table 3 Wastewater biorefinery reactor design requirements

#	Requirement	Comply?
Large volume		
1	Decouple hydraulic and solid retention times	✓
2	Continuous or semi-continuous (can't store flows)	✓
3	Think big! Commodity rather than niche	✓
Complex, variable		
4	Influence microbial community, non-sterile	✓
5	Give advantage to product: create ecological niche	✓
Environment		
6	Water released into environment eventually	✓
Downstream processing		
7	Product formation in different phase?	✓
8	Can product be recovered?	✓
9	Reactor design conducive to reducing DSP load?	✓

The same approach holds for the product. For cost and energy efficient downstream processes, the localised product in an accessible location with high apparent concentration is preferred. In dilute wastewater treatment, biomass retention is advantageous as conversion is limited by the amount of biomass present and retention allows the increase in biomass concentration. Importantly, biomass retention facilitates the effective decoupling of the hydraulic and solid retention time, required to improve reactor volumetric conversion capacity.

Conventional bioprocessing generally requires a homogenous, highly controlled environment, however, wastewaters tend to be more complex and heterogeneous (Verster et al. 2013). As the water is destined for eventual discharge to the environment, any additives to improve the characteristics of the stream need to be non-hazardous or removed within the process. The sheer volume of the stream also prevents extensive stream modification. Depending on the robustness of the organism, environmental regulation and social acceptability, the use of genetically modified organisms is also likely to be precluded. Further, sterilisation, either upstream or downstream, is typically not practical. This results in a limited scope to modify the microbial community in the manner currently favoured for industrial bioprocess applications. Instead, the most robust and resilient biological agents would occur in a mixed community well adapted to the physicochemical environment in which it exists, able to withstand shock loads and hostile environments,

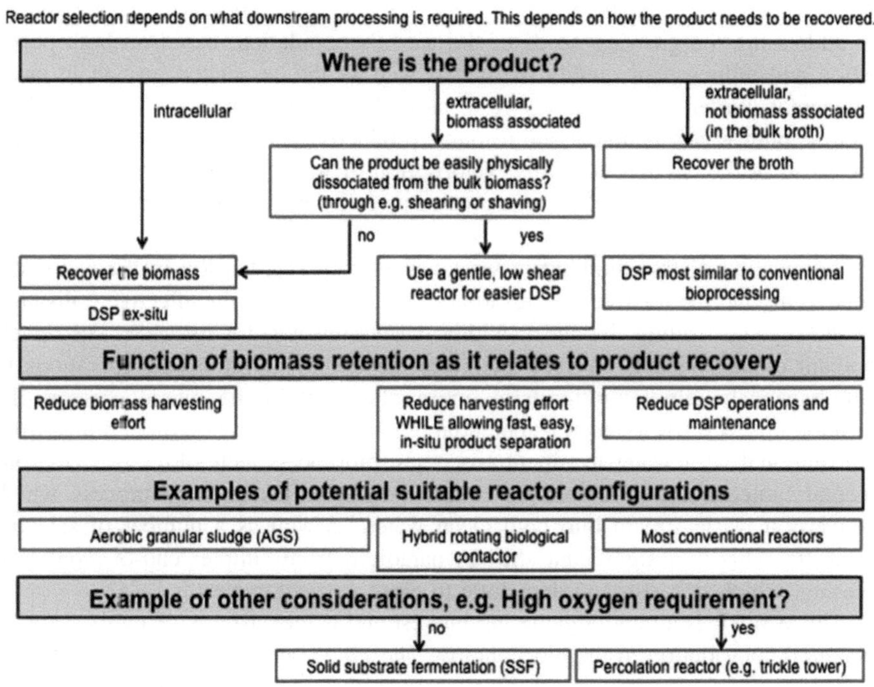

Fig. 3 Suggested guideline for WWBR reactor selection Reproduced with permission from Verster et al. (2013)

and preferentially selected for by the standard operational environment, rather than selected for externally (Kleerebezem et al. 2007).

Figure 3, reproduced from Verster et al. (2013), is an initial guideline for WWBR reactor selection. It gives a visual perspective on the pedagogy outlined above, which guides appropriate reactor choices. For a more comprehensive discussion of reactor selection, Verster et al. (2013) provide further insights into this aspect of WWBR development.

WWBRs are not suited to all products; the product needs to meet commodity market demand suited to the utilisation of organics from large stream flows, but also needs to serve an ecological function to the microorganisms to drive a competitive advantage within the reactor to ensure continued majority of the desired organism and product (Kleerebezem et al. 2007). Reactor design needs to complement this ecological niche to select for the desired products.

A fundamental consideration to make bioproducts from dilute streams viable from both an economic and environmental point of view is in the approach to the DSP (Richardson 2011). In a systems approach, the recovery and quality of both water and coproduct streams need to be considered.

5 Conclusions

In addressing the growing needs of the world's population, based on both population and affluence or quality of life, as well as the associated growth in environmental burden and required waste assimilation, new thinking is required to address both waste treatment and resource productivity. Strongly emerging themes are those of the circular economy; valorisation of waste as well as waste minimisation, i.e. use the value of the resource to its full potential before classifying any part of it as waste for disposal.

Application of this thinking to wastewaters creates a tension in the approach, centred around the relative prioritisation of clean water and the effective utilisation of the organic loading, N, and P within the wastewater, for instance. The development of the WWBR concept facilitates the use of multiple unit operations to simultaneously allow multi-criteria optimisation within the overall system. To develop this WWBR to reach its potential requires the integration of learnings from conventional wastewater treatment processes, bioprocess technology and environmental biotechnology. A solution to these systems results in a process which converts a waste stream into both clean water, as well as a number of saleable products. This is a significant shift in paradigm from simple 'end-of-pipe' processing, which is costly and does little to close nutrient cycles.

The WWBR is a rapidly gathering interest and is increasingly recognised for its potential contribution to the bioeconomy. The implementation of WWBRs moves industrial production towards closing resource cycles, by re-capturing those components of wastewaters which have value, and re-inserting them into economic circulation while at the same time remediating wastes and recovering clean, or 'fit for purpose' water as a product. This approach is consistent with the concepts of industrial ecology, cleaner production and the circular economy.

The pedagogy and design concepts around WWBRs are presented in this chapter, contributing to the thinking around WWBR development. Analysis of the grouping of waste streams to be treated in a similar manner is presented, showing that key considerations are (1) volume or volumetric flow rate, (2) concentration of dissolved solutes and (3) complexity. From this analysis, we can consider the nature of a potential WWBR flowsheet constructed around a waste stream. Further, consideration of the reactor design and selection is presented.

Acknowledgements The authors gratefully acknowledge the funding contribution of the South African Water Research Commission (WRC) through WRC K5/2380, as well as the technical input of the steering committee to this project.

References

Borole, A. P., & Hamilton, C. (2010). Energy production from food industry wastewaters using bioelectrochemical cells. In V. Shah (Ed.), *Emerging environmental technologies* (Vol. II, pp. 97–113). Dordrecht: Springer.

Burton, S., Harrison, S., Pather-Elias, S., Stafford, W., Van Hille, R., Von Blottnitz, H., et al. (2009). *Energy from wastewater—A feasibility study*. Essence report. Water Research Commission report TT 399-09. ISBN 978-1-77005-847-7.

Carey, D., Yang, Y., McNamara, P., & Mayer, B. (2016). Recovery of agricultural nutrients from biorefineries. *Bioresource Technology*. Available online March 2, 2016, ISSN 0960-8524.

Coetzee, B. (2012) Technical strategic support manager, utility services directorate, City of Cape Town. Personal communication.

Donofrio, J., Kuhn, Y., McWalter, K., & Windsor, M. (2009). Water sensitive urban design: An emerging model in sustainable design and comprehensive water cycle management. *Environmental Practice, 11*(3), 179–189.

Fava, F. (2012). *Biowaste biorefinery in Europe: Opportunities and R&D needs*. Barcelona: European Federation of Biotechnology.

Fisher-Jeffes, L., Carden, K., & Armitage, N. (2014). The future of urban water management in South Africa: Achieving water sensitivity. *Water Science and Technology, 14*(6), 1026–1034.

Graedel, T. E., & Allenby, B. R. (2009). *Industrial ecology and sustainable engineering*. Upper Saddle River: Prentice Hall.

Harding, K. G. (2009). *A generic approach to environmental assessment of microbial bioprocesses through life cycle assessment* (*LCA*) (Ph.D. thesis). University of Cape Town, South Africa.

Harding, K., Dennis, J. S., von Blottnitz, H., & Harrison, S. T. L. (2007). A life-cycle comparison between inorganic and biological catalysis for the production of biodiesel. *Journal of Cleaner Production, 16,* 1368–1378.

Harrison, S., Johnstone-Robertson, M., Pott, R., Verster, B., Rumjeet, S., Nkadimeng, L., et al. (2016). *Towards wastewater biorefineries: Integrated bioreactor and process design for combined water treatment and resource productivity*. Water Research Commission Report 2580/1/2016.

Heidrich, E., Curtis, T., & Dofling, J. (2011). Determination of the internal chemical energy of wastewater. *Environmental Science and Technology, 45*(2), 827–832.

Henze, M., Harremoes, P., Arvin, E., & LaCour Jansen, J. (2013). *Wastewater treatment: Biological and chemical processes*. Berlin: Springer.

Henze, M., Loodsdrecht, M., Ekama, G., & Brdjanovic, D. (2008). *Biological wastewater treatment: Principles, modelling and design*. London: IWA Publishing.

IEA Bioenergy. (2012). *IEA bioenergy task 42 biorefining*. [Online]. Available at: http://www.iea-bioenergy.task42-biorefineries.com/en/ieabiorefinery.htm. Accessed October 3, 2014.

Kamm, B., Gruber, P. R., & Kamm, M. (2007). Biorefineries—Industrial processes and products. In *Ullmann's encyclopedia of industrial chemistry*. Weinheim: Wiley-VCH.

Kleerebezem, R., & van Loosdrecht, M. (2007). Mixed culture biotechnology for bioenergy production. *Current Opinion in Biotechnology, 18*(3), 207–212.

Lettinga, G. (1995). Anaerobic digestion and wastewater treatment systems. *Antonie van Leeuwenhoek, 67*(1), 3–28.

Lobley, M., & Winter, M. (Eds.). (2009). *What is land for? The food, fuel and climate change debate*. New York: Routledge.

McCarty, P. L., Bae, J., & Kim, J. (2011). Domestic wastewater treatment as a net energy producer-can this be achieved? *Environmental Science and Technology, 45,* 7100–7106.

Pandey, A., Soccol, C., & Larroche, C. (Eds.). (2010). *Current developments in solid-state fermentation*. Dordrecht: Springer.

Rabelo, S., Carrere, H., Maciel Filho, R., & Costa, A. (2011). Production of bioethanol, methane and heat from sugarcane bagasse in a biorefinery concept. *Bioresource Technology, 102*(17), 7887–7895.

Ramaswamy, S., Ramarao, B., & Huang, H.-J. (2013). *Separation and purification technologies in biorefineries*. Chichester: Wiley.

Richardson, C. (2011). *Investigating the role of reactor design for maximum environmental benefit of algal oil for biodiesel*. MSc dissertation, University of Cape Town, South Africa.

Sheik, A., Muller, E., & Wilmes, P. (2014). A hundred years of activated sludge: Time for a rethink. *Frontiers in Microbiology, 5.*

Shizas, I., & Bagledy, D. (2004). Experimental determination of energy content of unknown organics in municipal wastewater streams. *Journal of Energy Engineering, 2*(45–53), 130.

Stafford, W., Cohen, B., Pather-Elias, S., von Blottnitz, H., van Hille, R., Harrison, S., et al. (2013). Technologies for recovery of energy from wastewaters: Applicability and potential in South Africa. *Journal of Energy in Southern Africa, 24*(1), 15–26.

Stuart, P., & El-Halwagi, M. M. (Eds.). (2012). *Integrated biorefineries: Design, analysis, and optimization*. Boca Raton: CRC Press.

Van den Berg, M. (2009). *The South African wastewater market—Business opportunities and export promotion for Dutch companies*. MSc thesis, University of Twente, Netherlands.

Verster, B., Madonsela, Z., Minnaar, S., Cohen, B., & Harrison, S. T. L. (2013). *Introducing the wastewater biorefinery concept: A scoping study of polyglutamic acid production from Bacillus rich mixed culture using municipal wastewater*. Pretoria: Water Research Commission Report TT587/13.

Verstraete, W., & Vlaeminck, S. E. (2011). Zero WasteWater: Short-cycling of wastewater resources for sustainable cities of the future. *International Journal of Sustainable Development and World Ecology, 18*(3), 253–264.

Part III
Climate Change

Small is Beautiful: Why a Club Approach is the Way to Go in Climate Change Mitigation

Rafael Leal-Arcas

Abstract This chapter argues that the Kyoto Protocol to the 1992 Framework Convention on Climate Change was doomed to face difficulties ab initio and tries to draw lessons from the international trading system's architecture for climate change mitigation negotiations. The UNFCCC has traditionally divided the world into developed countries (including countries undergoing the process of transition to a market economy) and developing countries. It places the responsibility of reducing emissions with developed countries as if they were the only sinners of climate change. A better (and arguably fairer) way to tackle the climate change issue today is by bringing together the major GHG emitters, irrespective of their GDP. Why so? Because seen retrospectively, rich-countries have been (and continue to be) the major polluters; they are responsible for most of the GHG emissions, and have the financial and technological means to tackle climate change. However, seen prospectively, it is a developing countries problem, as predictions indicate that, in the near future, developing countries will be the major polluters as well as the major victims of the consequences of climate change, especially countries near the equator. The longer we wait, the harder and more expensive it will become to deal with climate change. So the major GHG emitters (whether developed or developing countries), which are responsible for historic, current and future emissions, should, therefore, be the ones to take action.The chapter concludes that no breakthroughs will take place regarding climate change mitigation until there is more political maturity on the side of the U.S. and until rapidly emerging economies indicate that they are ready to play their part.

Keywords Bottom-up approach to climate change mitigation · Major economies forum · Variable geometry

---

Many ideas in this paper draw from Leal-Arcas, R. *Climate Change and International Trade*, Edward Elgar, 2013, Chap. 6.

R. Leal-Arcas (✉)
Centre for Commercial Law Studies, Queen Mary University of London, London, UK
e-mail: r.leal-arcas@qmul.ac.uk

W. Leal Filho and D. Surroop (eds.), *The Nexus: Energy, Environment and Climate Change*, Green Energy and Technology, https://doi.org/10.1007/978-3-319-63612-2_19

1 Introduction

This paper argues that the Kyoto Protocol to the 1992 UNFCCC was doomed to face serious difficulties ab initio because it places the responsibility of reducing GHG emissions only with developed countries (i.e. Annex I countries) as if they were the only sinners of climate change. If the international community is serious about GHG emissions reduction, a more plausible way to reduce GHG emissions (which is the whole purpose of the exercise) is to involve major GHG emitters, irrespective of their GDP.

International law is fragmented. The case of international trade law is a good illustration of the fact that international law, generally speaking, operates in a fragmented fashion (Cottier and Delimatsis 2011). At first, international trade agreements were bilateral. Then came the GATT 1947, which multilateralized bilateral trade agreements. Years later, international trade law saw the collapse of multilateralism in 1979, which broke down during the Tokyo Round of multilateral trade negotiations (World Trade Organization). A series of new plurilateral (or selectively multilateral) agreements were adopted during the Tokyo Round, which caused a fragmentation of the multilateral trading system (Leal-Arcas 2011a). In 1994, international trade law was again multilateralized with the World Trade Organization (WTO) Agreement (Leal-Arcas 2010).

The same thesis could be used for climate change law. Although not always easy, as demonstrated by the Doha Round, why not emulate the successful experience of multilateralization of international trade law from the bottom-up (which started with the GATT 1947) for the case of climate change law? If so, to what extent can the experience in international trade negotiations and the WTO framework as a bottom-up approach serve as an example? (Cottier 2006; Hong 1990) The trend of using bilateral or plurilateral trade agreements to build towards eventual trade multilateralization may be worth emulating for the case of climate change law (Cottier 2011).

The paper explores the architecture and operation of international instruments relating to climate change mitigation and their interaction with the international trading system. It examines the possibility of a complementary combination of top-down and bottom-up approaches to building a global climate landscape and the link of national and regional climate arrangements with the UNFCCC framework (Leal-Arcas 2011b; Rayner 2010; Pielke 2007). Although the UNFCCC's mandate is about mitigation and adaptation of climate change, the present paper focuses solely on the climate change mitigation dimension of the UNFCCC with its aim of achieving an international agreement for climate change mitigation post-2015. In this sense, the paper explains and justifies the bottom-up approach to climate change negotiations, and explores the possibility of variable geometry. It examines various *fora* for implementing a more effective GHG emissions reduction.

2 Paradigm Shift: Justifying a Bottom-Up Approach

The Kyoto Protocol is a highly centralized, top-down agreement on climate change, which has proven to be very rigid in the target of reducing GHG emissions (Kopp 2010). For the purposes of GHG emissions reduction, the UNFCCC divides the world into Annex I countries and developing countries, legally binding only Annex I countries to reducing their GHG emissions by a certain deadline.[1] Why so? Because seen retrospectively, rich countries have been the major polluters; they are responsible for most of the GHG emissions, and have the financial means and technological capability to tackle climate change.

However, instead of asking only Annex I countries to reduce GHG emissions, this chapter argues that a more effective (and presumably fairer) way to tackle climate change today is by bringing on board the major GHG emitters, irrespective of their GDP, and ask them to reduce their GHG emissions in an equitable manner without ignoring the historical responsibilities on the part of developed countries.[2] More than 20 years after the UNFCCC was created in the early 1990s, the climate change situation has changed. Why do we need to shift the paradigm? Because the Kyoto Protocol's stipulation that only Annex I countries reduce their GHG emissions does not reflect today's or tomorrow's climate change reality. In accordance with chapter 3 of the Kyoto Protocol, Annex I countries are asked to lead the way on GHG emissions reduction, for others to follow, but the U.S. refused. My argument is that it is not enough to ask only Annex I countries to reduce their GHG emissions (Bosetti 2009) if the aim is to solve the climate change issue. This is an example of the bottom-up approach proposed in this chapter. Moreover, asking major GHG emitters (both developed and developing countries) to commit to GHG reduction would have the advantage of avoiding a situation where a small group of industrialized countries would monopolize the regulation of GHG emissions reduction, something that many developing countries fear.

Major developing countries that are major GHG emitters should also be asked to reduce their GHG emissions. Why? Seen prospectively, climate change is a developing countries problem, since they have already surpassed the industrialized world in total GHG emissions (International Energy Agency 2011) and will account for more than 75% of GHG emissions growth in the next 25 years (Energy Information Administration 2007). Developing countries will also be the major victims of the consequences of climate change, especially countries near the equator (Farber 2008).

The US and other Annex I countries maintain that the terms of the Kyoto Protocol are unfair because they provide developing countries with inappropriate trade advantages and because the GHG emissions of leading developing countries

---

[1]Chapter 3 of the Kyoto Protocol.

[2]That said, the UNFCCC reflects the early 1990s situation, where developed countries decided to "take the lead in combating climate change". See chapter 3.1 of the UNFCCC. Twenty years later, the climate change situation has changed.

are growing rapidly without their being legally bound to reduce their GHG emissions.

The world has changed dramatically since the UNFCCC divided the world into two categories in 1992. For instance, approximately 50 non-Annex I countries now have higher per capita incomes than the poorest of the Annex I countries with commitments under the Kyoto Protocol (UNDP). In the same way, 40 non-Annex I countries ranked higher on the Human Development Index in 2007 than the lowest ranked Annex I country. The trend of developing countries catching up continues in the post-2008 financial crisis period, as developing countries are still growing macroeconomically, whereas many developed countries are in economic stagnation or recession. So major GHG emitters/economies (whether developed or developing countries), which are responsible for historic, current and future GHG emissions, should, therefore, be the ones to take action in an equitable manner.

As of 2000, the top 25 GHG emitters accounted for approximately 83% of global emissions (Baumert et al. 2005). Moreover, the top five GHG emitters today (China, US, the European Union (EU)—treated as a single entity—India and Russia) were responsible in 2000 for over 60% of global emissions. In contrast, most of the remaining countries contributed very little in absolute terms to GHGs in the atmosphere (i.e. the 140 least-pollutant countries were responsible for only 10% of global GHG emissions). These countries include the least-developed countries and many small island States.

To show the importance of rapid action in climate change mitigation, 15 out of the 195 UNFCCC members were responsible for approximately 80% of global GHG emissions in 2005, which means that, between 2000 and 2005, the level of GHG emissions coming from major emitters increased.[3] This figure means that the remaining UNFCCC membership was only responsible for around 20% of global GHG emissions. In other words, very many countries have contributed very little to climate change, but very few countries have contributed very much. This latter small group of countries should, therefore, be responsible for fixing the current situation, which would be easier and less complex to fix in a small club than among the entire UNFCCC membership.

3 The Underlying Rationale

Based on empirical observation, variable geometry (or a 'club' approach) seems both logical and fair as a mechanism to move the climate change agenda forward, given that a relatively small number of countries produces a large majority of GHG emissions. Moreover, from a practical viewpoint, it is easier and more manageable to negotiate amongst a small number of large players than amongst a large number of

[3]These numbers exclude emissions from land-use, land-use change, forestry, and international bunker fuels.

small players, which explains the creation of clubs. So bringing together a group of countries (i.e. major GHG emitters in the case of climate change, whatever the format may be, whether bilaterally or plurilaterally) seems to make sense, especially because there is more pressure to deliver when the group of countries is smaller (Abbot and Snidal 1998). Furthermore, less time is spent on procedural matters when dealing with a small group of countries. In addition, based on international negotiating experience from other fields, the only way to get any real business done is in small meetings (sometimes tête-à-tête meetings between key leaders) (Poteete et al. 2010).

If we accept this club approach, what *fora* may be used for formulating a global response to climate change? (Levi 2010) The G-20, the Major Economies Forum on Energy and Climate (MEF), the climate G-8,[4] (Stern and Antholis 2007) the G-3,[5] and regional groupings all seem plausible options to provide political leadership. They all have the shared vision that GHG emissions must be reduced, with targets for developed countries and actions from developing countries. As mentioned earlier, though, not every part of the world needs to be represented at the beginning. The global GHG contribution of the least-developed countries and small island developing States is so minimal, that it seems logical to start with the major GHG emitters and eventually have the rest of the world join in the quest for GHG emissions reduction. Once the major parties are grouping together, the chance of having other countries join increases (Barrett 2003). Previous experience shows that negotiating and decision-making resulting from such clubs has been valuable in *fora* such as the UN Convention on the Law of the Sea, the WTO or the creation of the Montreal Protocol (Kahler 1992). So there seems to be added value to formalizing negotiations in smaller groups.

There are, however, disadvantages to the variable geometry option. If one were to start with a smaller group of players, say the EU, the US and China, and, by virtue of doing so, one keeps, for instance, the most vulnerable countries out of the climate negotiations, how does one ensure sufficient ambition to reduce GHG emissions? There is a time window, optimistically, up to 2020 to reverse the trends in global GHG emissions and begin their sharp decline, at least if we are to keep global average temperature increases below 2 °C with reasonable certainty. In this sense, the Paris Agreement on Climate Change aims to keep the global average temperatures 'well below 2 °C above pre-industrial levels'[6] and also to 'pursue efforts to limit the temperature increase to 1.5 °C above pre-industrial levels, recognizing that this would significantly reduce the risks and impacts of climate change'.[7] Having only the major GHG emitters on the table may possibly translate

[4]The climate G-8 would be composed of China, the U.S., the EU, India, Brazil, South Africa, Japan, and Russia (currently suspended). This group represents the key economies in each region of the world and is responsible for 70 per cent of global GHG emissions. This number includes land use, land-use change, and forestry. For views on the climate G-8, see Stern and Antholis *Winter* 2007, pp. 43-49.

[5]I am referring to the G-2 (U.S. and China) plus the EU.

[6]chapter 2.1(a), Paris Agreement, Draft decision-/CP.21.

[7]Idem.

into a situation where these countries would indulge in a rather low GHG emission ambition (Sunstein 2008), agree quickly among themselves, and then sell the results as major achievements in climate diplomacy, whereas in fact they lock the world into a disastrous climate change scenario simply because of the very limited remaining carbon budget. This is in contrast to the negotiation of trade agreements, where delays may mean little more than a delay of economic benefits from such agreements, and ambition can be increased at any later stage.

Furthermore, why would countries engage in mini-lateral or even bilateral agreements that have any reasonable hope of meeting the 1.5–2 °C target? When negotiating trade agreements, parties have the interest to negotiate as they believe they will benefit from the agreement. By contrast, in climate change negotiations, most parties see mitigation as a burden to their economies, i.e. they negotiate to ensure that they do not have to do more than other parties in the negotiation. This would be different if there were a general belief that mitigation is beneficial for the economy and therefore needs an international agreement to maximize the parties' benefits. While there are voices that point to the many economic benefits of ambitious mitigation, this belief has yet to enter the minds of finance ministers, companies and society at large.

4 Forum Options

A collective action problem is a situation where those involved can choose between two options, and if they select the one that is individualistically rational (IR), the result for all parties concerned will be worse, in their own judgment, than if they were to choose the other option, i.e. the option that is not IR. Typically, in any collective action problem, the IR alternative is called defection and the non-IR alternative is referred to as cooperation (Olson 1965). By this token, then, climate change could be termed a collective action problem.

This section argues that polycentric systems can produce collective action more effectively than unified institutions such as the UNFCCC/Kyoto Protocol process (Ostrom 2010a, b). Moreover, it is also argued that climate governance should follow the examples of concentric circles in larger structures in other fields of global governance. For instance, just as the G-20 in the context of the International Monetary Fund or the Security Council in the context of the United Nations are examples of concentric circles for monetary and foreign policy, respectively, the Major Economies Forum on Energy and Climate may serve as a concentric circle for the UNFCCC in global climate governance. However, in the case of climate change negotiations, least-developed countries and small island States have constantly shown their preference for the UNFCCC as a negotiating platform because their voices can be heard (Durrant 2002). In spite of these objections, this book argues that alternative *fora* may be the most effective way to move forward. In this context, below are some non-exhaustive suggestions of plausible forum options to produce collective action in climate change mitigation and adaptation. The various selected concentric circles from smaller to larger are the G-2, the G-3, the MEF and the G-20, which does not reflect an order of preference. The list of possible concentric circles could go on and on, and there have been suggestions for other possible *fora*.[8]

4.1 The G-2

What is absurd is that the world's first and second largest CO_2 emitters—i.e. China and the US, respectively—are not bound by the Kyoto Protocol to reduce GHG emissions: China, because it is not an Annex I country, and the US, because it has not ratified the Kyoto Protocol (Seligsohn 2009; Stewart and Wiener 2003). Together, they (i.e. the G-2) account for over 42% of the world's total GHG emissions (United Nations Statistics Division). If we are serious about reducing GHG emissions, we must have both countries on board, without which it is difficult to continue effectively with climate change negotiations. The US can influence and push China's international climate change policy in three ways: first, setting an example; second, helping China reduce abatement costs; and third, promoting China's concern over the climate change issue. The US is a crucial country in climate change negotiations because it has both the technology and the financial capacity to reduce GHG emissions (Pacala and Socolow 2004). Having the US, China and the EU on board would certainly expedite the creation of a future global climate change agreement.

A way forward in climate change negotiations is the creation of bilateral deals between developed and developing countries, possibly (and desirably) involving the US These could include emissions allowances, a Kyoto-type Clean Development Mechanism, cash and non-climate change benefits in trade or other side payments or linkages, for instance, and may solve some of the equity problems among countries of who pays how much. The conclusion of bilateral deals is preferred among some countries (India, China and Brazil) because they offer more immediate returns. The international community should ideally aim at the multilateral coordination of these bilateral deals. Examples of bilateral deals are the Australia–China Natural Gas Technology Partnership Fund, the EU–China Partnership on Climate Change, the US–India Civil Nuclear Cooperation Initiative, the US–India Partnership to Advance Clean Energy, Energy Security and Climate Change and the EU–India energy cooperation.

---

[8]For instance, the G-8+5, which would be composed of the G-8 (i.e., the US, Canada, UK, Germany, France, Italy, Russia, and Japan) plus five leading developing countries (Brazil, India, China, Mexico, and South Africa).

4.2 The G-3

An agreement among a small group of major GHG emitters (for example, China, the US and the EU, i.e. the G-3) could provide a starting point for building new international emissions reduction commitments involving all major emitting countries. If this group of countries can agree to some meaningful measures, then the arrangement might be expanded to include Brazil, Japan, Australia, Canada, India, Indonesia, South Africa, possibly Russia and other major emitting countries. This major emitter 'club' could be built under the auspices of an existing international forum, such as the G-20 group of major developed and developing countries, or a new network organization, and eventually feed back into the UNFCCC, which would provide much more legitimacy to the exercise. On the other hand, major countries that are not at the table may object to a three-party initiative (such as the suggested one of China, the US and the EU), triggering a backlash that could impede progress on global GHG emissions reductions.

Linking the previous section with this section, the proposition of a G-2 or G-3 raises the following questions: is it wise to proceed with just two or three parties, rather than using the UNFCCC process or using a forum such as the G-20 or the MEF to negotiate new undertakings? Will a club approach promote or restrict broader progress?

4.3 The MEF

The Major Economies Forum on Energy and Climate (MEF) was initiated in 2007 by the Bush administration under the name 'Major Emitters Forum' and launched by the Obama administration on March 28, 2009. The MEF is intended to facilitate a candid dialogue among major developed and developing economies, help generate the political leadership necessary to achieve a successful outcome at future UN climate change conferences, and advance the exploration of concrete initiatives and joint ventures that increase the supply of clean energy while cutting GHG emissions. The MEF would provide for significant GHG limitations with credible assurances for compliance (Stewart 2011). The MEF partners include: Australia, Brazil, Canada, China, the EU, France, Germany, India, Indonesia, Italy, Japan, Korea, Mexico, Russia, South Africa, the UK and the US Bringing together these major emitters, which were responsible for around 75% of GHG emissions in the world as of 2009 (these numbers include land-use change) (Broder 2009), will increase the likelihood of reaching a climate change agreement, as the MEF is a more efficient negotiating forum than the UNFCCC (Oye 1985). Furthermore, an agreement amongst them would be almost as valuable as an agreement amongst all UNFCCC parties in terms of absolute GHG emissions reductions, since most GHGs come from the MEF partners.

The MEF is therefore a means to facilitate progress in the climate change negotiations. The MEF has a controversial relationship with the UNFCCC/Kyoto Protocol process and offers a substantially different means to respond to climate change. The Kyoto Protocol is universal in scope, whereas the MEF is based on small group negotiations among 17 parties; the Kyoto Protocol is legally binding, whereas the MEF stresses voluntary measures; the Kyoto Protocol focuses on GHG emissions reduction, whereas the MEF fosters technological innovation. To avoid the obstacles faced by the UNFCCC machinery, the MEF should focus on each member's economic weight, GHG emissions reduction responsibilities and the calculation of responsibility for GHG emissions such as sharing the burden equally between producers and consumers, in order to fairly decide who should reduce GHG emissions and by how much. For instance, most GHGs are emitted because countries do not have clean sources of energy. They have no choice but to use available technologies. If energy producing countries have to pay 50% of the cost, there would be a greater incentive to shift energy production from fossil fuels to clean energy.

4.4 The G-20

Most of the largest GHG emitters have large economies, large populations, or both. Given the direct link between climate change and the world economy, the G-20 could be a plausible forum for moving the climate change agenda forward. The G-20 'brings together important industrial and emerging-market countries from all regions of the world. Together, member countries represent around 90% of the global gross national product, 80% of world trade [including intra-EU trade] as well as two-thirds of the world's population. The G-20's economic weight and broad membership give it a high degree of legitimacy and influence over the management of the global economy and financial system'.[9] In 2008, the G-20 represented 66% of the world's population and produced over 80% of the world's GHG emissions (US Energy Information Administration 2009).

5 Conclusion

The international community should provide a burden-sharing element designed to produce a fair distribution of burdens across countries (whether developed or developing), taking into account the historical responsibilities of the developed world, while also giving priority to green economic development, addressing concerns about wealth inequality and achieving GHG emissions reductions

[9]http://www.g20.org/about_what_is_g20.aspx.

consistent with limiting the expected increase in global average temperature to 1.5–2 °C, without sacrificing economic growth. Nations should approach economic growth as part of the climate solution, not part of the problem.

However, there is no need to have a global solution/universal agreement to this global problem so long as the major GHG emitters reduce their GHG emissions locally, taking into account the historical responsibilities of the developed world in order for the process to be equitable. In other words, if China's (or any other major emitter's) GHG emissions decrease, it will benefit the rest of the world, whereas if China's (or any other major emitter's) GHG emissions increase, it will affect the rest of the world. A local solution can, therefore, have a global effect. So from an economic and environmental point of view, it is vital that China bring down its GHG emissions. The ambition to reduce GHG emissions to safe levels cannot be reached without equity in the process, similar to the argument of 'no peace without justice'.

Pragmatism should be the crucial element in moving the climate agenda forward: bilateral/plurilateral agreements in addition to a universal climate agreement, flexibility over rigidity and practical results over utopian ideals. Today, the UNFCCC is one platform among several in climate governance. The UNFCCC is a large structure with limited authority to move quickly and decisively. Ideally, the international community should preserve the successes of the global regime (top-down approach) in order to provide credibility to the multilateral system and, at the same time, move on regionally or with coalitions of the willing (bottom-up approach), which, in turn, will help expedite the goal of GHG emissions reduction. This would mean a complementary relation between the top-down and bottom-up approaches to climate change. The post-2015 multilateral framework on climate change action will most likely take place at domestic and regional levels.

In addition to such a global, universal agreement on climate change, it would make sense to explore, along with the current legal platform of UNFCCC/Kyoto Protocol, the 'clubs approach'—such as the MEF, the G-2, the G-3, or the G-20—given that the atmosphere does not care where emissions come from because GHG emissions mix globally in the atmosphere. This means that it does not matter where GHG emissions reduction takes place.

Regarding ways to move the climate change agenda forward, it is well known that equitable and efficient international cooperation is very difficult at the multilateral level. The geometries of power have fundamentally changed with the rising power of China and India. At the time of the signature of the Kyoto Protocol in 1997, very few people could envisage that China would grow macroeconomically so rapidly. With economic growth came higher GHG emissions. By 2007, China became the largest GHG emitter in the world, even if China's per capita emissions remain comparatively rather low so far.

There is increasing international pressure on developing countries to reduce their GHG emissions. No meaningful breakthroughs will take place regarding climate change mitigation until there is more political maturity and commitment on the side of the US and until rapidly emerging economies such as China and India indicate that they are ready to play their part in tackling climate change, since they are part of the solution. If the financial burden to reduce GHG emissions rests only on developed countries, it would mean a tremendous economic sacrifice for them. If, however, (major) developing countries are not given financial help by the developed world to mitigate climate change and to develop economically, it would be unfair on them too. This would imply a re-definition of the principle of common but differentiated responsibilities. Large emitters of GHGs need to be involved for climate change negotiations to come to a positive conclusion. Much progress is still needed until we reach a decarbonized world economy.

What is crucial is that the major GHG emitters reduce their GHG emissions. Every country in the world will benefit even in the case where only major GHG emitters reduce their emissions. *A* sensu *contrario*, the whole world will suffer if major GHG emitters increase their emissions, as climate change is a global problem.

Moreover, the fact that perhaps only a club of major GHG emitting countries may move the climate change agenda forward plurilaterally to limit GHG emissions —instead of the entire UNFCCC membership—is not as problematic as would be the case in the multilateral trading system, where issues of violation of the WTO law principle of non-discrimination would arise. Unlike the case of multilateral trade agreements, in the climate field, it may be preferable to have a mini-lateral climate change agreement (through clubs or coalitions of the willing) in addition to a global agreement, if that means making sure that the Earth's rising temperature is being addressed. This club approach proposition would be feasible through incentives for major GHG emitters to reduce GHG emissions and therefore to comply with such a mini-lateral climate change agreement. There are clear costs and risks to not mitigating climate change fast enough, analogous with gaining weight: the more weight one gains, the greater the chances of a serious cardiovascular pathology and the more expensive it will be to get back to being healthy. Therefore, a preventive approach to climate change mitigation is the safest and cheapest option.

References

Abbot, K., & Snidal, D. (1998). Why states act through formal international institutions. *Journal of Conflict Resolution, 41*(1), 1–32.

Barrett, S. (2003). *Environment and statecraft: The strategy of environmental treaty-making* (pp. 260–261). New York: Oxford University Press.

Baumert, K., Herzog, T., & Pershing, J. (2005). *Navigating the numbers: Greenhouse gas data and international climate policy, World Resources Institute, figure 2.1, p. 12. There are, however, various ways to measure emissions. Emissions per capita shall also be taken into account*; (these numbers exclude emissions from international bunker fuels, land use, land-use change, and forestry).

Bosetti, V., et al. (2009). Climate change Mitigation strategies in fast-growing countries: The benefits of early action. *Centre for Economic Policy Research* (See chapter 3 of the Kyoto Protocol, where Annex I countries were asked to lead the way on GHG emissions reduction, for others to follow, but the U.S. refused).

Broder, J. (2009). Clinton says U.S. is ready to lead on climate. *The New York Times*, 27 April 2009. http://nyti.ms/huEbYb.
Cottier, T. (2006). From progressive liberalization to progressive regulation in WTO law. *JIEL, 9* (4), 779–821.
Cottier, T. (2011). *Confidence-building for global challenges: The experience of international economic law and relations*, NCCR Trade Working Paper No. 2011/40. http://www.wti.org/fileadmin/user_upload/nccr-trade.ch/wp5/5.5a/International%20Economic%20Law%20Cottier%20final%200311%20%282%29.pdf.
Cottier, T., & Delimatsis, P. (Eds.). (2011). *The prospects of international trade regulation: From fragmentation to coherence*. New York: Cambridge University Press.
Durrant, N. (2002). *Guyana's participation in multilateral and regional trade negotiations and the United Nations Framework Convention on Climate Change (UNFCCC)*, Working Paper, Overseas Development Institute. http://www.odi.org.uk/resources/download/3614.pdf.
Energy Information Administration. (2007). *International energy outlook 2007*, Chapter 7.
Farber, D. (2008). The case for climate compensation: Justice to climate change victims in a complex world. *Utah Law Review, 2,* 377–412.
Hong, G. D. (1990). Linear controllable systems. *Nature, 135,* 18–27.
International Energy Agency. (2011). *CO_2 emissions from fuel combustion: Highlights* 7. See also Blanford, G., et al. Revised Emissions Projections for China: Why Post-Kyoto Climate Policy Must Look East. In Aldy, J. & Stavins, R. (eds.), *Post-Kyoto International Climate Policy: Implementing Architectures for Agreement*, New York: Cambridge University Press, 2009, pp. 822–856.
Kahler, M. (1992). Multilateralism with small and large numbers. *International Organization, 46*(3), 706.
Kopp, R. (2010). "The climate has changed—So must policy," *Resources for the Future*, Issue Brief 11-03, March 2011. See also Simon, N. "International Environmental Governance for the 21st century: Challenges, reform and options for action on the way to Rio 2012," SWP Research Paper, German Institute for International and Security Affairs, February 2011; Fauchald, O.K. "International Environmental Governance: A legal analysis of selected options," Fridtjof Nansen Institute, Report 16/2010.
Leal-Arcas, R. (2010). *International trade and investment law: Multilateral, regional and bilateral governance*. London: Edward Elgar.
Leal-Arcas, R. (2011a). The fragmentation of international trade law: Is now the time for variable geometry? *The Journal of World Investment and Trade, 12*(2), 145–195. Statement of the GATT director–general on the Tokyo Round, April 12, 1979, 18 I.L.M. 553.
Leal-Arcas, R. (2011b). Top-down versus bottom-up approaches for climate change negotiations: An analysis. *The IUP Journal of Governance and Public Policy, 6*(4), 7–52.
Levi, M. (2010) Beyond Copenhagen: Why less may be more in global climate talks. *Foreign Affairs*; see also Stavins, R. "Options for the institutional venue for international climate negotiations," *The Harvard Project on International Climate Agreements*, Issue Brief 2010-3, May 2010.
Olson, M. (1965). *The logic of collective action: Public goods and the theory of groups*. Cambridge, MA: Harvard University Press.
Ostrom, E. (2010a). Polycentric systems for coping with collective action and global environmental change. *Global Environmental Change, 20,* 550–557.
Ostrom, E. (2010b). Beyond markets and states: Polycentric governance of complex economic systems. *American Economic Review, 100*, 641–672, June 2010 [See for instance the work of Elinor Ostrom on the management of common pool resources and on global environmental change in Ostrom, E].
Oye, K. (1985). Explaining cooperation under anarchy. *World Politics, 38*(1), 21.
Pacala, S., & Socolow, R. (2004). Stabilization wedges: Solving the climate problems for the next 50 years with current technologies. *Science, 305*(13), 968–972.
Pielke, R. A., Jr. (2007). *Statement to the House Committee on Science and Technology of the United States House of Representatives. The State of Climate Change Science 2007: The*

findings of the fourth assessment report by the intergovernmental panel on climate change (IPCC), Working Group III: Mitigation of Climate Change.

Poteete, A., Janssen, M., & Ostrom, E. (2010). *Working together: Collective action, the commons, and multiple methods in practice*. Princeton: Princeton University Press.

Rayner, S. (2010). How to eat an elephant: A bottom-up approach to climate policy. *Climate Policy, 10*(6), 615–621.

Seligsohn, D., et al. (2009). *China, the United States, and the climate change challenge. World Resources Institute Policy Brief*, October 2009; Pew Center & Asia Society, "Common Challenge, Collaborative Response: A Roadmap for U.S.-China Cooperation on Energy and Climate Change".

Stern, T., & Antholis, W. (2007). Toolbox: Creating an E-8. *American Interest*, 2(3), 43–49 (The climate G-8 would be composed of China, the U.S., the EU, India, Brazil, South Africa, Japan, and Russia (currently suspended). This group represents the key economies in each region of the world and is responsible for 70 per cent of global GHG emissions. This number includes land use, land-use change, and forestry. For views on the climate G-8) (Winter).

Stewart, R., et al. (2011). *Strategic analytics for building a global climate regime bottom-up* (pp. 4–5), paper presented at the conference 'Reaching International Cooperation on Climate Change Mitigation', December 21–23, 2011 (Unpublished paper; on file with author).

Stewart, R., & Wiener, J. (2003). *Reconstructing climate policy: Beyond Kyoto.* Washington, DC: American Enterprise Institute [Chapter 3 and pp. 102–109 (on how to attract the participation of China and other major developing countries)].

Sunstein, C. (2008). The world vs. the United States and China? The complex climate change incentives of the leading greenhouse gas emitters. *UCLA Law Review, 1675*. See also Yang, Z. and Sirianni, P. "Balancing contemporary fairness and historical justice: A 'quasi-equitable' proposal for GHG mitigations," *Energy Economics* 32(5): 1121–1130, 2010.

UNDP, *International human development indicators.* http://hdr.undp.org/en/data/build/.

United Nations Statistics Division, *Millennium development goals indicators: Carbon dioxide emissions* (CO_2). http://mdgs.un.org/unsd/mdg/SeriesDetail.aspx?srid=749&crid.

U.S. Energy Information Administration. (2009). These numbers exclude land use, land-use change, and forestry. See U.S. Energy Information Administration, *International Energy Statistics 2009.*

World Trade Organization, *The GATT years: From Havana to Marrakesh.* http://www.wto.org/english/thewto_e/whatis_e/tif_e/fact4_e.htm.

Contextualizing of Organized Labour Position on Carbon Tax in South Africa Using a Qualitative System Dynamics Model

Martin Kaggwa

Abstract The South African government planned to introduce a carbon tax from 2013 as part of its efforts to reduce pollution levels emanating from domestic economic activities. The government sought buy-into the carbon tax policy from all stakeholders before its introduction. The position of organized labour has, however, ranged from outright rejection to requests for a delay in implementation of the policy until employment safeguards are put in place. Organized labour felt strongly that a carbon tax would have a negative impact on local employment. The objective of this study was to analyse the potential impact of a carbon tax on employment in South Africa, in the medium to long-term, with the aim of establishing the validity of organized labour's concerns on possible job losses. A qualitative system dynamics approach was used in the analysis, based on its usefulness in allowing holistic appraisal of effects of policy interventions. The paper highlights the fact that the motivation for firms to migrate towards the adoption of low emission production processes could be constrained by the cost of the clean technology. On the consumer side, the shift towards consumption of low emission products as a result of the imposition of a carbon tax could be constrained by increases in prices of the low emission products resulting from their increased demand. A case is made that despite being set at low levels, the carbon tax has potential to negatively affect competitiveness of firms in the long-term. The potential job losses from the carbon tax will not necessarily be a result of increases in the immediate operational costs of firms, but rather will be from the loss of firms' competitiveness in the long-term. It is recommended that the implementation of the carbon tax be put on hold until safeguards to local employment are put in place. One of the possible safeguards is to ensure that local manufacturers have access to low cost but clean production technologies that do not substitute the local labour force but rather supplement its productivity.

Keywords Carbon tax · Employment · Low emissions · Organized labour

M. Kaggwa (✉)
Sam Tambani Research Institute, Johannesburg, South Africa
e-mail: Kaggwam@satri.org.za

W. Leal Filho and D. Surroop (eds.), *The Nexus: Energy, Environment and Climate Change*, Green Energy and Technology, https://doi.org/10.1007/978-3-319-63612-2_20

1 Introduction

Global warming and the problems it poses to humankind are no longer contested by many people and governments (Stern 2006). What is at stake is how to deal with global warming given the other social and economic challenges that developing countries face.

The South African government recognizes that the country is vulnerable to effects of climate change. As such, the government is taking conscious steps to reduce adverse effects to the environment emanating from domestic economic activities. The imposition of a carbon tax is one of the policy interventions that the South African government planned to introduce in 2013. The effect of carbon tax on current and long-term carbon emissions, climate change and national economies remain ambiguous (Andrade de Sá and Daubanes 2016), yet its effect on employment in the high emission energy sector is almost certain. Benavente (2016) argues that carbon tax will have a contractionary effect on a country's economy; hence the South African government wanted a buy-in by all stakeholders before implementation of the policy.

The position of organized labour on the introduction of a carbon tax in South Africa has ranged from outright rejection of the tax to requests for a delay in its implementation until safeguards on employment are put in place (COSATU 2012). This paper analyses the potential impact of a carbon tax on employment in South Africa in the medium to long-term, with the aim of establishing the validity of organized labours' concerns on possible job losses. A qualitative Systems Dynamics (SD) approach is used in the assessment. The approach choice was due to its usefulness in assessing the effects of policy interventions in a holistic way, taking into consideration feedback effects, time lags and nonlinearity between key policy variables (Coyle 1996).

The chapter is organized as follows: after the introduction, a description of South Africa's carbon tax policy is presented in Sect. 2. The section also highlights key arguments in support of the tax as well as the basis of organized labour in qualifying their support for the policy. In Sect. 3, a qualitative system dynamics model of the tax is presented after a critic of the one-way casual thinking of the tax policy that has thus far been the basis of contemporary debates on the policy. Important policy insights emanating from the articulation of the carbon tax policy as a qualitative system dynamics model are discussed in Sect. 4. Section 5 concludes the paper, and also gives the recommendation.

2 The Carbon Tax Policy for South Africa

South Africa's carbon tax policy is aimed at businesses that use and produce high emission products or services. The tax is supposed to be a disincentive for such businesses from using high emission processes during production, and so reduce their carbon foot print.

The tax is set at very low levels to make sure that it does not negatively affect local productive activities on the one hand, and stifle trade on the other due to loss of competitiveness. A tax relief of 60% is provided across all sectors of the economy. Additional tax reliefs are provided for the agricultural and waste management sectors (National Treasury South Africa 2013). Proposed carbon tax thresholds reflective of sectors that may be adversely affected by the tax are presented in Table 1.

Table 1 Carbon tax thresholds for South Africa

Sector	Tax free threshold	Trade exposure additional allowance	Process emissions additional allowance	Maximum offset	Total tax avoided
	This % of a company's emissions will not be taxed at all	Extra exemption for sectors that may be disadvantaged in competing with foreign companies who are not carbon-taxed	Extra exemption for sectors that can do little about the emissions involved in production, without changing their business substantially	Companies can deduct offsets from their emissions, which reduces the emissions they will be taxed on	
Agriculture, forestry, land use	60%	–	40%	–	100%
Waste	60%	–	40%	–	100%
Iron and steel	60%	10%	10%	5%	85%
Aluminum	60%	10%	10%	5%	85%
Cement	60%	10%	10%	5%	85%
Glass and ceramics	60%	10%	10%	5%	85%
Chemicals	60%	10%	10%	5%	85%
Fugitive emissions: coal	60%	10%	10%	5%	85%
Petroleum (coal-to-liquid and oil refinery)	60%	10%	–	10%	80%
Pulp and paper	60%	10%	–	10%	80%
Sugar	60%	10%	–	10%	80%
Other	60%	10%	–	10%	80%
Electricity	60%	–	–	10%	70%

Source National Treasury South Africa

Proponents of the carbon tax policy for South Africa argue that the policy can contribute towards reducing emissions in the country. Moreover, this can be done without adversely affecting employment since the tax is set at levels too low to have any impact on jobs. They further argue that the tax creates a precedent for the wider argument that producers should take full responsibility for funding externalities emanating from their productive activities.

Sceptics of the policy, including organized labour, put forward compelling counter arguments. They acknowledge that firms may not take decisions to retrench workers because of the carbon tax given the relatively low level it is set at but, irrespective of the level, the tax could still adversely affect the competitiveness of the local firms due to increases in operational costs. Competitiveness is about having an edge over your competitor. Such an edge can easily be lost due to small changes in the firms' production costs. Although the tax may seem insignificant, it could still count when it comes to competitiveness. Loss of competitiveness leads to less sales, reduced production and subsequently lower demand of factor inputs, one of which is labour. Loss of employment may not happen instantaneously with the introduction of the tax but it is possible in the medium to long-term.

The setting of a good precedent to internalise externalities of production by firms is indeed a valid observation, too. However, considered against the potential loss of competitiveness as explained in the preceding section, the question becomes whether the use of the market-based disincentive (the carbon tax) is the best and most effective way to do this. Any use of market-based policy intervention assumes that the applicable market is perfect; so the transmission mechanism of policy interventions is not constrained in any way. But this is rarely the case. Markets are imperfect and many policy inputs cannot be directly mapped onto the outcomes. For this reason, direct rather than indirect regulation aimed at forcing producers to internalise all their production costs may be more appropriated.

For a successful implementation of the carbon tax in South Africa, convergence of opinions among stakeholders is important. This study, therefore, undertook to formalize the thinking behind the carbon tax policy, first as a one-way causal model and second as a qualitative system dynamics model. Formalising a policy into a model is useful in making its assumptions explicit and identifying flaws in the thinking behind the policy. These two aspects are important for improving any policy intervention and providing insight into the most probable outcomes of the policy intervention.

3 Methodology and Data Sources

The study used a system dynamics approach to model the policy and draw conclusions on South Africa's carbon tax. The system dynamics approach combines systems thinking, and qualitative and quantitative modelling of a problem at hand. Systems thinking allows a holistic articulation of the problem at hand. It brings to the fore causal-effect assumptions between related variables in a system, facilitating assessment and improvement of mental conceptualisation of issues (Cabrera et al. 2008).

System dynamics modelling on the other hand enables formal comprehensions and capturing of observed phenomenon, and in establishing consequences of different options available at a decision point (Randers 1980). For qualitative system dynamic modelling, Causal Loop Diagrams (CLD) are often used. The CLDs are a pictorial presentation of relationships from a systemic perspective. They are referred to as 'causal' because their key aim is to capture the causal relationships and feedbacks (Kaggwa 2008). Feedback is a central concept of system dynamics, yet a mental model seldom takes into cognisance feedback while determining dynamics of a system (Sterman 2000).

The data used was secondary and was drawn from the national treasury of South Africa. The national treasury data is often more reliable as it is normally triangulated from various government department sources.

4 Study Limitations

A key limitation of this research study is that it did not quantify the developed model. Qualitative articulation and modelling of policies is useful in revealing inconsistencies in policy design. However, it cannot reveal internal policy levers that can influence the performance of a system an aspect which is critical in policy improvement. Quantitative modelling and simulation of the carbon tax is outside the scope of the chapter but its importance is acknowledged.

5 One-Way Causal Model for South Africa's Carbon Tax

The articulation of the carbon tax model for South Africa has been based largely on a one-way causal model. It is assumed, for example, that the carbon tax will increase the cost of high emission production processes from the manufacturers' perspective which will, in turn, decrease their profit margins. The reduction in profit margins will force the manufacturers to change their production to low emission processes. In other words, there will be a one-way causal link from the carbon tax to low emission production and subsequently to a cleaner economy.

On the consumer side, again a one-way causal relationship between carbon tax and low emission production is assumed. The causal relationship is via increase in prices of high emission products, substitution of high emission products with low emission products and the subsequent increase in demand of low emission products. Increased demand for low emission products motivates their production.

Figure 1 presents the one-way causal model for South Africa's carbon tax upon which contemporary debates on the possible effects of the tax on the country are based.

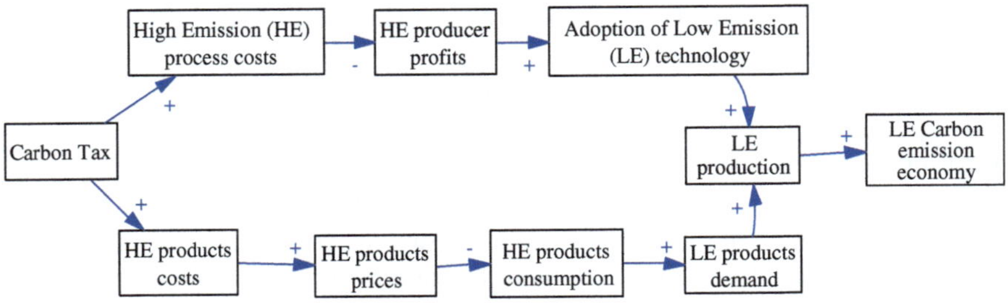

Fig. 1 The uni-direction of articulation of South Africa's carbon tax policy

There are fundamental omissions in the articulation of South Africa's carbon tax policy as presented in Fig. 1. These relate to lack of recognition that the direction of causality is not necessarily one way and that the causal effects are neither instantaneous nor linear overtime. By failing to recognize these aspects, key dynamics and effects of carbon tax are omitted. These omissions make it probable that erroneous conclusions are drawn on the potential effects of the carbon tax policy.

6 A Qualitative System Dynamics Model of South Africa's Carbon Tax Policy

System dynamics formalization and modelling of policy interventions help to make explicit the three aspects often omitted in the one-way causal articulation of policies, that is: double causality, time lags and nonlinearity of causal relations overtime (Sterman 2000).

SD modelling requires that model articulation takes into cognition, and explicitly include, feedback effects, time lags and nonlinearity of causal relationships (Richmond 2004). In the qualitative modelling, though, the element of nonlinearity is subtle. It only manifests itself in quantitative simulation of the model.

In line with the SD modelling protocol, four additional variables and two causal loops, relevant to the conceptualization of the carbon tax policy and the objective of the study, were identified and added to the model. The four additional variables included in the carbon tax SD model were as follows:

- Employment in high emission sectors
- Employment in the low emission sectors
- Low emission product prices
- Competitiveness of local firms.

The two feedback loops introduced related to the cost of low emission technology and the prices of low emission products from the producers' and consumers' perspectives, respectively.

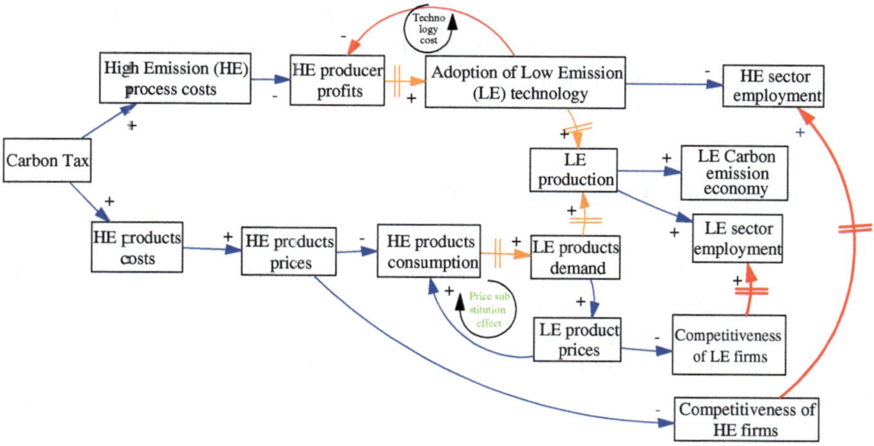

Fig. 2 Articulation of South Africa's carbon tax policy from a qualitative system dynamics perspective

South Africa's carbon tax policy model re-articulated as a qualitative system dynamics model is presented in Fig. 2.

A description of symbols used in Fig. 2 is given below:

	Causal relationship
	Causal relationship with time delay
	Anti-clockwise feedback loop
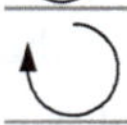	Clockwise feedback loop

A positive sign accompanying a causal relationship symbol indicates that the causal relationship between the two variables moves in the same direction. That is, if the 'causing' variable increases, the affected variable will also increase and vice versa. The negative sign indicates that the connected variables move in opposite directions.

7 Insights from the Articulation of South Africa's Carbon Tax Policy Into a System Dynamics Model

The articulation of South Africa's carbon tax as a system dynamics model brings to the fore assumptions that may not necessarily be true, and other omitted effects. The inaccurate assumptions and omitted effects weaken the key proposition that the carbon tax policy will reduce emissions in the domestic economy and will have minimal adverse effects on local employment. The following section considers some of the omitted effects and how their omission leads to inaccurate conclusions on the country's carbon tax policy.

8 Cost of Low Emission Technology

For producers using high emission processes, it is assumed that imposing carbon tax will reduce their profits from the high emission products. Driven by the desire to maintain or increase profits, such producers will adopt low emission technologies on which no or low carbon tax is payable. As result of this, low emission production will increase in the local economy and employment will be created in low emission sectors.

A fundamental assumption with this line of thought is that low emission technology adoption and use comes at a cost equal or less than that of high emission technology. If the cost of low emission technology is higher than that of the high emission technology, then the adoption of the former will not necessarily lead to higher profits even after imposing carbon tax. Without the higher profits being anticipated, producers will have no motivation to adopt low emission technologies despite the existence of the carbon tax.

Ideally, there is a two-way causal relationship between profits for high emission producers and the adoption of low emission technology that is determined by the relative cost of low emission and high emission technologies. This loop is named technology cost in the qualitative SD model of South Africa's carbon tax policy.

To the extent that low emission technology in South Africa is more expensive than high emission technology, the assumed effect of carbon tax motivating the adoption of low emission technologies will not happen. Subsequently, the anticipated increase in low emission production and increase in employment in low emission sectors will not take place either.

9 Price Substitution Effect Between Low Emission and High Emission Products

On the consumer side, it is assumed that the imposition of carbon tax will increase the cost of producing high emission products, and high emission product prices will increase as a result. Consumers of high emission products will find it more expensive to buy these products and will subsequently switch to buying low emission products. The increasing buying of low emission products will upsurge production levels of low emission sectors, subsequently creating more employment in these sectors.

An important effect that is not accounted for in the above logic is that an increase in the demand for low emission products will raise up their prices, according to the law of demand. This price increase will force consumers to substitute the low emission products with high emission products, according to the phenomenon of price substitution effect. If this happens, the causal link between carbon tax, low emission production and green job creation will cease to exist. The two-way causal relationship between the demand for low emission products and consumption of high emission products via the low emission product prices is the second loop in the carbon tax qualitative SD model. It is named price substitution effect in Fig. 2.

10 Competitiveness and Employment in the High and Low Emission Sectors

The focus of this study was to understand the potential effect of carbon tax on employment. One cannot talk about sustainable employment without considering the aspect of competitiveness. Competitiveness, defined as the degree to which a firm can—under free and fair market conditions—produce goods and services that meet the test of international markets, is critical in sustaining jobs (Clark and Guy 2000).

In the qualitative SD model of carbon tax, the causal relationship between low emission production, sector competitiveness and employment in both the low emission and high emission sectors was introduced. The following causal relationships were particularly important in tracking the potential effect of the tax on employment. First, there is a negative causal relationship between prices of products, whether high or low emission, with local firms' competitiveness. Lower market prices for the low emission products lead to competitiveness of low emission sectors. The same applies to a situation of low market prices for high emission products and the competitiveness of high emission sectors.

Carbon tax will increase the unit price of high emission products, subsequently making the high emission sectors less competitive. To the extent that competitiveness is a prerequisite for sustainable employment, jobs will most likely be lost in the high emission sectors in the long run. Based on the same logic, lower prices for low emission products will make low emission sectors of the economy more competitive and will create more jobs in these sectors.

From the qualitative model, it is clear that an increase in prices of high emission products is certain if a carbon tax is imposed. By implication, the likelihood of job losses in high emission sectors of the South African economy is very high in the long-term, holding other factors constant. On the other hand, the likelihood of a reduction in the prices of low emission products and, subsequently, the increase in the competitiveness of low emission sectors is uncertain. The uncertainty emanates from the low emission technology costs and the price substitution effect of low emission products with high emission products, as previously explained.

Whereas job loss in the high emission sectors of the country as a result of imposing a carbon tax is almost certain, the job creation potential of carbon tax in the low emission sector is doubtful and highly improbable. This observation supports organized labours' position that the introduction of a carbon tax should either be stopped or at least be delayed until safeguards are put in place to mitigate its effect on local employment.

11 Conclusions and Recommendation

There are so many factors that will influence the ultimate impact of the carbon tax on employment in the low emission and high emission sectors of the South African economy. After accounting for time lags and feedback effects using a qualitative system dynamics model, the scale of probability indicates that jobs will be lost in the high emission sectors of the South Africa economy if a carbon tax is introduced without employment safeguards. The job loss aspect is of major concern to the country since the high emission sectors, like the coal-energy generation sector, employ a significant number of citizens.

The modelling exercise validates, to a reasonable extent, labour concerns on unconditional implementation of the carbon tax in the country (despite being set at low levels). The tax has a potential to negatively affect competitiveness of firms in the long-term. The potential of job losses as a result of the policy will not necessarily be an outcome of increases in the immediate operational costs of firms, but rather a result of the loss of firms' competitiveness in the long-term. This fact is not limited to South Africa only, since fiscal rules of the tax and the effects are largely the same in any economy.

It is recommended that the implementation of carbon tax in South Africa be put on hold until safeguards to local employment are put in place. One of the safeguards is to ensure, beforehand, that local manufacturers have access to low cost clean technology that does not substitute the local labour force but supplements its efficiency. Again this recommendation is applicable to any developing country that has a significant non-renewable and high emission energy sector.

References

Andrade de Sá, S., & Daubanes, J. (2016). Limit pricing and the (in) effectiveness of the carbon tax. *Journal of Public Economics, 139,* 28–39.

Benavente, J. M. G. (2016). Impact of a carbon tax on the Chilean economy: A computable general Equilibrium analysis. *Energy Economics, 57,* 106–127.

Cabrera, D., Colosi, L., & Lobdell, C. (2008). *Systems thinking: evaluation and program planning* (vol. 31, pp. 299–310). Kluwer Publishers: New York.

Clark, J., & Guy, K. (2000). Innovation and competitiveness. *Technology Analysis & Strategic Management, 10,* 363–394.

COSATU (Congress of South Africa Trade Unions). (2012). *A carbon tax for South Africa*. South Africa: Braamfontein.
Coyle, R. G. (1996). *System dynamics modelling*. London: Chapman & Hall.
Kaggwa, M. (2008) *Modelling South Africa automotive incentives: The case of the productive asset allowance* (Thesis). University of Pretoria.
National Treasury South Africa. (2013). *Reducing Greenhouse gas emissions and facilitating the transition to a green economy*. South Africa: Pretoria.
Randers, J. (1980). *'Guidelines for model conceptualisation', in elements of the system dynamics method*. Cambridge, MA: MIT Press.
Richmond, B. (2004). *An introduction to systems thinking*. ISEE Systems. pp.6
Stern, N. (2006) *Stern review: The economics of climate change* United Kingdom: Government of the United Kingdom. Accessed February 13, 2016. http://www.hm-treasury.gov.uk/media/8A8/C1/Summary_of_Conclusions.pdf.
Sterman, J. D. (2000). *Business dynamics: Systems thinking and modelling for a complex world*. Irwin, NY: McGraw-Hill.

A Step-by-Step Calculation of the Combined Margin Grid Emission Factor of the National Electricity System of Mauritius

Prakash N.K. Deenapanray and Sanjay Sookhraz

Abstract A step-by-step calculation of the combined margin grid emission factor for the national electricity system of the island of Mauritius using the Clean Development Mechanism (CDM) Methodological Tool 07—"Tool to calculate the emission factor for an electricity system (Version 04.0)" is provided. Using statistical data for electricity generation and fuel consumption for the years 2012, 2013, and 2014, the operating margin emission factor has been calculated as 1.0170 tCO_2/MWh. while the build margin emission factor has been found to be 0.8135 tCO_2/MWh. These values give a combined margin emission factor of 0.9661 tCO_2/MWh for grid electricity generated from wind power or solar PV. For any other types of grid-tied renewable electricity generation or energy efficiency measures that displace grid electricity, the combined margin grid emission factor is 0.9152 tCO_2/MWh. The CDM methodological tool has been applied to develop a standardized greenhouse gas emission baseline for the national grid of Mauritius that will be valid until January 2019. A comparison of the grid emission factor of Mauritius to those of selected African countries is carried out, and the comparative advantages of attracting climate finance are discussed.

Keywords Grid emission factor · Operating margin · Build margin · Combined margin · Climate change mitigation · Mauritius

1 Introduction

Climate change has emerged as arguably the biggest threat to human development in the twenty-first century. The science is categorical that climate change is driven predominantly by the accumulation of greenhouse gases (GHG) through anthropogenic emissions (IPCC 2013). Since the atmosphere is a limited sink for GHG

P.N.K. Deenapanray (✉)
ELIA—Ecological Living in Action Ltd, La Gaulette, Mauritius
e-mail: sanju@ecolivinginaction.com

S. Sookhraz
Central Electricity Board, Ebene, Mauritius

W. Leal Filho and D. Surroop (eds.), *The Nexus: Energy, Environment and Climate Change*, Green Energy and Technology, https://doi.org/10.1007/978-3-319-63612-2_21

emissions as far as irreversible perturbations in the planet's climate are concerned, the United Nations Framework Convention on Climate Change (UNFCCC) has the goal of preventing "dangerous" human interference with the climate system. Given that the atmosphere has a fixed carbon budget, the 5th Assessment Report (AR5) of the Intergovernmental Panel on Climate Change (IPCC) implies that all countries will have to carry out reductions in GHGs to achieve net zero emissions by 2080 in order to reach the 450–530 ppm CO_{2e} by 2100 (IPCC 2014). According to the UNFCCC, the national GHG emission reductions need arrangements for measurement, verification, and reporting (MRV).[1] Concerning GHG emissions reductions related to grid electricity, the grid emission factor can form part of a national MRV system. The UNFCCC Secretariat has streamlined the process for countries to develop standardized baselines to facilitate the calculation of emission reduction and removals and/or determination of additionality for CDM project activities, while providing assistance for assuring environmental integrity (UNDP 2013). Standardized baselines for national grids can be developed using the appropriate methodological tool developed by the Clean Development Mechanism (CDM). This approach has a number of advantages including (UNDP 2013):

(a) A standard national grid emission factor approved for use in all relevant CDM projects (and also applicable for voluntary carbon offset schemes);
(b) Increased efficiency once the project-by-project approach is eliminated;
(c) Expedited development of CDM (and voluntary carbon offset) projects in the country;
(d) Accurate and up-to-date data available as needed; and
(e) Increased public sector understanding of the requirements of CDM projects (since the standardized baseline is developed and submitted for approval by the Designated National Authority—a public institution).

This paper outlines the step-by-step calculation of the grid emission factor of the electric grid of Mauritius. It also compares the grid emission factor with that of other African countries and it discusses the comparative advantages of attracting climate finance.

2 Methodology

The grid emission factor for the national electricity system has been calculated using the CDM Methodological Tool 07—i.e., "Tool to calculate the emission factor for an electricity system (Version 04.0)".[2] It uses statistical data on electricity

[1]Since the Bali Action Plan in 2007 (Decision 1/CP.13), all mitigation actions by Parties to the UNFCCC in the context of sustainable development are expected to be carried out in a measurable, reportable, and verifiable manner [paragraph 1(b) (ii)] (UNFCCC 2008). All decisions at subsequent Conference of parties have reinforced the need to robust MRV mechanisms.

[2]http://cdm.unfccc.int/methodologies/PAmethodologies/tools/am-tool-07-v4.0.pdf—accessed 3 April 2015.

generation and power plant fuel consumption for 2012, 2013, and 2014 that was provided by the Central Electricity Board (CEB). The CEB has an internal data quality control and quality assurance system. The data are obtained through calibrated meter and verified internally at the Production Department of the CEB. They are then circulated to other CEB departments on a monthly basis in the form of a Generation Progress Report (GPR) and Operation Report (OR). Verification of the data is also carried out on a daily basis at the system control section of the CEB. The Independent Power Producers (IPPs) submit their data to the CEB as part of their obligations under the Power Purchase Agreements (PPAs) signed with the CEB. All the data related to fuel consumption and electricity generation for the national grid is provided to Statistics Mauritius for onward dissemination.[3] Similar data are also published in CEB's Annual Report. Hence, the data used to calculate the combined margin grid emission factor are publicly available.

The combined margin grid emission factor is calculated as a weighted average of the operating margin emission factor and the build margin emission factor through a series of six steps. The grid emission factor is calculated from the perspective of a project activity that will either produce grid-connected renewable electricity or reduce the quantity of grid electricity consumed through energy efficiency measures or reduce transmission and distribution losses.

3 Results

3.1 Step 1—Identifying the Relevant Electricity System

The electrical grid of the island of Mauritius is interconnected—i.e., power plants are physically connected through transmission and distribution lines to the project activity (Fig. 1). Being a remote island, the electrical grid is not connected to any other country's grid. Therefore, the relevant electric power system is the national grid, which is managed by the CEB.

3.2 Step 2—Choosing Whether to Include Off-grid Power Plants in the Project Electricity System

The methodological tool allows selecting one of the following two options to calculate the operating margin (OM) and build margin (BM) emission factor:

[3]Statistics Mauritius is mandated to provide coherent, timely, relevant and reliable statistics, consistent with international principles and standards, for effective policy and decision-making, and for monitoring national development processes. Statistical data for the energy sector can be found at http://statsmauritius.govmu.org/English/StatsbySubj/Pages/Energy-and-Water.aspx—accessed 4 May 2016.

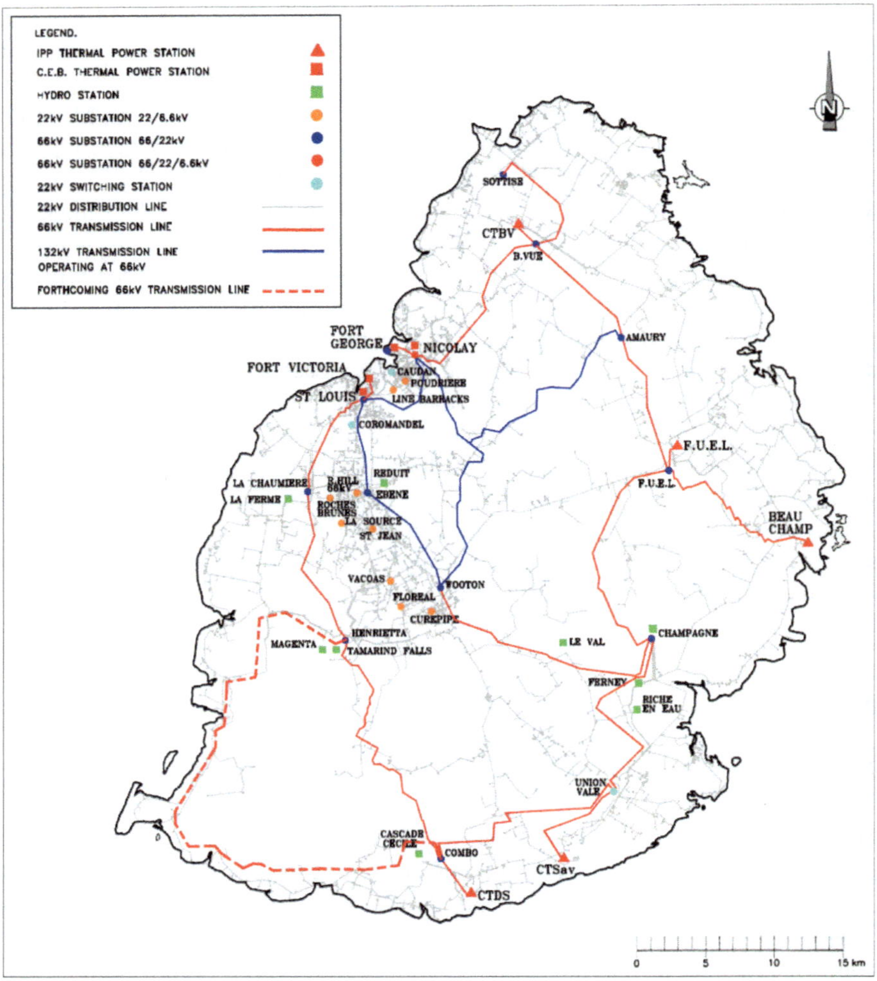

Fig. 1 Map of the Mauritian electrical grid (as of 06.08.08)

Option 1: Only grid power plants are included in the calculation; or
Option 2: Both grid power plants and off-grid power plants are included in the calculation.

Since there are no off-grid power plants on the island of Mauritius, **Option 1** is selected for the calculation of both the OM and BM emission factors.

3.3 Step 3—Select a Method to Determine the Operating Margin (OM)

The calculation of the operating margin emission factor ($EFgrid,OM,y$) is based on one of the following methods:

(a) Simple OM, or
(b) Simple adjusted OM, or
(c) Dispatch Data Analysis OM, or
(d) Average OM.

Method (c) requires the detailed operation and hourly dispatch data of power plants in the grid. To date, there is no publicly available (i.e., either reported by the CEB in its annual report or website, or published and disseminated by Statistics Mauritius) dispatch data with that level of details for the grid on the island of Mauritius. Method (c) is therefore not applicable. Method (b) needs the annual load duration curve of the grid that is also not available publicly. Therefore, method (b) is also not applicable. In order to choose between method (a) or method (d), the share of low-cost/must-run in the total power generation has to be calculated. Method (a) [or (d)] is used when low-cost/must-run resources constitute less [or more] than 50% of the total amount of power generation on the grid. The share (%) of low-cost/must-run renewable electricity in the national electricity grid of the island of Mauritius has been calculated for the 2010–2014 period (Table 1). The data in brackets are the equivalent electricity generated in GWh. Data for 2010–2013 were taken from the Digest of Water and Energy Statistics (Statistics Mauritius 2014), while electricity generation data for 2014 was provided directly by the CEB.

The Simple OM method [i.e., Method (a)] is chosen since low-cost/must-run resources constituted on average 17.7% (<50%) of the total amount of the power generation on the grid between 2010 and 2014.

Table 1 Share (%) of low-cost/must-run renewable electricity in the national grid of the island of Mauritius

Renewable source	2010	2011	2012	2013	2014
Hydropower	4.2 (100.7)	2.3 (56.5)	3.0 (74.1)	(94.8)	(90.8)
Bagasse	14.4 (342.8)	14.5 (352.6)	13.8 (344.0)	(346.5)	(328.3)
Photovoltaic	–	–	0.0(0.3)	(1.24)	(20.4)
Landfill gas	–	0.1 (3.1)	0.7 (17.8)	(20.0)	(21.3)
Total RE	18.6 (443.5)	16.9 (412.2)	17.5 (436.8)	18.0 (462.5)	17.5 (460.8)
Total grid electricity	(2376.1)	(2433.2)	(2495.5)	(2575.5)	(2633.2)

3.4 Step 4—Calculate the Operating Margin Emission Factor According to the Selected Method

According to the tool, the simple OM emission factor (*EFOM,simple,y*) is calculated as the generation-weighted average CO_2 emissions per unit net electricity generation (tCO_2/MWh) of all generating power plants serving the system, not including low-cost/must-run power plants/units. The simple OM may be calculated in two ways:

(a) **Option A**: Based on the net electricity generation and a CO_2 emission factor of each power unit; or
(b) **Option B**: Based on the total net electricity generation of all power plants serving the system and the fuel types and total fuel consumption of the project electricity system.

According to the tool, Option B can only be used if:

(a) The necessary data for Option A is not available; and
(b) Only nuclear and renewable power generation is considered as low-cost/must-run power sources and the quantity of electricity supplied to the grid by these sources is known; and
(c) Off-grid power plants are not included in the calculation (i.e., if Option 1—only grid power plants are included in the calculation—has been chosen in **Step 2**).

Data for Option A are not available in official statistics as can be gathered from official sources (Statistics Mauritius 2014). Further, power generated from bagasse, hydro, PV, and LFG is considered low-cost/must-run and the quantity of electricity supplied to the grid by these sources is known (Statistics Mauritius 2014). Finally, as discussed under Steps 1 and 2 above, there are no off-grid power plants on the island of Mauritius. Given these conditions, Option B is used to calculate the OM emission factor. Under this option, the simple OM emission factor is calculated based on the net electricity supplied to the grid by all power plants serving the system, excluding low-cost/must-run power plants/units, and based on the fuel type (s) and total fuel consumption of the project electricity system, as follows:

$$EF_{grid,OMsimple,y} = \frac{\sum_i FC_{i,y} \times NCV_{i,y} \times EF_{CO_2,i,y}}{EG_y}, \tag{1}$$

where $EF_{grid,OMsimple,y}$, Simple operating margin CO_2 emission factor in year y (tCO_2/MWh), $FC_{i,y}$, Amount of fuel type i consumed in the project electricity system in year y (mass or volume unit), $NCV_{i,y}$, Net calorific value (energy content) of fuel type i in year y (GJ/mass or volume unit), $EF_{CO_2,i,y}$, CO_2 emission factor of fuel type i in year y (tCO_2/GJ), EG_y, Net electricity generated and delivered to the grid by all power sources serving the system, not including low-cost/must-run power plants/units, in year y (MWh), i, All fuel types combusted in power sources

Table 2 Data for fuel consumption and electricity delivered to grid from different fuel sources, 2012–2014

Fuel source	Fuel consumption (FC, t)			Electricity delivered to grid (EG, MWh)		
	2012	2013	2014	2012	2013	2014
Coal	649,157	683,207	696,038	1,021,400	1,067,200	1,125,400
HFO[a]	207,874	210,055	215,615	1,026,823	1,043,976	1,045,001
Kerosene	3437	645	676	10,984	1668	1991

[a]Includes diesel inputs

Table 3 Net calorific value and emission factor of fuel sources

Fuel Source	NCV [GJ/t(fuel)]	EF (tCO_2/TJ)	EF [tCO_2/t(fuel)][a]
Coal	25.5	87.3	2.226
HFO	40.19	75.5	3.034
Kerosene	43.4	69.7	3.025

[a]EF [tCO_2/t(fuel)] = [NCV (GJ/t(fuel)) × EF (tCO_2/TJ)]/1000

in the project electricity system in year y, y, The relevant year as per the data vintage chosen in Step 3.

For the calculation of the OM emission factor, the consumption data for each fossil fuel used to power the different power plants were obtained from the CEB. The calculation of the OM is based on data for the years 2012, 2013, and 2014. Local values of NCV_i (Tier 2) and IPCC (IPCC 2006) default values (Tier 1) of $EF_{CO_2,i}$ are used. Data for 2012 and 2013 were taken from officially published statistics (Statistics Mauritius 2014), while data for 2014 were provided by the CEB (Tables 2 and 3).

The emission factors for the fossil combustibles were taken from Table 2.2—Chapter 2—Stationary Combustion, 2006 IPCC Guidelines for National Greenhouse Gas Inventories.[4] The values correspond to the lower bound default emission factors for CO_2.

Using the data in Tables 2 and 3, the calculated OM is $\mathbf{EF_{OM} = 1.0170\ tCO_2/MWh}$ (Tables 4, 5 and 6).

3.4.1 Calculation of the 3-Year Average Fuel Consumption, ton (t)

First, the average consumption of fossil combustibles to produce electricity is calculated for the last 3 years of vintage data (Table 4).

[4]http://www.ipcc-nggip.iges.or.jp/public/2006gl/pdf/2_Volume2/V2_2_Ch2_Stationary_Combustion.pdf—accessed 4 May 2016.

Table 4 3-Year average fuel consumption, t

Fuel source	Fuel consumption (FC, t)			3-Year average fuel consumption (t)
	2012	2013	2014	
Coal	649,157	683,207	696,038	(649,157 + 683,207 + 696,038)/3 = 676,134.0
HFO	207,874	210,055	215,615	(207,874 + 210,055 + 215,615)/3 = 211,181.3
Kerosene	3437	645	676	(3437 + 645 + 676)/3 = 1586.0

Table 5 3-Year average of electricity generated, MWh

Fuel source	Electricity delivered to grid (EG, MWh)			3-Year average electricity generated (MWh)
	2012	2013	2014	
Coal	1,021,400	1,067,200	1,125,400	(1,021,400 + 1,067,200 + 1,125,400)/3 = 1,071,333.3
HFO	1,026,823	1,043,976	1,045,001	(1,026,823 + 1,043,976 + 1,045,001)/3 = 1,038,600.0
Kerosene	10,984	1668	1991	(10,984 + 1668 + 1991)/3 = 4881.0

Table 6 Set of five most recently built power units, $AEG_{SET\text{-}5\text{-}units}$

Power unit	Power plant	Date commissioned	Fuel type	Installed capacity (MW)	Net electricity to grid in 2014 (MWh) [fossil + renewables]	Fuel consumption in 2014 (excluding renewables) (t)
N/A	Sarako PV plant	February 2014	Solar	15	20,406	N/A
N/A	Midlands Dam	14 March 2013	Hydro	0.35	1310	N/A
G6	Fort Victoria	May 2012	HFO	15	154,633	31,653
G5		Apr 2012		15		
G4		Apr 2012		15		
G3		Apr 2012		15		
G2	Fort Victoria	Sept 2010	HFO	15	75,327	15,435
G1		Sept 2010		15		
N/A	La Nicolière	2010	Hydro	0.35	797	N/A
Total					252,473	47,088

3.4.2 Sum of the 3-Year Average Electricity Generated, MWh

Similarly, the average electricity generated from the combustion of each fossil fuel source is calculated (Table 5).

The total 3-year average electricity generated = (1,071,333,3 + 1,038,600.0 + 4881.0) MWh = **2,114,814.3 MWh.**

3.4.3 Calculating the OM Emission Factor, EF_{OM}

The Operating Margin emission factor is calculated as the net CO_2 emissions emanating from the combustion of fossil fuels (using data from Tables 3 and 4) divided by the 3-year average electricity generated from fossil fuel sources (Table 5) as:

$$\begin{aligned} \mathbf{EF_{OM}} &= [(\mathbf{676{,}134 \times 2.226}) + (\mathbf{211{,}181.3 \times 3.034}) + (\mathbf{1586 \times 3.025})](\mathbf{tCO_2}) \\ &\quad /\mathbf{2{,}114{,}814.3} \\ (\mathbf{MWh}) &= \mathbf{1.0170\, tCO_2/MWh}. \end{aligned} \tag{2}$$

3.5 *Calculate the Build Margin (BM) Emission Factor*

According to the *tool to calculate emissions factor for an electricity system*, project participants should use the set of power units that comprise the larger annual generation. The build margin consists of either:

(a) **The set of five power units ($SET_{5\text{-units}}$) that have been built most recently (excluding registered CDM projects), or**
(b) **The set of power capacity additions in the electricity system that comprises 20% of the system generation (in MWh) and that has been built most recently ($SET_{\geq 20\%}$).**

If electricity generated by $SET_{5\text{-units}}$ is larger than electricity generated by $SET_{\geq 20\%}$, then the sample set of plants used to calculate the BM emission factor, $SET_{sample} = SET_{5\text{-units}}$. Else, $SET_{sample} = SET_{\geq 20\%}$.

The methodological tool provides guidance regarding the selection of the sample group of power units that is needed to calculate the build margin (Fig. 2).

3.5.1 Determining the Electricity Generated (MWh) by the 5 Most Recently Built Power Units, $AEG_{SET\text{-}5\text{-}units}$

Using the list of electricity generated (MWh) and year of commissioning of power units provided by the CEB (not shown), the details of the five most recently built

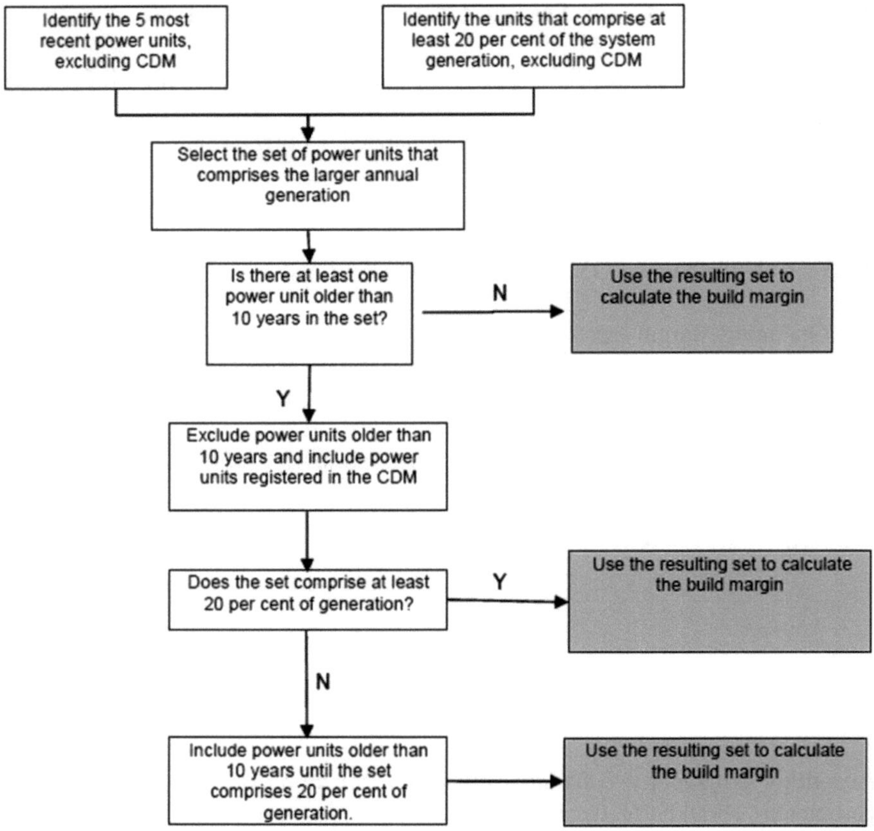

Fig. 2 Procedure to determine the sample group of power units used to calculate the build margin

power units $SET_{5\text{-}units}$ (excluding the registered CDM landfill gas project at Mare Chicose that started power generation in August 2011) were identified (Table 6). In this case, $AEG_{SET\text{-}5\text{-}units}$ = 252,473 MWh.

A power plant/unit is a facility that generates electric power. Several power units at one site comprise one power plant, whereas a power unit is characterized by the fact that it can operate independently from other power units at the same site. Where several identical power units (i.e., with the same capacity, age and efficiency) are installed at one site, they are considered as one single power unit.

3.5.2 Determining the Set of Power Units that Generate 20% of Total Grid Electricity, and That Have Been Most Recently Built, $SET_{\geq 20\%}$

The total grid electricity generated in 2014 was 2633.2 GWh (Table 1). Twenty per cent of this value is 526,650 MWh. Hence, additional power plants than the five most recent plants (Table 6) have to be identified. Since a fraction of a power plant/power unit cannot be used for additional generation, the 20% of total grid electricity is met through a larger set of power plants (Table 7).

A comparison between results (Tables 6 and 7) reveals that $SET_{sample} = SET_{\geq 20\%}$. It is pointed out that none of the power plants/units needed to reach at least 20% of total grid electricity (Table 7) is older than 10 years.

The build margin emissions factor is the generation-weighted average emission factor (tCO_2/MWh) of all power units m during the most recent year y for which electricity generation data is available (2014 in the present case). It is calculated as follows:

$$EF_{grid,BM,y} = \frac{\sum_m EG_{m,y} \times EF_{EL,m,y}}{\sum_m EG_{m,y}}, \tag{3}$$

where, $EF_{grid,BM,y}$, Build margin CO_2 emission factor in year y (tCO_2/MWh); $EG_{m,y}$, Net quantity of electricity generated and delivered to the grid by power unit m in year y (MWh); $EF_{EL,m,y}$, CO_2 emission factor of power unit m in year y (tCO_2/MWh); m, Power units included in the build margin; y, Most recent historical year for which electricity generation data is available.

Using data given in Tables 3 and 7, the BM emission factor has been calculated as **0.8135 tCO_2/MWh**. The detail of the calculations is as follows:

$$\begin{aligned}\mathbf{EF_{BM}} &= [(\mathbf{31{,}653} \times \mathbf{3.034}) + (\mathbf{15{,}435} \times \mathbf{3.034}) + (\mathbf{208{,}429} \times \mathbf{2.226})](\mathbf{tCO_2}) \\ &\quad /\mathbf{746{,}032}\,(\mathbf{MWh}) \\ &= \mathbf{0.8135\,tCO_2/MWh}.\end{aligned} \tag{4}$$

3.6 *Step 6—Calculate the Combined Margin (CM) Emissions Factor*

The combined margin emission factor is calculated as follows:

$$EF_{grid,CM,y} = EF_{grid,OM,y} \times w_{OM} + EF_{grid,BM,y} \times w_{BM}, \tag{5}$$

where

$EF_{grid,BM,y}$ = Build margin CO_2 emission factor in year y (tCO_2/MWh);
$EF_{grid,OM,y}$ = Operating margin CO_2 emission factor in year y (tCO_2/MWh);
w_{OM} = Weighting of operating margin emissions factor (%); and
w_{BM} = Weighting of build margin emissions factor (%).

The following default values are used for w_{OM} and w_{BM}:

Table 7 Set of 5 most recently built power units that generate at least 20% of total grid electricity, $SET_{\geq 20\%}$

Power unit	Power plant	Date commissioned	Fuel type	Installed capacity (MW)	Net electricity to grid in 2014 (MWh) [fossil + renewables]	Fuel consumption in 2014 (excluding renewables) (t)
N/A[a]	Sarako PV plant	February 2014	Solar	15	20,406	N/A
N/A	Midlands Dam	14 March 2013	Hydro	0.35	1310	N/A
G6	Fort Victoria	May 2012	HFO	15	154,633	31,653
G5		Apr 2012		15		
G4		Apr 2012		15		
G3		Apr 2012		15		
G2	Fort Victoria	Sept 2010	HFO	15	75,327	15,435
G1		Sept 2010		15		
N/A	La Nicolière	2010	Hydro	0.35	797	N/A
G1	CTSav	October 2007	Coal/bagasse	45	359,561	208,429
G2		April 2007		45	133,998	N/A
Total					746,032	

[a] *N/A* not applicable

Table 8 Combined margin grid emission factor for the national electricity system of the island of Mauritius

Type of project	OM (tCO_2/MWh)	BM (tCO_2/MWh)	CM (tCO_2/MWh)
PV and wind	1.0170	0.8135	0.9661 [(1.0170 × 0.75) + (0.8135 × 0.25)]
All other projects	1.0170	0.8135	0.9152 [(1.0170 × 0.5) + (0.8135 × 0.5)]

(a) **Wind and solar power generation** project activities: $w_{OM} = 0.75$ and $w_{BM} = 0.25$ (owing to their intermittent and non-dispatchable nature) for the first crediting period and for subsequent crediting periods;
(b) **All other projects**: $w_{OM} = 0.5$ and $w_{BM} = 0.5$ for the first crediting period, and $w_{OM} = 0.25$ and $w_{BM} = 0.75$ for the second and third crediting period, unless otherwise specified in the approved methodology which refers to this tool.

The combined margin grid emission factors for the island of Mauritius have been calculated using Eq. (5) (Table 8).

4 Discussion

As mentioned above, the methodological tool has been used to develop a standardized baseline for the national electricity system of the island of Mauritius. The application of the methodology through the above six steps has been reviewed by the UNFCCC Secretariat and CDM Methodology Panel, and the combined margin emission factor was adopted by the CDM Executive Board as the official standardized baseline for the national electricity system of the island of Mauritius on January 7, 2016.[5]

The combined margin emission factor of Mauritius can be compared with the officially reported grid emission factors for selected African countries that can be found in the public domain. The values used for comparison (Table 9) are derived from the Institute for Global Environmental Strategies (IGES 2015), and, given the geographical location of the island of Mauritius, they are limited to sub-Saharan African countries. The information is extracted from publicly available sources on the UNFCCC website, and the combined margin emission factors are registered for CDM projects on April 3, 2015. For the purpose of the comparative analysis, the average values of the grid emission factors are used.

[5]https://cdm.unfccc.int/filestorage/e/x/t/extfile-20160308100109783-ASB0019_.pdf/ASB0019_.pdf?t=b2t8b3V2bDJmfDCVTDSX-DEw9i9QQolOQ8Z_—accessed 4 May 2016.

Table 9 Average combined margin emission factor (tCO_2/MWh) for the national grid of selected African countries

Country	CM emission factor	Country	CM emission factor
South Africa	0.9545	Nigeria	0.5733
Angola	0.8408	Madagascar	0.5520
Libya	0.7937	Tanzania	0.5290
Rwanda	0.6540	Sierra Leone	0.4020
Kenya	0.6032	Sudan	0.3050

The combined margin grid emission factor of Mauritius is larger than any of these average values (Table 9). The countries that have the lowest emission factors typically have a large share of hydroelectric power in their national grid. Others like Nigeria and Kenya generate electricity from natural gas (cleaner than other thermally generated electricity from fossil combustibles like fuel oil or coal) and geothermal, respectively. Electricity in South Africa is generated mainly from coal. There are two pertinent observations that can be drawn here: (1) despite having a relatively large share of renewable electricity in its power mix (Table 1), Mauritius has a very high grid emission factor showing that the national grid is not as "clean" as might be expected; and (2) given its relatively high emission factor, Mauritius may be seen as being well positioned to develop project activities that can leverage climate (or carbon) finance.

However, Mauritius has only one registered CDM project to date that is delivering certified emission reductions. This is because Mauritius lacks from the scale of economies that other African countries may enjoy despite their lower grid emission factors. The constraint of scale can be overcome by developing carbon finance projects at the sectoral level for which the nationally appropriate mitigation action approach may be appropriate. The technical capacity for African countries to be able to calculate and update their grid emission factors on an annual basis is necessary to capitalize on post-2020 mitigation opportunities, as those formulated in their intended nationally determined contributions (INDCs) and reported to COP21 in 2015.[6]

5 Conclusion

The combined margin grid emission factor for the national electricity system of the island of Mauritius has been calculated using the Clean Development Mechanism Methodological Tool 07—"Tool to calculate the emission factor for an electricity system (Version 04.0)". Using statistical data for electricity generation and fuel

[6]http://www4.unfccc.int/submissions/indc/Submission%20Pages/submissions.aspx—accessed 4 May 2016.

consumption for the years 2012, 2013, and 2014, the operating margin and build margin emission factors have been calculated as 1.0170 tCO_2/MWh and 0.8135 tCO_2/MWh, respectively. The combined margin emission factor is 0.9661 tCO_2/MWh for grid electricity generated from wind power or solar PV. For other types of grid-tied renewable electricity generation or energy efficiency measures that displace grid electricity, the combined margin grid emission factor is 0.9152 tCO_2/MWh. A comparative analysis of the grid emission factor of Mauritius has been carried out relative to that of selected sub-Saharan African countries. Mauritius has a comparatively high grid emission factor despite having a relatively large share of renewable electricity in its power mix. This reveals that the national grid may not be as "clean" as might be expected from a grid with close to 20% of renewables in the mix. Nevertheless, a high grid emission factor provides an opportunity to develop project activities that can leverage climate (or carbon) finance. The lack of projects making recourse to carbon finance is related to the lack of scale of economies that is an inherent vulnerability of a small island developing state. However, Mauritius has the opportunity develop larger scale initiatives that reduce GHG emissions at the sectoral scale, such as those communicated in its INDC to the UNFCCC.

Acknowledgements The combined margin grid emission factor was developed under the CD4CDM project implemented by the Ministry of Environment, Sustainable Development, Disasters and Beach Management through technical assistance from the UNEP DTU Partnership.

References

IGES. (2015) *List of grid emission factor*. Available online at http://pub.iges.or.jp/modules/envirolib/view.php?docid=2136. Accessed June 1, 2015.

IPCC. (2006). In H. S. Eggleston, L. Buendia, K. Miwa, T. Ngara, & K. Tanabe (Eds.), *2006 IPCC guidelines for national greenhouse gas inventories* (Vol. 2). Japan: IGES.

IPCC. (2013). Summary for policymakers. In T. F. Stocker, D. Qin, G.-K. Plattner, M. Tignor, S. K. Allen, J. Boschung, A. Nauels, Y. Xia, V. Bex, & P. M. Midgley (Eds.), *Climate change 2013: The physical science basis. Contribution of working group i to the fifth assessment report of the intergovernmental panel on climate change* (pp. 2–17). Cambridge: Cambridge University Press.

IPCC. (2014). *Technical summary of working group III—Mitigation of climate change AR5*. Cambridge: Cambridge University Press.

Statistics Mauritius. (2014). *Digest of energy and water statistics—2013*. Port Louis: Statistics Mauritius.

UNDP. (2013). *Guidance note—Standardized baselines*. http://www.undp.org/content/dam/undp/library/Environment%20and%20Energy/MDG%20Carbon%20Facility/Guidance%20note_SB_20%2012%202013.pdf. Accessed May 3, 2016.

UNFCCC. (2008). *Part 2: Action taken by the conference of the parties at its thirteenth session*. Report of the conference of the parties on its thirteenth session, held in Bali from 3 to 15 December 2007. FCCC/CP/2007/6/Add.1 (pp. 3–6).

Author Biographies

Dr. Prakash Deenapanray holds a Bachelor of Systems Engineering (Interdisciplinary Program) from the Australian National University (ANU), M.Sc. in Physics from the University of Pretoria (South Africa), Ph.D. in Semiconductor Physics, ANU, and MBA in Technology Management from LaTrobe University (Australia). He has published over 130 research articles, technical and policy papers, and generated patents while developing novel thin-film silicon solar cells technologies for photovoltaic applications. He has supported several sub-Saharan and MENA countries in accessing climate finance for mitigation and adaptation. He has been a member of UNDESA's Technical Advisory Group supporting the development of the Vulnerability-Resilience Profile (VRP) methodology for SIDS. He has been a member of the UNEP's Expert Group for developing an Inclusive Green Growth Index. He served as a member of the Mauritius Stock Exchange of Mauritius Sustainability Index Supervisory Committee. He is currently a member of the High Level Task Force on Sustainable Development under the Prime Minister's Office.

Sanjay Sookhraz holds a Bachelor Degree in Chemical Engineering from the National Institute of Technology Durgapur, India, and Master's Degree in Water Resources Engineering from Katholieke University/Vrije University Belgium. He is currently studying for a Master's degree in Sustainable Energy Engineering with the Royal Institute Technology Sweden, and an MBA from University of Bradford. He currently works as Environmental Affairs Officer at the Central Electricity Board (CEB), a para-statal organization that is responsible for the generation and distribution of electricity in Mauritius. His responsibility is to ensure that CEB's activities are meeting the requirements of the prevailing environmental regulation, and that all activities are carried out to effectively minimize environmental risks. Mr. Sookhraz has been extensively involved in the climate change activities on a national level, especially in the preparation of the National Communications under the UNFCCC. He has been involved in the preparation of national policies on energy, environment, and climate change.

Greenhouse Gas Emission Reductions from Solid Waste Management: Prognosis of Related Issues

Sunjaya Leelah and Ackmez Mudhoo

Abstract The implementation of cleaner waste treatment technologies in developing countries is accompanied by myriads of limitations and obstacles. This study reports that greenhouse gas (GHG) emission reductions that would not have happened with business-as-usual scenarios are achievable in the waste management system under the Clean Development Mechanism (CDM). The grounds for CDM failures and successes were reviewed by reflecting on the past and current episodes in developing countries. It was found that economic development was directly proportional to increase in waste generation rate and influenced waste composition. The need to decouple waste production from Gross Domestic Product (GDP) and linking the prevalence of more low-carbon emissions waste treatment options were found essential to ensure sustainable development and mitigate climate change. Based on the present evaluation, Brazil and Thailand have had the largest number of CDM projects and mostly small-scale projects with the highest maximum emission reductions in Brazil (751, 148 Mt CO_{2e}) followed by Vietnam (158, 727 Mt CO_{2e}) and India (158, 077 Mt CO_{2e}). It was found that small-scale projects were more feasible as technology transfer and financial status were hindering the implementation of large-scale CDM projects. Moreover, emissions trading through CDM were accompanied by technical obstructions in terms of carbon leakage, permit allowances and unfair carbon price with irregular international policies and regulations to control GHGs. This study highlights the potential of a carbon trading scheme through CDM waste management projects designed to aid highly polluting developed countries achieve their target baselines for carbon emissions. The increasing GHG emissions in developing and least developed countries could be tackled at the early stages of economic growth through financial and technological assistance from industrialized countries.

S. Leelah (✉)
Department of Agriculture and Food Science, Faculty of Agriculture, University of Mauritius, Reduit 80837, Mauritius
e-mail: s.leelah@uom.ac.mu

A. Mudhoo
Department of Chemical and Environmental Engineering, Faculty of Engineering, University of Mauritius, Reduit 80837, Mauritius

W. Leal Filho and D. Surroop (eds.), *The Nexus: Energy, Environment and Climate Change*, Green Energy and Technology, https://doi.org/10.1007/978-3-319-63612-2_22

Keywords Greenhouse gas · Clean development mechanism · Waste management · Carbon emissions trading

1 Introduction

The waste sector contributes about 3–4% to the total global anthropogenic GHG emissions (Bogner et al. 2008; Hoa and Matsuoka 2017) and the most important greenhouse gas contribution is methane. The impact of methane is 34 times than that of carbon dioxide for one ton of GHG emitted considering a 100-year time horizon and 86 times when considering a 20-year time horizon as reported in the Fifth Assessment Report of the IPCC (Myhre et al. 2013). It was the 20-year horizon that was relevant to global efforts to prevent mean global temperature from exceeding either the limit of 2 °C above the pre-industrial average as agreed in Copenhagen in 2009 or 1.5 °C as endorsed in the 2015 Paris agreement (Fearnside 2015).

Alternatively, waste management also potentially reduces impacts indirectly by recovering materials and by producing energy that otherwise would be produced from fossil fuels (Couth and Trois 2012). The recovery potential for energy from major waste management technologies has been considerable and this was achieved through the Clean Development Mechanism (CDM) under the 1997 Kyoto Protocol (Rong et al. 2012). The heat and power recovery from incineration and landfill gas collection and utilization proved to be beneficial in China during 2008–2011 (Wang et al. 2016). In 2008, about 30×10^6 tCO_{2e} was recovered globally from CDM projects in developing countries through methane recovery (Monni et al. 2006). For the year 2020, an increase in recycling could save −207,056 tons of CO_{2e} and the introduction of anaerobic digestion with about −157,682 tons of CO_{2e}. However, the highest overall GHG savings of −630,867 tons of CO_{2e} will be possible if a landfill gas collection and utilization system for the Buffelsdraai landfill site is introduced in Durban (Friedrich and Trois 2016).

China, which is among the largest GHG emitters, had very high ratio of CER revenues during the period of 2008–2011 (Wang et al. 2016). In March 2010, over 50% of the 2062 projects which were registered at the United Nations Framework Convention on Climate Change (UNFCCC) as CDM activities were based on renewable energy and one quarter on methane avoidance (Fenhann 2010). The Nejapa landfill project allowed the government of Luxembourg to decrease its carbon footprint by purchasing the certified emission reductions (CERs) produced in the first year of the gas-flaring operation (Martinez and Bowen 2012). More than 400 studies have been undertaken since 1997 on the CDM and with some 3000 more projects awaiting registration, the CDM was expected to generate nearly 3 billion CERs by 2012 (Sreekanth et al. 2012), when the first Kyoto commitment period ended.

1.1 Limitations in GHG Emissions Accounting

GHG inventories from the waste sector were distinguished by methodology constraints, lack of data, poor reliability of available information, and extreme uncertainty levels (Pikon and Gaska 2010; Friedrich and Trois 2016). The overall methane gases emitted from landfills across the world were 40–50% lower or 20–25% higher than those existing in reality (Lou and Nair 2009). The main causes were the adoption of assumptions and extrapolations during data acquisition (Lim et al. 2016; Monni et al. 2004). The 2006 IPCC Guidelines indicated that uncertainties for global emissions from waste could be as high as 10–30% for developed countries to +60% for developing countries (DCs) that did not have annual data (IPCC 2006). This was a challenging issue for developing countries since higher accuracy methods were more data intensive (Friedrich and Trois 2011) whereby most of the time default values were used.

1.2 Discrepancies from Emissions Trading

In response to complaints about inequalities produced by free allocation, a vast majority of trading firms (84%) agreed that it was important for firms to reduce their carbon footprint over time and not only rely on purchased allowances (Olsson et al. 2016). A possible unintentional environmental consequence was the potential for carbon leakage (Clo 2008; Huang et al. 2016). The aim of this study was to trace the main limitations, barriers, and technical obstacles encountered prior and subsequent to the implementation of cleaner waste management technologies for GHG emission reductions in developing countries to achieve targeted benchmarks. The main objectives were to assess (i) first, the factors affecting waste generation and GHG concentrations in the atmosphere; (ii) second, the uncertainties in the process of quantification of GHG emissions; and (iii) third, the discrepancies encountered before, during, and after the implementation of CDM projects; (iv) lastly, limitations and strengths of carbon trading routes across the globe.

1.3 Approach to Study

An exploratory research was carried out to locate the key limitations, barriers, and obstacles that have already been encountered in previous studies in developing countries along with CDM failures and successes followed by an analytical research to analyze and review the causes for CDM failures and successes to find potential solutions and create opportunities. This study employed a three-pronged approach to achieve its objectives; (i) analysis of the opportunities of the waste sector for emission reductions through cleaner technologies, (ii) assessment of the factors affecting implementation of CDM projects and trading of CERs in developing countries and finally, (iii) reviewing the discrepancies from greenhouse gas accounting and acquisition of data to improve accuracy of inventories in developing countries through secondary data collections. Literature was analyzed and reviewed extensively. Figures were collected and analyzed to make comparisons and predictions. Illustrations were presented to demonstrate the past and future trends.

1.4 Limitations

Developing countries and developed countries were chosen for the study and least developed countries were omitted due to inaccessibility of correct data. The choice of the countries from each continent depended on (i) the availability of data on waste generation, composition and treatment, (ii) the amount of GHGs emissions, (iii) the largest economy in the world with the highest industrialization, and (iv) the highest population density.

2 Results and Discussions

2.1 Waste Generation

Current global municipal solid waste (MSW) generation levels are approximately 1.3 billion tons per year, and were expected to increase to approximately 2.2 billion tons per year by 2025 (Hoornweg and Bhada-Tata 2012). The drivers for increasing global waste generation have been associated with population size and economic growth of a country (Uddin et al. 2016). The gross domestic product (GDP) growth rate measures the affluence and income level of a country (Alam 2016; Bhattacharya et al. 2016).

Figure 1 depicts the relationship between economic growth, population growth, and municipal waste generation in selected developing and developed countries. Overall, there was an increasing trend for all three indicators. The population of developing countries (e.g., Ethiopia, Egypt, Algeria South Africa, Malaysia, and India) was higher than developed countries (Denmark, United States, New Zealand, France, and Australia) in Fig. 1. It involved a rise in material consumption, food production, and resource depletion in terms of quantity of consumables needed per capita to sustain life (Kamran et al. 2015).

Inevitably the waste generation rate increased with rising GDP for most countries and was highest in Ethiopia, Egypt, China, India, Indonesia, and South Africa. India experienced an average GDP growth of 7%, and estimated municipal waste increased from 48 to 70 million tons during the same period (Sharholy et al. 2008). Figure 1 clearly depicts the correlation of GDP with MSW generation and has

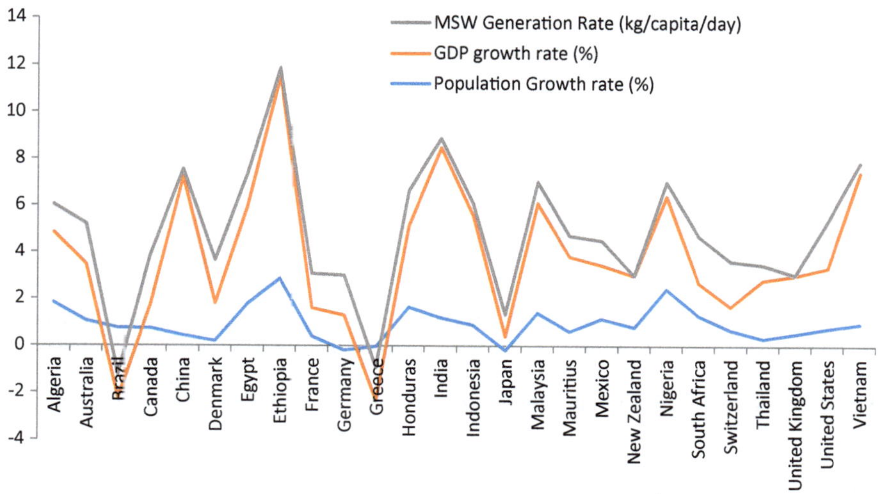

Fig. 1 Relationship between population growth (http://data.worldbank.org/indicator/SP.POPO.GROW), economic growth (http://data.worldbank.org/indicator/NY.GDP.MKTP.KD.ZG_) and municipal waste generation rate (Kawai and Tasaki 2016) for the year 2015

always been the case since 1999 (Couth and Trois 2011a, b). By 1999, people were already exceeding the Earth's capacity to regenerate the resources used in 1 year by roughly 20% (Wackernagel et al. 2002). The economic growth trend is an existing driver for GHG emission reduction whereas the introduction of renewable energies and energy efficiency are premeditated strategies by the government of Spain to fulfill its Kyoto Commitments (Martinez de Alegria et al. 2016). It was observed that as affluence of a country rises, the demand for development in ecological improvement will increase due to high financial investment facilities (Alam 2016). When waste production data are not available, historical disposal of waste was estimated proportional to GDP and was used as the driver in the Tier 1 method (IPCC 2006).

2.2 *Waste Composition*

Organic waste comprised the bulk of MSW, following paper, metal, plastic, glass, and other wastes as shown in Fig. 2. Low- and middle-income countries had a high percentage of organic matter in the urban waste stream, ranging from 50 to 65%. Paper, plastic, glass, and metal fractions increased in the waste stream of upper middle- and high-income countries. Organics made up 64% of the MSW stream for low-income countries and paper only 5%, whereas in high-income countries it was 28 and 31%, respectively. Globally organic waste maintained a high fraction with 46% followed by 17% paper, 10% plastic, 5% glass, and 4% metal. Consumption

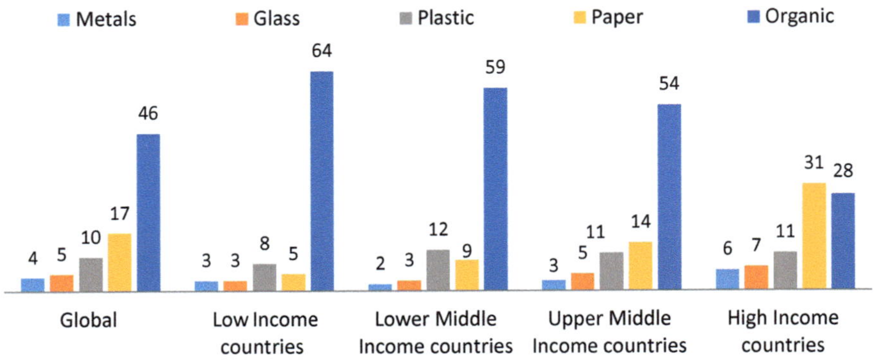

Fig. 2 Waste composition variation with regard to income level in percentage

pattern may differ due to higher income level in developed countries. The trend would shift towards more inorganic waste compared to developing countries as shown in Fig. 2. Recent and accurate data on the composition of MSW help to choose the right technology for management of the waste streams (Kamran et al. 2015).

2.3 *Biogenic Carbon Versus Fossil Carbon in MSW*

Biogenic carbon accounts for 49–90% of the organic waste composition (food, paper, cardboard, and wood) of MSW and is responsible for carbon sequestration (Eggleston et al. 2006). $CO_{2biogenic}$ and $CO_{2fossil}$ are calculated separately from MSW such that the former has a zero GWP and the latter has a GWP equivalent to 1 (Wiloso et al. 2016; Christensen et al. 2009). The emission ranking of waste technologies remains same regardless whether $CO_{2biogenic}$ has a neutral GWP (Muñoz and Schimidt 2016). The neutrality of biogenic carbon is assumed when no CO_2 is added into the atmosphere according to the carbon cycle (Christensen et al. 2009). Fossil carbon is nondegradable fraction found in plastic and synthetic fibers. It remains in the landfill for 100 years by forming stable humus compounds. Hence, landfills can provide carbon sinks for fossil carbon (Galik et al. 2016).

Degradable organic carbon (DOC) is one of the main parameters affecting the CH_4 emissions from solid waste disposal and varies for different waste fractions (IPCC 2006). Highest percentage of fossil carbon is found in plastics (Table 1). The DOC is considered as biogenic carbon and it contains organic carbon. It may be noted that metal and glass contain very little carbon and no carbon, respectively (Table 1). Munoz and Schimidt (2016) concluded that there is a discrimination of 2.75 kg $CO_{2\text{-eq}}$. regarding the molecular weights of CO_2 and methane and have therefore recommended to rectify the GWP and GTP values for methane.

Table 1 Default values of DOC in different waste types (IPCC 2006)

MSW component	DOC % content of wet waste	DOC % of dry waste	Total Carbon content	Fossil carbon fraction
Paper/cardboard	36–45	40–50	42–50	0.5
Textiles	20–40	25–50	25–50	0–50
Food wastes	8–20	20–50	20–50	–
Wood	39–46	46–54	46–54	–
Nappies	18–32	44–80	54–90	10
Plastics	–	–	67–85	95–100
Metal	–	–	NA	NA
Glass	–	–	NA	NA

2.4 *Decoupling: Economic Growth and Waste Management*

It has always been nearly impossible for an absolute decoupling waste generation from GDP to ensure sustainable use of the world's resources, with consequent climate benefits whereby the waste quantities stabilize or decrease while GDP still increased (Mazzanti and Zoboli 2008). In its place, researchers have shown the prevalence of cleaner waste treatment technologies with high GDP with an impact decoupling process in which economic growth was increasing while the associated environmental impacts were reduced simultaneously (UNEP 2011; Mazzanti and Zoboli 2008). Several studies have been carried out to find a relationship between GDP and waste intensities (Johnstone and Labonne 2004; Barton et al. 2008; Mazzanti 2008; Sjostrom and Ostblom 2010; Fazeli et al. 2016). In this context, environmental degradation could be replaced by low-carbon management strategies. A strong positive relationship between waste paper recovery rates and income for rich countries was obtained (Berglund and Söderholm 1997). Absolute decoupling from economic growth demanded significant reductions of the waste intensities linked to three driving forces: technical change, economic growth, and household consumption (Yang et al. 2016; Uddin et al. 2016).

2.5 *Waste Management Strategies*

As shown in Fig. 3, landfilling remained the main option for waste disposal in most developing countries with minimal amount of composting and anaerobic digestion. However, incineration was highest in China and Denmark. Recycling was remarkable in Nigeria, Honduras, and Mexico. The potential emission from a conventional landfill from MSW Australian default quantified values was approximately 1.287 ton CO_{2e} for 1 ton of waste (Lou and Nair 2009).

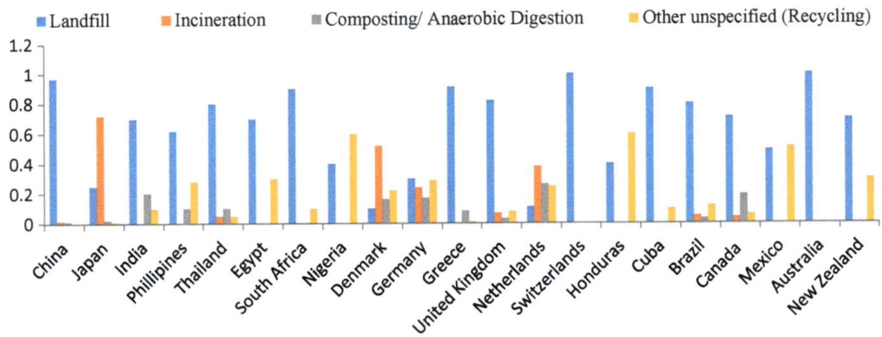

Fig. 3 Type of waste treatment option by country

The most feasible waste management options for energy recovery are landfill gas flaring or utilization and heat recovery from incineration (Yu and Zhang 2016).

2.6 *Clean Development Mechanism (CDM)*

The CDM is a trading scheme under the 1997 Kyoto Protocol and it allows countries under certain criteria to fulfill their GHG reduction targets by implementing green projects with proven and significantly reduced carbon footprint with respect to baseline scenarios. The CDM Executive Board (EB) requires a monitoring report from the Designated Operational Entity, already prepared by the project participants prior to CER issuance. CERs are emission reductions which have been validated and approved after scrutinizing implemented project performance. The monitoring report gives an estimation of emission reductions throughout the crediting period. However, uncertainties in accounting were affecting the emission reductions estimates.

2.7 *Greenhouse Gas Accounting Uncertainties*

The waste management sector in developing countries had to consider a series of parameters to provide real and transparent outcomes. Methane emissions were calculated using many simple and complex models with different orders of kinetics affecting reproducibility and consistency of inventories. The main criticism was their lack of accuracy and validation (Thompson et al. 2009; Couth and Trois 2011a, b). Unavailability and poor quality of annual data on waste type, quantity, and management practices resulted in inaccurate outcomes (Bogner et al. 2008). Subsequent assumptions on decomposition rates, methane generation rates, and oxidation rates added error and ambiguity to the calculations (Mohareb et al. 2011) although activity data were acceptable. The errors due to assumptions ranged from 15 to 20% for 2001 for which real data existed. Double counting was inevitable when the same source of emissions was accounted in two different sectors (e.g., waste and energy sectors during waste to energy projects (Schneider et al. 2015). The uncertainty of the dry matter content ranged between ±10% up to ±50% every time wet weight converted to dry weight (Gentil et al. 2009). The overall rate of emissions for landfilling was influenced by operational intrusions like waste compactions, leachate recirculation, or aerobic landfilling (Lou and Nair 2009).

It was difficult to gauge the accuracy of current estimates of the climate impact of waste activities both at national and global scale. Global predictions of GHG emissions with and without the CDM project depended on the assumed rates of waste generation, composition and disposal, and uncertainties due to emission factors and activity data (Lim and Boileau 1999). Transparency of data compilation would be reduced only if assumptions, estimations, and extrapolations were clearly disclosed and referenced at both national and international level (Kawai and Tasaki 2016). The assumption of biogenic carbon neutrality can render the calculated values of GWP considerably superior or inferior than the true inventories when diverse system margins, types of GHGs, and waste composition are accounted (Wiloso et al. 2016).

2.8 Small- and Large-Scale CDM Projects

Most of the capable host countries were in Latin America and Asia as shown in Fig. 4. The eligibility of African host countries has been relatively low with the exception of Nigeria and South Africa with a high GDP value. The highest number of registered CDM projects in the waste handling and disposal sector in the first quarter of 2015 were found in Brazil followed by Indonesia and Thailand with a total of 40, 32, and 28 large and small-scale projects, respectively. South Africa has not been left apart with a new large-scale project registered in May 2014 for the Manufacture and utilization of bio-coal briquettes.[1] The CDM projects in Fig. 5 were mainly methane capture and landfill gas recovery and some combustion, composting/anaerobic digestion, biogas system with energy generation and wastewater treatment practices.

The realization of programmatic CDM would have been useful for developing countries as it would have allowed bundling of small-scale projects to be evaluated in one common large-scale process and facilitating the CDM registration. It would allow emerging economies to meet the challenge of emission reduction efforts. Switzerland, Netherlands, and Japan were the most active countries in validating these CDM projects acting as Designated Operational Entity (DOE). Technology

[1]United Nations Framework Convention on Climate Change. https://cdm.unfccc.int/Projects/projsearch.html. Last Updated 2015.

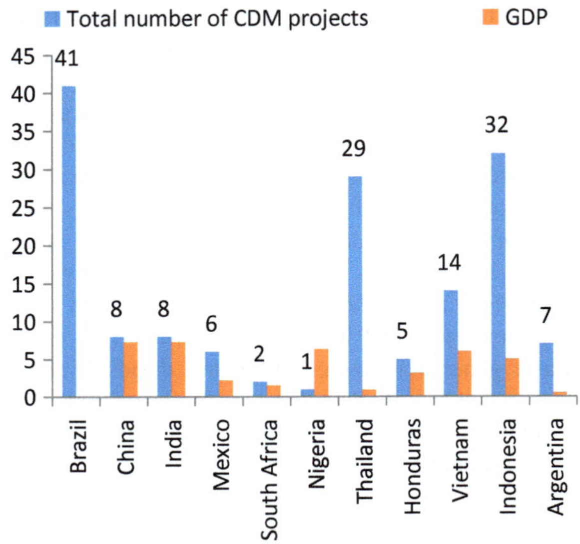

Fig. 4 Relationship between GDP and prevalence of CDM projects

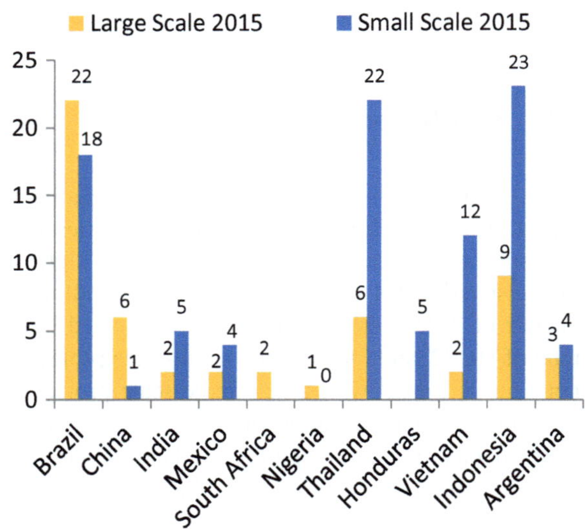

Fig. 5 Number of CDM large and small projects in the waste sector

and know-how for CDM projects were likely to come from a relatively small set of countries such as Germany, USA, Denmark, Japan, and China as the top suppliers. The "rent-sharing approach" was introduced to pledge significant involvement of developing countries without challenging their economic development (Viguier 2004). They had the opportunity to enter the emission markets and benefit from financial transfers such as China may be better off while accepting a significant CO_2 reduction effort (Streimikiene and Girdzijauskas 2009). The critical issues addressed by developing countries were Additionality and establishing a baseline emission and crediting period.

2.9 Recovery of Emission Reductions from Solid Waste Management

Greenhouse gas emissions from municipal solid waste are affected by changes in composition and disposal methods (Yu and Zhang 2016). A projection was made from 2015 until 2031; the potential CO_2 emission reduction would be 42,18 CO_2 Mt accounting for a reduction of 63,72 CO_2 Mt within the National Solid Waste Policy (NSWP) in Brazil (Colling et al. 2016). The conversion of food waste (food waste generation is 1080 tons daily in Hong Kong) into biogas for its use as vehicle fuel has reduced 1.9% of greenhouse gas emissions in the transport sectors from 2005 to 2010 (Woon et al. 2016).

A yield of 77 m^3 of methane was recovered from anaerobic digestion to generate electricity (Barton et al. 2008). A potential CO_2 reduction of about 500,000 tons per year was achievable by co-incineration of waste in Austrian industry (Hackl and Mauschitz 2008). Significant amount of GHGs has been avoided through controlled composting with emissions savings of 0.9 tCO_{2e} per ton of waste composted (Boldrin et al. 2009) and by state-of-the-art incineration with 0.48 tCO_{2e} per ton of waste incinerated (Astrup et al. 2009). Substitution of fertilizer by compost and substitution of peat were estimated to save about 0.008 tCO_{2e}/t of waste and 0.081 tCO_{2e}/t of waste, respectively (Boldrin et al. 2009). A key factor for obtaining the full synergetic benefits of this energy recovery is the presence of local heat distribution infrastructures, without which no large-scale recovery and utilization of excess heat are possible (Persson and Münster 2016). The above studies showed the potential of GHG emission reductions from the optimization of waste management practices for climate benefits.

Figure 6 represents the amount of certified emission reductions recovered from solid waste management in some host countries. The highest maximum emission reduction was noted in Brazil (751, 148 Mt CO_{2e}) followed by Vietnam (158, 727 Mt CO_{2e}) and India (158, 077 Mt CO_{2e}) as shown in Fig. 6. Amazingly, South Africa recovered 282, 349 Mt CO_{2e}. The lowest emission reduction recovered was in Honduras with small-scale projects.

2.10 CDM Failures

The CDM Executive Board was answerable for the validation and registration of any project prior to the Additionality and eligibility. Any project was rejected in accordance with "Procedures for review of requests for registration". However, rejected project activities were resubmitted for validation and registration provided they complied with the above requirements, in accordance with the CDM Modalities and Procedures (UNFCCC PDD 2010). The CDM Ruling Notes have brought out all the projects which have been rejected by the DOE in its latest 2015

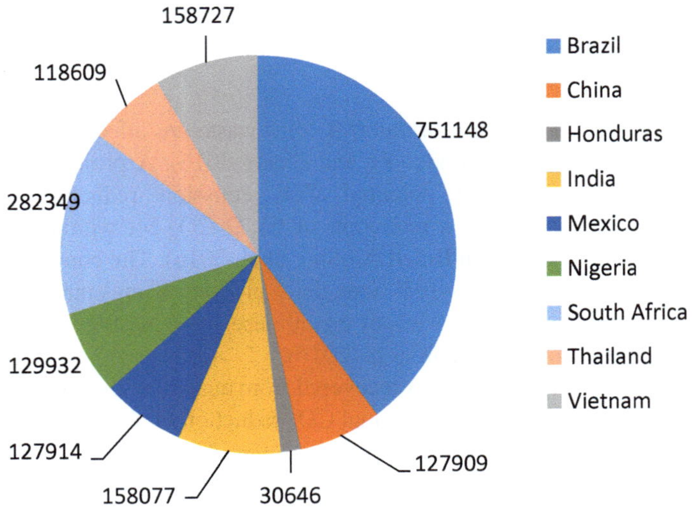

Fig. 6 Volume of CERs recovered from CDM projects in (Mt CO_{2e})

report.[2] From September 2013 to November 2014, 16 projects have been rejected while 5 projects were in the waste sector involving methane recovery, wastewater treatment, and waste heat recovery. All the projects were either rejected at the stages of Validation and Registration or Monitoring and Issuance correspondingly. Additionality, proper baseline quantification, and accounting methodologies were the most important requirements.

2.11 Barriers to CDM Implementation

The difficulty in establishing Additionality was the subject of dynamic debate (Lederer 2010). Benchmarking and CDM Baselines were crediting too many free-riders while being too lenient (Lazarus et al. 1999; Pasour 1981). The abatement potential of host countries was not enough substantial to attract investors for potential sustainable CDM projects promotion (Burian et al. 2011). CDM project implementation was constrained due to high investment costs even higher than expected carbon revenues (Schneider et al. 2008). Countries which lacked existing national policies and mitigation actions, not interested in climate change avoidance, were unlikely to attract CDM projects. The host country had to be proactive to the promotion of the project implementation with easy access to foreign investors to obtain external support (Lederer 2010). A lack of resources, for example, low

[2]United Nations Framework Convention on Climate Change. Available from https://cdm.unfccc.int/Reference/Rulings/index.html. Last updated 2015.

electrification rate, was obstructing possible CDM projects (Burian et al. 2011). The amenability of the waste sector was governed by political preferences where certain countries were privileged due to poverty (Newell 2009). The setting up of potential CDM projects was hindered by the undersized financial status of Least Developed Countries (LCDs) which were eligible for CDM (Okubo and Michaelowa 2010). High transaction costs and lack of applicable methodology have been the most important barriers for China and Japan (Huang et al. 2016).

CDM projects should involve technology transfer to facilitate its implementation and its importance should not be devalued as developing countries become more established (Lema and Lema 2016). Technological support in individual host countries for adaptation actions occurs as capital inputs, policy guidance, and research and development of updated technologies. China could offset GHG emissions to meet the target of global warming to 2% before 2025 through technological advancement and more stringent policies (Lin et al. 2015). The nature of technology transfer changes over time in host countries due to different levels of development of developing countries. Lower middle-income countries have greater needs for technical and industrial aptitudes whereas supportive activities may be proper for upper middle-income countries that already have facilities to deal with climate change (Lema and Lema 2016).

2.12 Trading of GHG Emissions

It was found that carbon emissions decreased around 1.58% per year since Emission Trading Schemes implementation. Around 23.43% of CO_2 reduction was reached after 10 years of their existence (Villoria-Sáez et al. 2016). The process of determining an equitable exchange for CERs was shown to be confounded by a number of factors: data availability, carbon leakage, carbon price, motivation for trading, and permit allocations as pointed out in Fig. 7.

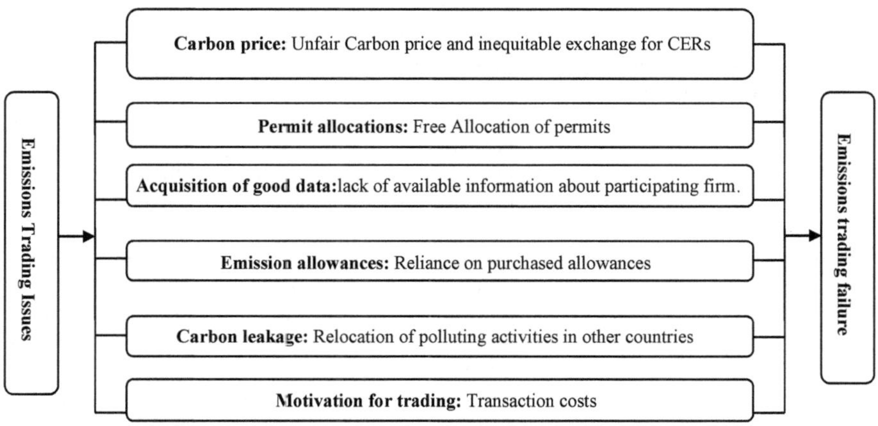

Fig. 7 Technical issues and obstacles leading to carbon trading failures

The allocations of permits were so generous that a number of Europe's biggest polluters had made huge profits by selling on permits that they did not need (Ellerman and Joskow 2008). There was a risk of over allocation with the total emission value of allocated permits being more than the benchmarked level of emissions. Strict environmental regulations in one country resulted in a company choosing to relocate their industrial activities to another country with less stringent regulations so as to avoid large increases in production costs (Parker and Blodgett 2008). Moreover, the price of carbon had always been so low that it was always cheaper for companies to buy permits than to start paying for more expensive infrastructural changes (Hodes and Kamel 2007). Transactional costs were difficult to internalize and occurred at different phases of trading (before entering any market it is essential to learn the rules, demands and required permits of transactions), thus increased the costs for gathering information on the market (Lohmann 2009). Firms are commonly not keen to reveal information about their workings to the regulating body. Regulating bodies have difficulty to report for individual firms with effective

Table 2 Potential solutions to carbon trading [data compiled from Clifton (2009) and Hepburn (2009)]

Solutions	Expected changes
Setting a low enough cap	Setting a cap on emitters covered by carbon trading schemes to ensure that reductions in emissions in developed countries take place in line with climate science and an equitable reduction of emissions globally based on historic responsibility
Removal of all forms of offsetting	Stopping the inclusion of all forms of offset credits in emissions trading schemes including the CDM
Global regulation of emissions trading	Agreement on global rules and standards for emissions trading
Driving real innovation	Address the perverse incentives created by carbon trading for lock-into high-carbon infrastructure by the industries for which emissions cuts are hardest and most expensive to make
Favoring emission reductions rather than purchasing credits	Incentivizing highly polluting heavy industries, for which emissions cuts are expensive, to shift to cleaner technologies rather than simply opting out through the purchasing of credits would require either the removal of their trading option altogether, or the setting of clean technology standards alongside the emissions trading mechanisms
Auctioning for permit allocation	This problem of allocating and pricing permits can be addressed by using market-based systems, such as auctioning to efficiently distribute and accurately price permits. With auctions, there is less of a rent seeking behavior
Setting a price collar or symmetric safety valve	A potentially useful form of market reform would introduce mechanisms to insure that prices neither fall too low nor rise too high

recommendations and proposals (Evans and Bertraw 2009). The appeal for carbon offsets is heavily subjective to carbon trading programme factors and is specific to the type of trading scheme (Dormady and Englander 2016).

2.13 Overcoming Carbon Trading Obstacles

There are a variety of views on whether or not the problems with carbon trading can be overcome. Some more optimistic critics believed that it was within the realms of possibility that the fundamental problems with carbon trading could be addressed (Table 2).

Table 3 Issues and critiques of carbon trading in developing countries with data compiled from Ellerman and Joskow (2008), Evans and Bertraw (2009) and Gillenwater (2008)

Issues	Critiques
Corruption	Corruption was approached in a nuanced way because carbon trading provided new and different ways of profiting illegitimately at the expense of a deteriorating climate
Property rights	The carbon market was criticized for creating unjustified property rights by co-modifying air, a common resource. It placed a disproportionate burden of responsibility on developing countries, which were expected to host carbon trading projects
Carbon offsets	The offsetting mechanism was it actually led to an increase in emissions, allowed corporations and countries to simply transfer responsibility for their emissions reductions to the developing world
Fraud	Companies were profiting from the sale of credits where emission reductions were overestimated
Conflicts of interest	If a carbon trading company was owned by an individual sitting on the CDM Executive Board (one who designed rules for projects) would likely benefit from those rules
Revolving door	Whereby people leave positions of influence to take up others where they can make use of information gathered or connections gained to benefit themselves (when a CDM Executive Board member who quits and becomes a carbon broker)
Perverse incentive	Very often the implementation of a carbon trading project which was meant to result in the reduction of pollution, but rather encouraged polluters to emit more
Confidence trickery	E.g., when a confident carbon trading consultant who knows how to profit from the system by technically meeting the guidelines drawn up by the CDM Executive board, but knew in reality that the project in question was not viable and did not actually reduce emissions
Erosion of rule of law	When a law to control pollution is put in place but a company that previously failed to comply with it now develops a project to control pollution so that it can gain from carbon credits

2.14 Carbon Trading and Developing Countries

Most developing countries were lacking the financial resources, technological know-how, and institutional capacity to deploy such strategies at a rate appropriate with the urgency of the climate challenge. Failure to honor long-standing commitments of international support in those three areas remained the single biggest obstacle to meeting the challenge. The political, social, environmental, and ethical issues in line with the carbon trading system were illustrated in Table 3.

Carbon emissions do not appear to be relevant to it due to the lack of a framework to regulate them effectively. CDM is a specific form of carbon governance that might need re-regulation. Carbon governance might take the form of policy instruments (Mazzanti and Zoboli 2008).

3 Conclusions

Waste management technologies have the ability of reducing GHG emissions to a great extent through energy recovery. Economic growth is a constant change and there are still many least developing countries which are on the path of development. However, as the GDP increases, the ability to invest in cleaner technologies is amplified. The investigation into the role of carbon trading remains an emerging area with many avenues to explore so as to eradicate barriers and obstacles faced by developing countries. Despite the shortcomings of the carbon market, it continues to expand, and is becoming increasingly entrenched as a way of reducing GHG emissions. Carbon trading is, therefore, an excellent opportunity for developing countries in need of funds to spur sustainable development.

This study has pointed out the factors hindering the implementation of CDM projects and CER trading. The potential solutions have brought to light facilitating the task of national and international policy makers. Decoupling should be explored to dissociate waste generation from economic growth and link the chances for cleaner technologies to the rising GDP annual percentage. Countries would become more attractive to CDM projects by becoming more accessible to financial resources. Programmatic CDM in developing countries would increase their participation in the climate change actions. The voluntary cancelation of CERs would avoid the illegal and misuse of emission reductions. To date, the UNFCCC has canceled about 20% of CERs from the Waste handling and disposal sector representing true emission reductions. Policymakers would have to undertake important decisions concerning the role of Annex 1 GHG cuts towards Non-Annex 1 countries with green investment plans and concrete visions.

References

Alam, M. J. (2016). Critical evaluation of impact on economic growth of India due to the environmental degradation. *Asian Journal of Multidisciplinary Studies, 4*(4), 98–103.

Astrup, T., Moller, J., & Fruergaard, T. (2009). Incineration and co-combustion of waste accounting of GHG and global warming contributions. *Waste Management and Research, 27,* 789–799.

Barton, J. R., Issaias, I., & Stentiford, E. I. (2008). Carbon-making the right choice for waste management in developing countries. *Waste Management, 28,* 690–698.

Berglund, C., & Söderholm, P. (1997). An econometric analysis of global waste paper recovery and utilization. *Environmental and Resources Economics, 26*(2), 429–456.

Bhattacharya, M., Paramati, S. R., Ozturk, I., & Bhattacharya, S. (2016). The effect of renewable energy consumption on economic growth: Evidence from top 38 countries. *Applied Energy, 162,* 733–741.

Bogner, J., Pipatti, R., Hashimoto, S., Diaz, C., Mareckova, K., Diaz, L., et al. (2008). Mitigation of global greenhouse gas emissions from waste: Conclusions and strategies from the Intergovernmental Panel on Climate Change (IPCC) Fourth Assessment Report. Working Group III (Mitigation). *Waste Management and Research, 26*(1), 11–32.

Boldrin, A., Andersen, J. K., Christensen, T. H., & Favoino, E. (2009). Composting and compost utilization: Accounting for GHG and global warming contributions. *Waste Management, 28,* 8000–8012.

Burian, M., Arens, C., Sterk, W., & Helmreich, H. W. (2011). *Integrating Africa's least developed countries into global carbon market: Analysing CDM implementation barriers.* Wuppertal Institute for Climate, Environmental and Energy. Available from http://www.jiko-bmu.de/english/service/host_country_information/doc/1096.php. Accessed May 23, 2013.

Christensen, T. H., Gentil, E., Boldrin, A., Larsen, A. W., Weidema, D. C., & Hauschild, M. (2009). C balance, carbon dioxide emissions and global warming potentials in LCA-modelling of waste management system. *Waste Management and Research, 27,* 707–715.

Clifton, S. J. (2009). *A dangerous obsession: The evidence against carbon trading and for real solutions to avoid a climate crunch.* Research report. Available from www.foe.co.uk/resource/reports/dangerous_obsession.pdf. Accessed August 25, 2013.

Clo, S. (2008) The effectiveness of the EU emissions trading scheme. *Climate Policy, 9*(3), 227–241.

Colling, A. V., Oliveira, L. B., Reis, M. M., da Cruz, N. T., & Hunt, J. D. (2016). Brazilian recycling potential: Energy consumption and Green House Gases reduction. *Renewable and Sustainable Energy Reviews, 59,* 544–549.

Couth, R., & Trois, C. (2011a). Waste management activities and carbon emissions in Africa. *Waste Management, 31,* 131–137.

Couth, R., & Trois, C. (2011b). Waste management activities and carbon emissions in Africa. *Waste Management, 31,* 131–137.

Couth, R., & Trois, C. (2012). Sustainable waste management in Africa through CDM projects. *Waste Management, 32,* 2115–2125.

Dormady, N. C., & Englander, G. (2016). Carbon allowances and the demand for offsets: A comprehensive assessment of imperfect substitutes. *Journal of Public Policy, 36*(1), 139–167.

Eggleston, S., Buendia, L., Miwa, K., Ngara, T., & Tanabe, K. (2006). *IPCC guidelines for national greenhouse gas inventories.* Hayama: Waste Institute for Global Environmental Strategies (IGES). Available from http://www.ipccnggip.iges.or.jp/public/2006gl/pdf/0_overview/v0_0_cover.pdf. Accessed January 6, 2013.

Ellerman, A. D., & Joskow, P. (2008). *The European Union's emissions trading system in perspective.* Pew Center on Global Climate Change. Available from http://www.pewclimate.org/docUploads/EU-ETS-In-Perspective-Report.pdf. Accessed July 1, 2013.

Evans, D. A., & Burtraw, D. (2009). Tradable rights to emit air pollution. *The Australian Journal of Agricultural and Resource Economics, 53,* 59–84.

Fazeli, A., Bakhtvar, F., Jahanshaloo, L., Sidik, N. A. C., & Bayat, A. E. (2016). Malaysia's stand on municipal solid waste conversion to energy: A review. *Renewable and Sustainable Energy Reviews, 58,* 1007–1016.

Fearnside, P. M. (2015). Emissions from tropical hydropower and the IPCC. *Environmental Science Policy, 50,* 225–239.

Fenhann, J. (2010). *CDM pipeline overview.* United Nations Environment Program Risoe Centre on Energy, Climate and Sustainable Development (URC). Available from http://cdmpipeline.org/. Accessed August 9, 2013.

Friedrich, E., & Trois, C. (2016). Current and future greenhouse gas (GHG) emissions from the management of municipal solid waste in the eThekwini Municipality–South Africa. *Journal of Cleaner Production, 112*(5), 4071–4083.

Fischer-K. M., Swilling, M., Weizsäcker, E. U., Ren, Y., Moriguchi, Y., Crane, W., et al. (2013). http://uncsd.iisd.org/news/unep-releases-report-on-decoupling-resource-use-and-economic-growth/. Accessed July 23, 2013.

Galik, G. S., Murray, B. C., Mitchell, S., & Cottle, P. (2016). Alternative approaches for addressing non-permanence in carbon projects: An application to afforestation and reforestation under the Clean Development Mechanism. *Mitigation and Adaptation Strategies for Global Change, 21*(1), 101–118.

Gentil, E., Christensen, T. H., & Aoustin, E. (2009). GHG accounting and waste management. *Waste Management and Research, 27,* 696–706.

Gillenwater, M. (2008). Forgotten carbon: Indirect CO_2 in greenhouse gas mission inventories. *Environmental Science and Policy, 11*(3), 195–203.

Hackl, A., & Mauschitz, G. (2008). Role of waste management with regard to climate protection: A case study. *Waste Management and Research, 26*(1), 5–10.

Hepburn, C. (2009). International carbon finance and the Clean Development Mechanism. In D. Helm & C. Hepburn (Eds.), *The economics and politics of climate change* (pp. 409–432). Oxford: Oxford University Press.

Hoa, N.T., & Matsuoka, Y. (2017). The analysis of greenhouse gas emissions/reductions in waste sector in Vietnam. *Mitigation and Adaptation Strategies for Global Change, 22*(3), 427–446.

Hodes, G., & Kamel, S. (2007). *Equal exchange: Determining a fair price for carbon.* UNEP RISOE Centre, Capacity Development for CDM (CD4CDM) Project. Available from http://www.unep.org/pdf/dtie/FairPriceCarbon.pdf. Accessed August 25, 2013.

Hoornweg, D., & Bhada-Tata, P., (2012). What a waste: A global review of solid waste management. *Urban Development Series Knowledge Papers, 15,* 1–98.

Huang, B., Mauerhofer, V., & Geng, Y. (2016). Analysis of existing building energy saving policies in Japan and China. *Journal of Cleaner Production, 112*(2), 1510–1518.

IPCC. (2006). *IPCC guidelines for national greenhouse gas inventories.* Hayama: Institute for Global Environmental Strategies (IGES) for the IPCC. Available from www.ipcc-nggip.iges.or.jp/support/Primer_2006GLs.pdf. Accessed January 28, 2013.

Johnstone, N., & Labonne, J. (2004). Generation of household solid waste in OECD countries: An empirical analysis using macroeconomic data. *Land Economics, 80*(4), 529–538.

Kamran, A., Chaudhry, M. N., & Batool, S. A. (2015). Effects of socio-economic status and seasonal variation on municipal solid waste composition: A baseline study for future planning and development. *Environmental Sciences Europe, 27*(1), 16. Available from https://link.springer.com/article/10.1186/s12302-015-0050-9. Accessed March 10, 2016.

Kawai, K., & Tasaki, T. (2016). Revisiting estimates of municipal solid waste generation per capita and their reliability. *Journal of Material Cycles and Waste Management, 18*(1), 1–13.

Lazarus, M., Kartha, S., Ruth, M., & Bernow, S. (1999). *Evaluation ofbenchmarking as an approach for establishing Clean Development Mechanism baselines.* Available from http://sei-us.org/Publications_PDF/SEI-EvaluationBenchmarkingCleanDev-99.pdf. Accessed August 25, 2013.

Lederer, M. (2010). Evaluating carbon governance: The Clean Development Mechanism from an emerging economy perspective. *Journal of Energy Markets, 3*(2), 3–25.

Lema, A., & Lema, R. (2016). Low-carbon innovation and technology transfer in latecomer countries: Insights from solar PV in the clean development mechanism. *Technological Forecasting and Social Change, 104,* 223–236.

Lim, B., & Boileau, P. (1999). Methods for assessment of inventory data quality: Issues for an IPCC expert meeting. *Environmental Science and Policy, 2*(3), 221–227.

Lim, S. L., Hwee Lee, L., & Wu, T. Y. (2016). Sustainability of using composting and vermicomposting technologies for organic solid waste biotransformation: Recent overview, greenhouse gases emissions and economic analysis. *Journal of Cleaner Production, 111*, 262–278.

Lin, E., Jiang, K., Hu, X., Zuo, J., Li, M., & Ju, H. (2015). Climate change mitigation and adaptation: Technology and policy options. In *Climate and environmental change in China: 1951–2012* (pp. 107–127). Berlin: Springer.

Lohmann, L. (2009). Regulatory challenges for financial and carbon markets. *Carbon and Climate Law Review, 3*(2), 161–171.

Lou, X. F., & Nair, J. (2009). The impact of land filling and composting on greenhouse gas emissions—A review. *Bioresource Technology, 100*(16), 3792–3798.

Martinez, C. A., & Bowen, J. D. (2012). The Clean Development Mechanism in the solid waste management sector: Sustainable for whom? *Ecological Economics, 82*, 123–125.

Martínez de Alegría, I., Basañez, A., Díaz de Basurtoa, A., & Fernández-Sainz, A. (2016). Spain's fulfillment of its Kyoto commitments and its fundamental greenhouse gas (GHG) emission reduction drivers. *Renewable and Sustainable Energy Reviews, 59*, 858–867.

Mazzanti, M. (2008). Is waste generation de-linking from economic growth? Empirical evidence for Europe. *Applied Economics Letters, 15*, 287–291.

Mazzanti, M., & Zoboli, R. (2008). Waste generation, waste disposal and policy effectiveness: Evidence on decoupling from the European Union. *Resources, Conservation and Recycling, 52*, 1221–1234.

Mohareb, E. A., Maclean, H. L., & Kennedy, C. A. (2011). Greenhouse gas emissions from waste management—assessment of quantification methods. *Journal of the Air and Waste Management Association, 61*(5), 480–493.

Monni, S., Pipatti, R., Lehtilla, A., Savolainen, I., & Syri, S. (2006). *Global climate change scenarios for solid waste management Espoo 2006*. VTT Publications 603. Available from www.vtt.fi/inf/pdf/publications/2006/P603.pdf. Accessed June 15, 2013.

Monni, S., Syri, S., & Savolainen, I. (2004). Uncertainties in the finnish greenhouse gas emission inventory. *Environmental Science and Policy, 7*(2), 87–98.

Muñoz, I., & Schmidt, J. H. (2016). Methane oxidation, biogenic carbon, and the IPCC's emission metrics. Proposal for a consistent greenhouse-gas accounting. *The International Journal of Life Cycle Assessment, 21*(8), 1069–1075.

Myhre, G., et al. (2013). Anthropogenic and natural radiative forcing. In T. F. Stocker et al. (Eds.), *Climate change 2013: The physical science basis. Contribution of working group 1st to the 5th assessment report of the intergovernmental panel on climate change* (pp. 661–740). Cambridge: Cambridge University Press.

Newell, P. (2009). Varities of CDM governance: Some reflections. *The Journal of Environment and Development, 18*(4), 425–435.

Okubo, Y., & Michaelowa, A. (2010). Effectiveness of subsidies for the Clean Development Mechanism: Past experiences with capacity building in Africa and LDCs. *Climate and Development, 2*(1), 30–49.

Olsson, A., Grönkvist, S., Lind, M., & Yan, J. (2016). The elephant in the room—A comparative study of uncertainties in carbon offsets. *Environmental Science and Policy, 56*, 32–38.

Parker, L., & Blodgett, J. (2008). *Carbon leakage and trade: Issues and approaches*. Congressional Research Service. Available from www.crs.gov. Accessed June 15, 2013.

Pasour, E. C. (1981). The free rider as a basis for government intervention. *Journal of Libratarian Studies, 5*(4), 453–464.

Persson, U., & Münster, M. (2016). Current and future prospects for heat recovery from waste in European district heating systems: A literature and data review. *Energy, 110*, 116–128.

Pikon, K., & Gaska, K. (2010). Greenhouse gas emission mitigation relevant to changes in municipal solid waste management system. *Journal of the Air and Waste Management Association, 60*(7), 782–788.

Rong, F., Lan Y., Zeng, S., & Yu, H. (2012). Waste recovery CDM projects in China: Status, challenges and suggestions. *Climate Policy, 12*(1), 98–114.

Schneider, M., Holzer, A., & Hoffmann, V. H. (2008). Understanding the CDM's contribution to technology transfer. *Energy Policy, 36*(8), 2930–2938.

Schneider, L., Kollmuss, A., & Lazarus, M. (2015). Addressing the risk of double counting emission reductions under the UNFCCC. *Climatic Change, 131*(4), 473–486.

Sharholy, M., Ahmad, K., Mahmood, G., & Trivedi, R. C. (2008). Municipal solid waste management in Indian cities—A review. *Waste Management, 28,* 459–467.

Sjöström, M., & Östblom, G. (2010). Decoupling waste generation from economic growth—A CGE analysis of the Swedish case. *Ecological Economics, 69,* 1545–1552.

Sreekanth, K. J., Sudarsan, N., & Jayaraj, S. (2012). Clean Development Mechanism as a solution to the present world energy problems and a new world over: A review. *International Journal of Sustainable Energy, 33*(1), 1–27.

Streimikiene, D., & Girdzijauskas, S. (2009). Assessment of post-Kyoto climate change mitigation regimes impact on sustainable development. *Renewable and Sustainable Energy Reviews, 13,* 129–141.

Thompson, S., Sawer, J., Bonam, R., & Valdivia, J. E. (2009). Building a better methane generation model: Validating models with methane recovery rates from 35 Canadian landfills. *Waste Management, 29,* 2085–2091.

Uddin, G. A., Alam, K., & Gow, G. (2016). Estimating the major contributors to environmental impacts in Australia.*International Journal of Ecological Economics and Statistics, 37*(1), 1–14.

UNEP. (2011). *Decoupling natural resource use and environmental impacts from economic growth.* A report of the working group on decoupling to the international resource panel.

UNFCCC PDD. (2010). *Omdurman landfill municipal solid waste composting project.* Available from http://cdm.unfccc.int/filestorage/LQK9O8I2XS6HB4V037TPYCGEMN5DZW/Project_Design_Document.pdf?t=aXZ8MTI5Njk3NTI2Ni4yNw==|zAptLt540aM-lvnQb0Ka0Vmzy3c=. Accessed August 9, 2013.

Viguier, L. L. (2004). A proposal to increase developing country participation in international climate policy. *Environmental Science and Policy, 7*(3), 195–204.

Villoria-Sáez, P., Tam, V. W., del Río Merino, M., Arrebola, C. V., & Wang, X. (2016). Effectiveness of greenhouse-gas emission trading schemes implementation: A review on legislations. *Journal of Cleaner Production, 127,* 49–58.

Wackernagel, M., Schulz, M. B., & Deumlin, D. (2002). Tracking the ecological overshoot of the human economy. *Proceedings of the National Academy of Science, 99,* 9266–9271.

Wang, Y., Geng, S., Zhao, P., Du, H., He, Y., & Crittende, J. (2016). Cost–benefit analysis of GHG emission reduction in waste to energy projects of China under clean development mechanism. *Resources, Conservation and Recycling, 109,* 90–95.

Wiloso, E. I., Heijungs, R., Huppes, G., & Fang, K. (2016). Effect of biogenic carbon inventory on the life cycle assessment of bioenergy: Challenges to the neutrality assumption. *Journal of Cleaner Production, 125,* 78–85.

Woon, K. S., Lo, I. M. C., Chiu, S. L. H., & Yan, D. Y. (2016). Environmental assessment of food waste valorization in producing biogas for various types of energy use based on LCA approach. *Waste Management, 50,* 290–299.

Yang, J., Fujiwara, T., Matsuoka, Y., & Wang, W. (2016). A systematic approach to projecting industrial solid waste generation by industrial sector in Shanghai. *Journal of Material Cycles and Waste Management, 18*(1), 81–92.

Yu, Y., & Zhang, W. (2016). Greenhouse gas emissions from solid waste in Beijing: The rising trend and the mitigation effects by management improvements. *Waste Management & Research, 34*(4), 368–377.

Assessing the Implementation of Eco-Driving in Mauritius—A Climate Change Mitigation Measure

Joanna Doorghen-Gorden, Manta Devi Nowbuth and Virendra Proag

Abstract Mauritius depends on imported fuel for close to 82% of its energy needs. The carbon dioxide (CO_2) emissions associated with the burning of fossil fuels are on the rise due to growing electricity and transportation demand. In 2010, transportation accounted to 24.9% of total CO_2 emissions and the number of vehicles registered in Mauritius has increased from 265,841 in 2002 to 384,115 in 2010 (Statistics Mauritius in Digest of environment statistics 2011, Mauritius, 2012). This is a highly unsustainable trend for both energy security and climate-related reasons. Therefore, an analysis of the approach to encourage and implement eco-driving, an important climate change mitigation measure, is a necessity for the island. This study aimed at developing an in-depth understanding of eco-driving, as a driver education initiative and as a climate change mitigating measure. A questionnaire was designed with the objectives of highlighting the key parameters that should be addressed in order to promote the concept of eco-driving in Mauritius. The findings of the survey first gave an insight of the willingness of drivers to adopt an eco-driving behaviour and identified the barriers that should be overcome so as to successfully implement eco-driving in Mauritius. In addition, a cost–benefit analysis was carried out to determine the extent to which implementing eco-driving will be cost-effective. An action plan of measures needed for implementing eco-driving in a timely manner was then proposed. The survey revealed that 68.1% respondents are willing to adopt eco-driving practices into their driving habits and only 45% respondents are willing to purchase and use an eco-driving device in their vehicles. It was observed that most drivers have little knowledge of

J. Doorghen-Gorden (✉)
United Nations Development Programme/Ministry of Environment, Port Louis, Republic of Mauritius
e-mail: d.joanna@hotmail.com

M.D. Nowbuth
Department of Ocean Engineering and ICT, Faculty of Ocean Studies, University of Mauritius, Moka, Republic of Mauritius

V. Proag
Department of Civil Engineering, Faculty of Engineering, University of Mauritius, Moka, Republic of Mauritius

W. Leal Filho and D. Surroop (eds.), *The Nexus: Energy, Environment and Climate Change*, Green Energy and Technology, https://doi.org/10.1007/978-3-319-63612-2_23

eco-driving and they do not practice eco-driving frequently. Hypotheses were tested and it was shown that drivers are willing to adopt eco-driving behaviours but they are not willing to purchase an eco-driving device; at a confidence interval of 95%. Possible barriers to the implementation of eco-driving were highlighted from the survey. From the barrier analysis, a cost–benefit analysis was carried out to assess the cost-effectiveness of specific measures. The study identified measures, like conducting awareness campaigns on eco-driving, providing eco-driving training to different groups of drivers, giving subsidy to drivers for purchasing eco-driving feedback devices, investing in an eco-driving reference project and designing road layouts so as to encourage eco-driving. A benefit/cost ratio of 1.06 was obtained which means that it is worthwhile to go ahead with the project of implementing eco-driving in Mauritius. Besides, additional measures were recommended over a long-term basis, such as encouraging Public–Private Partnership to secure financial support and optimising traffic lights to enable eco-driving. The outcomes of this research work provide reassurance to the government that investing money for the implementation of eco-driving in Mauritius can reduce the country's dependence on imported fuel and consequently, lowering the CO_2 emissions from transport sector. In addition, eco-driving is associated with social benefits by enabling a safer and pleasant driving; ensuring community welfare.

Keywords Energy · Environment · Climate change · Energy efficiency Transport · Eco-driving

1 Introduction

Mauritius has experienced a drastic increase in the number of motor vehicles because of rapid industrialisation. This is thus leading to consequential impacts on the environment and on the country's energy needs. It has been found that the transport sector is the heaviest energy consumer, accounting for 48% of total energy imports. In 2008, some 407 ktoe of energy were used for transport; diesel oil accounted for 37.8%, aviation fuel 33.7%, gasoline 27.0% and liquefied petroleum gas 1.5%. From 2007 to 2008, the consumption of diesel oil rose by 0.6% and gasoline by 2.8% while that of aviation fuel fell by 4.9% (Ministry of Environment and Sustainable Development 2011).

There are important opportunities to increase the operating efficiencies of transport vehicles. Road vehicle efficiency might be improved by 5–20% through strategies such as eco-driving styles, increased load factors, improved maintenance, in-vehicle technological aids, more efficient replacement tyres, reduced idling and better traffic management and route choice (Kahn Ribeiro et al. 2007).

In France, La Poste Group has implemented ongoing training on eco-driving and early test results have shown a reduction of 8% in carbon dioxide (CO_2) emissions (La Poste 2011). Japan achieved its 2010 goal of reducing CO_2 emissions by 31 million tonnes below 2001 levels by encouraging drivers to use their vehicles more efficiently through eco-driving (Rakatonirainy et al. 2011). According to Barkenbus (2010), eco-driving can reduce fuel consumption by 10%, on average and over time, thereby reducing CO_2 emissions from driving by an equivalent percentage (Barkenbus 2010).

In this research work, we carried out a survey study amongst Light Duty Vehicles (LDV) users of Mauritius to determine their awareness on CO_2 emission from vehicles and the impact on the climate. Their level of understanding with respect to eco-driving and their willingness to adopt such a driving behaviour were evaluated; and the potential barriers to the adoption of an eco-driving behaviour were investigated as well. Based on the barriers and measures identified in the survey results, a cost–benefit analysis of the measures for adopting eco-driving in Mauritius was conducted.

2 Literature Review

The transport sector relies almost exclusively on oil products (gasoline and diesel) for its energy supply. Since the first oil crisis of 1973, this dependence is considered to be problematic; this is the 'energy problem' of transport. High retail prices for fossil fuels and/or fuel economy regulations have been justified, and used, as instruments to reduce energy dependence. Since the 1990s, awareness has risen of greenhouse gas emissions associated with the use of fossil fuels, and the transport sector's contributions to climate change are now perceived as a major problem, indeed sometimes the major policy challenge for transport (Proost and Dender 2012).

The Intergovernmental Panel on Climate Change (IPCC) warns that a continued increase of GHGs in the atmosphere is one of the leading causes of rising global temperatures today, adding that unless measures are put in place to reduce emissions, the earth may experience a further rise in average temperatures of up to 6 °C in the coming decades, and may have adverse effect on water supplies, food production and human health.

Many countries have promoted eco-driving as a key element of national strategies to reduce CO_2 emissions. European Union regulations already require eco-driving to be taught to novice drivers. Japan achieved its 2010 goal of reducing CO_2 emissions by 31 million tonnes below 2001 levels by encouraging drivers to use their vehicles more efficiently through eco-driving. The claimed advantages of the eco-driving approach are that it can apply to vehicles of any age or size, it can take effect across the entire fleet of vehicles immediately at low cost, and that it can result in immediate savings to individuals from greater fuel efficiency, better safety and perhaps lower insurance rates (Rakatonirainy et al. 2011).

Eco-driving attempts to change drivers' behaviour, so as to reduce fuel consumption, through advice such as driving more smoothly by anticipating changes in the traffic, shifting gear sooner, eliminating excessive idling, operating the vehicle within an optimum range of engine revolutions, avoiding jerky braking/acceleration, reducing frequency of stops (faster acceleration is acceptable if it means that a stop is avoided) and avoiding traffic congestion.

Avoiding rapid starts and stops; maintaining a steady speed when travelling on highways; keeping rolling in traffic; and using the highest gear possible are key safe driving behaviours, which also form the basis of eco-driving (Rakatonirainy et al. 2011).

Eco-driving brings benefits beyond reducing emissions—it also creates more fluid, more aware and safer drivers. By thinking ahead and avoiding sharp acceleration and deceleration, eco-drivers stop far less often (13% less often, on average), driving more fluidly overall (FIAT 2010).

3 The Research Study

This section describes the methodology adopted to carry out the research.

3.1 Survey Questionnaire

With the use of a well-structured questionnaire and specific communication methods, like face-to-face interview and online questionnaire, data was collected from a sample of the vehicle users' population. To achieve this, on-site visits were effected at the two vehicle fitness centres of Mauritius for interviewing the vehicle users, and online questionnaire was sent to friends, family and colleagues.

The questions aimed at collecting primary data from drivers so as to extract pertinent information relevant to the objectives and aim of the study. The questionnaire was divided into five sections, to gather the following information:

- Technical data:
  Technical features of the drivers' cars.
- Behavioural data:
  Reasons that causes people to drive, driving styles, concern about fuel consumption monitoring, frequency of adopting alternative transport modes (public, bicycle, footing) and whether drivers practice eco-driving or not. Factors considered by drivers when buying a car and their willingness to buy an eco-driving device. The drivers' choice of eco-driving devices/interfaces was also determined.
- Awareness on eco-driving:
  Awareness and perception with regards to eco-driving. Perceptions concerning barriers to eco-driving implementation and measures that can be adopted to overcome those barriers.

- Awareness on climate change:
  Level of awareness of drivers with respect to climate change and CO_2 emissions.
- Personal data:
  Age, level of education and gender of respondents.

4 Pre-testing of Questionnaire

Once the questions have been prepared, a pre-testing of the questionnaire was done (face-to-face interview and via email) with five vehicle users to identify any weaknesses in the questionnaire, to provide information about field procedures and to help organising the full scale survey. Through the pre-testing stage, it was observed that some questions were not properly structured causing confusion to the respondents. The question on the choice of eco-driving devices required a visual explanation. The questions were then refined and the questionnaire was amended accordingly.

5 Target Population and Sample Size

- Determining target population

For the purpose of this study, the targeted population was taken as the number of registered Light Duty Vehicles (LDVs) in Mauritius. According to the United Nations Environment Programme Partnership for Clean Fuels and Vehicles (UNEP-PCFV), this is the group of vehicles with a gross weight of less than 3.5 tonne and includes the following:

- Passenger cars;
- Small buses with a gross weight of more than 3.5 tonne, usually carrying up to 14 persons (including driver); and
- Lorries, vans and pickups designed to carry goods with a gross vehicle weight of less than 3.5 tonne.

LDVs do not include:

- Motor cycles and mopeds (2 or 3 wheels);
- Large buses with a gross weight of more than 3.5 tonne, usually carrying more than 14 persons; and
- Heavy trucks (gross weight over 3.5 tonne).

For Mauritius, the number of registered LDVs (population size) for the year 2011 has been obtained from Statistics Mauritius and is tabulated as shown in Table 1.

Table 1 Light Duty Vehicles registered in Mauritius (Statistics Mauritius 2012)

Type of vehicle	Description	Number
Car (of which taxi car)	Passenger cars	136,225 (6907)
Dual purpose vehicle	Includes pickups designed to carry goods with a gross vehicle weight of less than 3.5 tonne	49,132
Heavy motor car	Small buses, with a gross weight of more than 3.5 tonne, usually carrying up to 14 persons	1230
Van	Designed to carry goods with a gross vehicle weight of less than 3.5 tonne	26,090
Total		212,677

Therefore, the population size amounts to 212,677.

- Determining sample size

Given that the population size is unusually large, a sample is mostly desired. Hence, the sample size for the survey was calculated by assuming a population proportion (p) of 50% (worst scenario) where $q = 1 - p$.

The sample size was determined using a confidence interval of 95% and a margin of error of 5%, and the following statistical formula was used to obtain the sample size, n.

$$n = \frac{pqZ^2}{B^2},$$

where p is the proportion in the population, q is $(1 - p)$, Z is the Z score for the confidence interval and B is the margin of error.

In our case, $pq = 0.25$, $Z = 1.96$ at a 95% degree of confidence and $B = 0.05$ (5%). Hence, n is equal to 384. Therefore, a sample size of 384 drivers had to be surveyed for this study.

Drivers were chosen at random for the survey; whereby every unit in the population has an equal and non-zero chance of being selected to ensure that adequate number of responses is obtained since it is quite difficult to get 100% participation in a survey. The survey was administered to the sample population over a period of three months (April till June 2013).

During the month of April, face-to-face interviews were carried out once a week with drivers who came for vehicle examination at Plaine Lauzun vehicle fitness centre. Similarly, in May interviews were done once a week with drivers coming at Curepipe vehicle fitness centre. Drivers were selected at random, after they completed their vehicle examination, depending on their willingness and time available to respond to the questionnaire. In June, the online questionnaire was sent to those family members, colleagues and friends who have a driving licence. Drivers were chosen irrespective of their age, gender, age of their vehicles and driving experience.

6 Hypothesis Testing

Statistical methods were applied to assess any association between variables of the survey. Specific questions have been formulated when conducting the survey, to test hypotheses set for this study, namely:

From the survey results, 68.1% respondents were willing to adopt eco-driving practices into their driving habits, and 45% respondents were willing to purchase and use an eco-driving device in their vehicles.

The following two hypotheses were then tested:

Hypothesis Test I: Willingness of drivers to adopt eco-driving behaviour

H_o: Drivers are not willing to change their driving habits
H_a: Drivers are willing to adopt eco-driving habits

Based on the Hypothesis Test results, the null hypothesis is rejected. Therefore, there is sufficient evidence that, at a confidence interval of 95%, drivers are willing to adopt eco-driving behaviours.

Hypothesis Test II: Willingness of drivers to purchase and use an eco-driving device

H_o: Drivers are not willing to purchase an eco-driving device for use in their cars
H_a: Drivers are willing to purchase an eco-driving device for use in their cars

Based on the Hypothesis Test results, we fail to reject the null hypothesis. Therefore, at a confidence interval of 95%, there is insufficient evidence to conclude that drivers are willing to purchase an eco-driving device to be used in their vehicles.

7 Barrier Analysis

The questionnaire has allowed the classification of potential barriers to eco-driving implementation, in order of significance, as expressed by the survey respondents.

Based on the results obtained, an in-depth barrier analysis was conducted using the tool Logical Problem Analysis. This technique aims at arranging observed problems (barriers) into a hierarchy of causes and effects as a basis for preparing a concrete and realistic action plan. Each barrier is linked to causes and effects, with direct causes below and direct effects above, so that multi-level cause-and-effect paths are created to form the problem tree.

Due to the complexity of the barrier analysis for eco-driving, a separate problem tree has been worked out for each barrier, whereby the effects for not adopting eco-driving are similar in all the problem trees. Outcomes of the barrier analysis are summarised in Table 2.

Table 2 Outcomes of barrier analysis

Economic and financial	Market	Policy, legal and regulatory	Social, cultural and behavioural	Information and awareness	Training and skills	Physical infrastructure
1. Financially not viable • High resource cost (labour to give training on eco-driving, investing in eco-driving devices) • High modification and implementation *costs* (redesign roads) **2. High transaction cost** • Technology acquisition and implementation (investing in satellite navigation system and intelligent traffic lights)	**1. Restricted access to technology** • Eco-driving technology is limited on the Mauritian market **2. Lack of reference projects in country** **3. Inadequate sources of increasing returns** • Low ability or willingness to pay by drivers • Market size for eco-driving technology is small	**1. Insufficient legal and regulatory framework** • Absence of laws on climate change, town planning to facilitate eco-driving • Absence of plans and programmes to implement eco-driving behaviour **2. Inefficient enforcement** • Missing executive and regulatory body to enforce eco-driving legislation • Insufficient willingness or ability to enforce laws and regulations	**1. Traditions and habits** • Resistance to change (not willing to change driving behaviours) **2. Lack of confidence in new technology (eco-driving device)** • Unknown product (due to lack of information on the device) **3. Routine driver behaviour** • Inefficient trip planning and routine choice by the driver	**1. Inadequate information** • Poor dissemination of information to users (on eco-driving behaviour and devices) • Lack of knowledge (on implementation requirements, on how to drive efficiently) **2. Feedback mechanism lacking or inadequate** • Benefits of eco-driving not visible	**1. Lack of trainers** • Lack of skilled personnel to provide training on eco-driving **2. Inadequate training facilities** • Educational system or driving school curriculum do not include eco-driving **3. Lack of finance to support training needs**	**1. Roads layout** • Roads are not designed in an eco-friendly manner to facilitate eco-driving **2. Traffic lights** • No intelligent traffic management, e.g. intelligent traffic lights

8 Measures to Overcome Barriers

Similarly, the Logical Problem Analysis tool was used to identify measures to be adopted to overcome the identified barriers. In this case, the barriers are reformulated into solutions (objectives) to the problem and the cause–effect relations of the problem tree are converted into measure-result relations, producing what is called an objective tree. The latter is a logically organised presentation of objectives.

In principle, by implementing measures (at the root of the tree), all the objectives above in the tree should automatically be achieved. A summary of the measures is provided in Table 3.

9 Cost–Benefit Analysis

To assess the economic feasibility of implementing eco-driving in Mauritius, a cost–benefit analysis of the measures being proposed for adopting eco-driving in Mauritius was conducted based on the barriers and measures identified.

A benefit-to-cost (*B*/*C*) ratio was then calculated as such:

$$B/C \text{ ratio} = \frac{\text{Present worth of total benefits}}{\text{Present worth of total costs}}$$

If the *B*/*C* ratio is greater than 1, the project is worthwhile and vice versa. A *B*/*C* ratio of exactly 1 means that the project will break even at some point in time (Tang 2003).

Three main benefits have been identified for the project:

A. Lower CO_2 emissions in the transport sector;
B. Running cost of vehicle is lowered; and
C. Reduced expenditure on imported fuel.

Barkenbus (2010) stated that eco-driving can reduce fuel consumption by 10%, on average and over time, thereby reducing CO_2 emissions from driving by an equivalent percentage. Therefore, for the purpose of this study, a 10% reduction in terms of CO_2 emissions and fuel consumption is assumed for calculating the cost savings if eco-driving is implemented in Mauritius.

By taking the year 2010 as baseline, the following information has been obtained from Statistics Mauritius (2012):

- CO_2 emissions contributed by the transport sector = 912 kton
- Fuel consumption by land transport (LPG, gasoline, diesel oil) = 279,378 tonne
- Expenditure on imported diesel = Rs. 6,945,099,000.

Table 3 Measures for overcoming barriers to eco-driving

Barriers	Information and awareness	Training/skills	Social, cultural and behavioural	Policy, legal and regulatory	Enabling Market	Economic and financial	Physical infrastructure
Measures	1. Conduct mass media sensitisation campaign on benefits of eco-driving, and provide tips to drivers 2. Include principles of eco-driving in school curricula, driving school and training centres 3. Include eco-driving concepts in traffic codes and in driving licensing system	1. Train driving instructors on eco-driving principles 2. Provide eco-driving training to apprentice drivers 3. Organise training sessions on eco-driving for existing drivers	1. Provide incentive to encourage eco-driving (people will be less resistant to change) e.g. subsidy on eco-driving devices 2. Communicate the benefits and saving obtained from eco-driving to drivers	1. Eco-driving to be included in town planning legislation 2. Include non eco-driving behaviour as an offence in road traffic laws 3. Include eco-driving in government programmes and action plans	1. Give loan facilities to wholesaler/re-tailers of eco-driving device 2. Invest in an eco-driving project (that would become a reference project for Mauritius) 3. Government to provide subsidies on eco-driving devices	1. Make eco-driving a government priority, to secure funds for eco-driving 2. Ask support from other countries (expertise, training and financial) 3. Encourage Public–Private Partnership for implementing eco-driving	1. To design road layouts/traffic lights in such a way so as to encourage eco-driving

We assumed that the expenditure on imported diesel would be reduced by 10% on a yearly basis if eco-driving is implemented in Mauritius. This represents a saving of Rs. 694,509,900 per annum.

The following measures are considered for the cost–benefit analysis:

i. Conduct sensitisation campaigns to raise awareness on eco-driving and to communicate its benefits to the Mauritian population at large;
ii. Provide training to driving instructors, apprentice and existing drivers on how to practice eco-driving and how to monitor their vehicles' fuel consumption;
iii. Provide subsidy to those drivers who purchase eco-driving feedback devices;
iv. Invest in an eco-driving project, which would become a reference project for Mauritius; and
v. Design road layouts in such a way so as to encourage eco-driving.

Cost estimates for the above measures were obtained from different sources of information namely; Mauritius Broadcasting Corporation, Traffic Road Management Safety Unit and Ministry of Public Infrastructure, National Development Unit, Land Transport and Shipping, and assumptions were made whenever necessary.

A *B*/*C* ratio of 1.08 was obtained by calculating the Net Present Values (NPV) for the discounted net cash flows over a project life of 10 years, using appropriate assumptions.

10 Results of Analysis

Findings from the quantitative analysis of the survey and from the economic analysis are provided hereunder.

10.1 Behaviours of Drivers in Mauritius

It is clear that in Mauritius, drivers tend to monitor their fuel consumption with a view to minimise the running cost of their cars. Even if they are not really aware of the impacts that CO_2 emission has on climate, they are conscious that fuel savings will ensure costs savings for their own benefits.

Since the majority of drivers have a careful driving style, it would not be difficult to sensitise them on adopting safe and eco-driving driving habits. Furthermore, 75% of the sample drivers mentioned that they do use alternative modes of transport like bicycle and public transport. This demonstrates that drivers in Mauritius do not depend solely on cars for mobility purposes.

10.2 Awareness Level Towards Climate Change and Eco-Driving

Most of the drivers stated that they are aware of the reasons why the climate is changing. This is so because all the respondents are educated, represented as 76.8% being degree-holders. It reveals that there is sufficient awareness on climate change in Mauritius, amongst graduates. However, awareness campaigns are needed to reach the whole population.

A large portion of the sample (69.6%) has little awareness on eco-driving. Therefore, eco-driving is not a totally new concept to drivers. However, knowledge dissemination would be required to raise awareness on eco-driving since 18.8% of sample does not know about eco-driving at all.

10.3 Perception Towards Eco-Driving and Eco-Friendly Vehicles

Majority of respondents are agreeable that eco-driving saves fuel and money and above 40% respondents are aware that eco-driving protects the environment. Though 68.1% respondents do practice eco-driving, there are 31.9% that do not. Most probably this 31.9% of drivers are either not willing to practice eco-driving or they do not know how to practice eco-driving.

Also, there is sufficient evidence, as per Hypothesis Test I, that drivers are willing to adopt eco-driving behaviour. It is to be noted that even if drivers are aware of the benefits of eco-driving, they are not practising eco-driving frequently. This could probably be due to the fact that they are not aware of how to drive in an eco-friendly manner.

In general, drivers consider the cost aspect when buying and driving a car. They agree to choose an eco-friendly car so as to save money on running cost. However, they will opt for a fuel efficient vehicle only if it is not more expensive to buy than any other car. Moreover, the study reveals that people will only change their energy-consuming habits if they are forced to do so.

10.4 Use of Eco-Driving Devices

Drivers prefer the Eco-Matic type device since the latter automatically reduces fuel consumption by adjusting various parameters within the car. Other devices like Eco-Advisor and Eco-Display are not favoured because they require the attention of drivers, requesting them to perform specific tasks while driving, which could be inconvenient and not user-friendly to the drivers.

The survey showed that only 45% respondents are willing to purchase and use an eco-driving device in their vehicles. Also, Hypothesis Test II has proved that drivers are not willing to purchase an eco-driving device to be used in their vehicles. This could be due to the fact that the drivers are not aware of the benefits of the devices, which therefore represent an additional cost for them.

10.5 Barriers to Eco-Driving

Drivers have pointed out that the primary barriers to eco-driving implementation in Mauritius are the lack of training on how to adopt an eco-driving behaviour, the lack of knowledge on the concept of eco-driving and the roads in Mauritius are not properly designed to facilitate eco-driving. Indeed, 90% of the sample survey had not attended any such training, though majority of respondents were graduates.

Other barriers such as lack of legal provision to regulate eco-driving, lack of eco-driving devices on the local market and inefficient trip planning have been given less priority by the sample drivers.

10.6 Measures Needed in Mauritius to Implement Eco-Driving

Similarly, measures considered as a priority by the interviewed drivers were:

- To include eco-driving in the government's transport and environmental policies so as to secure funding for eco-driving projects;
- To provide continuous training to drivers on how to adopt eco-driving behaviour; and
- To raise awareness on eco-driving through a mass media sensitisation campaign.

11 Benefit/Cost Ratio

The economic analysis has revealed a benefit/cost (*B*/*C*) ratio of 1.08 for the project. This implies that the government can go ahead with implementing eco-driving in Mauritius since a *B*/*C* ratio greater than 1 means that the project is worthwhile.

For the purpose of this study, five specific measures have been considered for the economic analysis. Nevertheless there are additional measures, as detailed in Table 3, which could be implemented.

12 Limitations of the Survey

a. Low response rate:
   A total of 400 respondents were targeted to answer the questionnaire via online questionnaire, via email and face-to-face interview. However, only 70 questionnaires were filled-in resulting in a response rate of only 17.5%.
b. Unavailability of respondents:
   The main problem encountered at the fitness centres was that most respondents were not available and willing to fill-in the questionnaire. Drivers were in a hurry to go back to work or home after conducting their fitness tests.
c. Resource constraints:
   Financial resources were needed relating to travel expenses to access the two fitness centres which are found in two different regions.
d. Time constraints:
   Due to time constraints, the survey was conducted over 3 months only, thus limiting the number of visits at fitness centres for interviewing drivers. This has contributed to the generation of a small sample size and consequently a higher margin of error of 12%.

13 Conclusions

We have observed that most drivers are aware of the reasons why the climate is changing, bearing in mind that all the respondents are educated, 76.8% being degree-holders. However, awareness on eco-driving is insufficient in Mauritius and drivers are not practising eco-driving frequently.

Therefore, an action plan of measures is being proposed, for implementing eco-driving in a timely manner, as follows:

Short-term measures:

- Including eco-driving in government programmes, action plans and legislation;
- Raising awareness on benefits and principles of eco-driving;
- Provision of training on eco-driving to different groups of drivers;

Long-term measures:

- Providing incentives/facilities to drivers and wholesalers of eco-driving devices;
- Encouraging Public–Private Partnership to secure financial support; and
- Designing road layouts and optimising traffic lights to enable eco-driving.

As an overall conclusion, this study is able to reassure government decision-makers that new spending for eco-driving will result in a net increase in national welfare and meet government's public policy and fiscal objectives. Besides, eco-driving leads to social benefits which are intangible, such as safer and pleasant driving; meaning less stress while driving and higher comfort for drivers.

Acknowledgements The authors gratefully acknowledge the contribution of the Ministry of Environment, Statistics Mauritius, National Transport Authority and the University of Mauritius in this project.

References

Barkenbus, J. N. (2010). Energy policy. In *Eco-driving: An overlooked climate change initiative* (pp. 762–769).

FIAT. (2010). *Eco-driving uncovered. The benefits and challenges of eco-driving based on the first study using real journey data.*

Kahn Ribeiro, S., Kobayashi, S., Beuthe, M., Gasca, J., Greene, D., Lee, D. S. et al. (2007). *Transport and its infrastructure.* UK.

La Poste. (2011). *La Poste hands workers keys to 10,000 electric vehicles.* France.

Ministry of Environment and Sustainable Development. (2011). *Maurice Ile Durable—Working Group I on Energy.* Mauritius.

Proost, S., & Dender, K. V. (2012). Energy and environment challenges in the transport sector. In *Economics of transportation* (pp. 1–11).

Rakatonirainy, A., Haworth, N., Saint-Pierre, G., & Delhomme, P. (2011). Research issues in Eco-driving. In *22nd International Technical Conference on the Enhanced Safety of Vehicles*, Washington.

Statistics Mauritius. (2012). *Digest of Environment Statistics 2011.* Mauritius.

Tang, S. L. (2003). *Economic feasibility of projects: Managerial and engineering practice.* Hong Kong.

Author Biography

Mrs. Joanna Doorghen-Gorden has a bachelor degree in Chemical and Environmental Engineering and a master degree in Project Management from the University of Mauritius. In addition, she holds a National Diploma in Occupational Safety and Health from the National Examination Board in Occupational Safety and Health (NEBOSH). She works as Environment Officer at the Ministry of Environment where she monitors industrial waste procedures and liaises with the community to ensure compliance to environmental laws, through sensitization and enforcement mechanisms. As a member of the Ozone Officers Network for English-speaking Africa, she is committed to assist Mauritius in complying with the Montreal Protocol (environmental agreement for phasing out Ozone Depleting Substances). She has working experience in project management at the United Nations Development Programme, where she also works as Project Manager on environmental projects. She is also a member of the Young African Leaders Initiative (YALI) Network, a member of the Lions Club of Ebene and she is a Mandela Washington Fellow 2016.

The Nexus Energy, Environment and Climate Change: Future Trends

Walter Leal Filho

Abstract This chapter offers a short overview of some of the matters concerning the nexus environment, environment and climate change, and outlines some of the areas where future action is needed.

Keywords Climate change · Nexus · Water · Energy

1 Introduction

The individual challenges seen in the areas of energy, environment and climate change, including water access and use, are significant. Indeed, the goal of ensuring water, energy and food security is a daunting task due to the complexities inherence to each area. As far as **water** is concerned, estimates by UN Water indicate that:

- almost 2.5 billion do not have access to adequate sanitation
- 783 million people do not have access to clean water
- 6–8 million people die annually from the consequences of disasters and water-related diseases (UN Water 2013).

Water is known to be essential to life on Earth and its wide use in areas as diverse as industry, personal consumption, sanitation, agriculture, and as energy source, means that water shortages are bound to affect societies and economies in a significant way. The problems related to water use are not only related to its scarcity or wastage in many parts of the world. UN Water estimates that up to 90% of the wastewater in developing countries flows untreated into rivers, lakes and highly productive coastal zones, threatening health, food security and access to safe drinking and bathing water (UN Water 2013).

W. Leal Filho (✉)
Faculty of Life Sciences, Hamburg University of Applied Sciences, Ulmeliet 20, 21033 Hamburg, Germany
e-mail: walter.leal@haw-hamburg.de

W. Leal Filho and D. Surroop (eds.), *The Nexus: Energy, Environment and Climate Change*, Green Energy and Technology, https://doi.org/10.1007/978-3-319-63612-2_24

The OECD Environmental Outlook to 2050 projects that global demand for energy and water will increase by 80 and 55% by 2050, respectively (OECD 2012). Current projections for the population increase in developing countries indicate that higher food demands over the same period will have to be expected and planned for. It is no secret to anyone that elements such as population growth in the developing world, poor economic development prospects and the pressures posed by climate change will exacerbate the current competition for food, water and energy.

Farming communities across the world have realised the graveness of the problem (Knowler and Bradshaw 2007), and the literature documents examples of adaptation in Sub-Saharan Africa (e.g. Juana et al. 2013). In addition studies such as the one performed by Nhemachena (2009) on agriculture and future climate dynamics in Africa have tried to understand the problem which is related to climate change (Paavola 2008; Pant 2009) and negatively influences agriculture, hence threatening the livelihoods of millions of people throughout the developing world, often with severe health implications, which have been considered by the World Health Organisation (WHO 2004).

In terms of **energy**, the World Energy Outlook 2016 has shown that 1.2 billion people are without access to electricity and more than 2.7 billion people rely on the traditional use of biomass for cooking, which is associated with the approximately 3.5 million deaths annually from indoor air pollution (IEA 2016).

As far as energy is concerned, a growing emphasis is being given across the world to two main areas:

1. the promotion of energy efficiency
2. the development of renewable energy.

In respect to the latter, many European Union countries are playing a key role and the fact that the European Union has set its 20–20–20 goals, i.e. 20% in reductions of emission, a 20% increase in the use of renewable energy and a 20% increase in energy efficiency by the year 2020, means the Nexus energy–environment–climate change in Europe is seriously taken.

This however is not as wide a trend, as it should be. In the United States and across the developing world, the emphasis is still on the use of fossil fuels. Along with coal and oil, shale gas is also playing a prominent role as an energy source, despite the many risks associated with it. The European Union is still relying much on natural gas for around 25% of its electricity production, and imports a substantial proportion of it, especially from Norway and Russia.

Yet, both in Europe and in the developing world, a great dependency on conventional fossil fuels and the use of hydrocarbons as energy sources are known to pose a risk to the environment, since the emissions deriving from them are acknowledged as being a potential threat to human and environmental health. Moreover, the use of fossil fuels is known to be associated with the release of greenhouse gases, which exacerbate climate change (IPCC 2014).

Bearing in mind that the use of fossil fuels as energy sources may lead to damages to the physical environment (especially in respect of extraction, transport as well their consumption in vehicles or factories) and contribute to global warming, it is fair to say that without a proper emphasis to the nexus of these three areas, little progress can be achieved in addressing the problems each one is associated with.

In rural areas in Africa, Latin America or Asia for instance, harsh environmental conditions and the sheer lack of options virtually force rural communities to rely on biomass as their primary energy sources. The fact that in many rural areas in Sub-Saharan Africa biomass is no longer so easy to access, means that their use worsens already difficult living conditions. The nexus approach can help to foster an understanding of the interplays and support synergies across sectors.

2 Action Needed

In order to address the current and emerging threats to energy and food security, there is a need for concerted efforts and more strategic planning, at two levels:

(a) Level 1 National: national governments should be encouraged to engage in concrete action aimed at improving the efficiency with which water and energy resources are used, and to diversify their energy sources. One first step towards achieving this is by ensuring a greater coherence between national, regional and local planning and priorities. Also, even though its involvement so far has not been so high, it is very appropriate to identify ways to catalyse private sector investment in the energy and food security nexus, especially in the highly profitable field of energy. There are no reasons why African countries should not be leading the world in respect of renewable energy generation and use, hence alleviating the pressure on the use of biomass, and opening up the new opportunities for the social, economic and even educational benefits (e.g. open schools and community centres in the evenings) energy access can lead to.

(b) Level 2 International: the many multinational humanitarian operations to handle food, energy and water scarcity should not be limited to food or water donations. Rather, they should encourage more join efforts and information sharing between the various actors, and publicise innovative and successful efforts to improve climate resilience, and foster water and energy efficiency.

Unfortunately, the past and current problems on the energy, food and water nexus are indicative of the lack of priority many countries have decided to afford to water and energy security. This needs to be changed. Across the developing world, we need a more active participation from government agencies on the one hand, but also local companies, NGOs and academics. They all need to work harder in designing and testing new water and energy efficiency technologies, or otherwise find ways to optimise their use. More information dissemination, more networking, further communication and exchanges about successful local-level experiments and initiatives are also needed, so that those who are innovative can share and learn from one another more effectively.

One example which can be provided, is related to **rainwater harvesting**. This is a term used to describe systems via which rainwater can be collected and stored, for future use. Despite the fact that it is rather simple and requires little investments, rainwater harvesting has proven to be an efficient, low cost and time-tested technology for improving water security. The project "ACP-EU Technology-Transfer Network on Rainwater Harvesting Irrigation Management for Sustainable Dryland Agriculture, Food Security and Poverty Alleviation in sub-Saharan Africa" (AFRHINET), undertaken from 2014 to 2017 and funded by the European Union ACP Programme, was a project aimed at the implementation of integrated theoretical and practical capacity-building, and the development of technology-transfer and demonstration projects in the field of rainwater harvesting irrigation (RWHI). The rationale behind the project is that the knowledge and use of RWHI management for small-scale irrigation needs to be enhanced in rural dryland areas of sub-Saharan Africa, so as to encourage its further use. Figure 1 shows the project's web site (http://afrhinet.eu/).

In addition, through the development of research and technology-transfer centres, and a transnational network, a platform for co-operation and the exchange of experience in RWHI management will be created. The network comprises micro-enterprises, non-governmental and public actors, academic/scientific institutions, and rural dryland local communities, especially farmers, women and youth groups (see http://afrhinet.eu/). Table 1 outlines the many advantages of using rainwater harvesting for irrigation purposes, as part of subsistence farming practices.

Fig. 1 Web site of the project AFRHINET

Table 1 Advantages of rainwater harvesting in subsistence farming

Item	Relevance
Support to yields and prevention of crop failure	Watering can reduce the chances of low yields and help to prevent crop failure
Flexibility of methods	The use of ponds, bawaries, step wells, tanks, etc. can be flexibly made, according to the site
Low technology	Use of methods entail no complex technologies
Low costs	Costs associated with rainwater harvesting methods are low
Durability	The chosen methods can be used over many years
Easy maintenance	Maintenance simple and effective

Perhaps one of the main advantages of rainwater harvesting for irrigation purposes is that people can be responsible for the maintenance and optimal use of water by themselves, ensuring the water use is the one which fulfils their needs, without wastages.

We do not only need a better understanding of the interdependencies between water, energy and food security. We also need to identify ways to reach synergies and trade-offs, involving the different actors and economic sectors, in a way that the communities which are so badly in need of water and energy security, can access it.

3 Conclusions

This paper has outlined some of the main features of the energy, environment and climate change nexus. But the progresses achieved to date should not make us overlook the fact that there are two main problems in promoting and in consolidating the energy, environment and climate change nexus:

1. Each theme is traditionally handled in separate, i.e. their interrelations are often ignored, and quite often only sectoral approaches are pursued. The lack of integration means that duplications of initiatives frequently take place and that substantial resources are therefore wasted.
2. Even though the combination of energy and climate strategies and policies would maximise their effectiveness and lead to real impacts, their pursuance in separate means that progress is modest and impacts are on a small scale.

In order to overcome the current conflicts, there is a need for a diversification of rural livelihood systems (Niehof 2004) and to improve policy responses (Osano 2012). Ultimately, new governance systems are needed (Knieling and Leal Filho 2012) so that benefits can be yielded in the long term.

The above state of affairs illustrates the fact that a new thinking is needed, which can cater for the integration of energy and climate issues in a more consistent way. Due to the fact that substantial amounts of money are spent in both areas, there is an urgent need to change the current state of affairs and seek more effective and long-term solutions, which may allow water, climate and energy issues to handled in a combined way, and in way which is commensurate with their importance.

As the contributions amassed on this book have shown, there have been various technological advances over the years, which have set a sound basis upon which the Nexus energy, environment and climate change can be further developed. The benefits such an approach would yield, namely a greater integration of two important and mutually complementary challenges (energy–climate change) and the corresponding synergies in the use of resources, means that a substantial contribution would be provided towards addressing climate change and sustainable energy use, two of the most important challenges of modern times.

References

Intergovernmental Panel on Climate Change (IPCC). (2014). *Fifth assessment report—Synthesis report*. Geneva: IPCC.

International Energy Agency. (2016). *World energy outlook*. Paris: International Energy Agency.

Juana, J. S., Kahaka, Z., & Okurut, F. N. (2013). Farmers' perceptions and adaptations to climate change in Sub-Sahara Africa: A synthesis of empirical studies and implications for public policy in African agriculture. *Journal of Agricultural Science, 5*(4), 121–135.

Knieling, J., & Leal Filho, W. (Eds.). (2012). *Climate change governance*. Berlin: Springer.

Knowler, D., & Bradshaw, B. (2007). Farmers' adoption of conservation agriculture: A review and synthesis of recent research. *Food Policy, 32*(1), 25–48.

Nhemachena, C. (2009). *Agriculture and future climate dynamics in Africa: Impacts and adaptation options*. Pretoria: University of Pretoria.

Niehof, A. (2004). The significance of diversification for rural livelihood systems. *Food Policy, 29,* 321–338.

Organisation for Economic Cooperation and Development (OECD). (2012). *OECD environmental outlook to 2050*. Paris: OECD.

Osano, P. (2012). Improving policy responses and finding sustainable solutions to drought in Kenya. *Africa Portal, 33,* 1–7.

Paavola, J. (2008). Livelihoods, vulnerability and adaptation to climate change in Morogoro, Tanzania. *Environmental Science & Policy, 11,* 642–654.

Pant, K. P. (2009). Effects of agriculture on climate change: A cross country study of factors affecting carbon emissions. *Journal of Agriculture and Environment, 10,* 72–88.

Water, U. N. (2013). *Water facts*. Rome: UN Water/FAO.

World Health Organization. (2004). *Health and global environmental change, series no. 2*. Geneva: WHO.

MIX
Papier aus verantwortungsvollen Quellen
Paper from responsible sources
FSC® C105338

If you have any concerns about our products,
you can contact us on
ProductSafety@springernature.com

In case Publisher is established outside the EU,
the EU authorized representative is:
Springer Nature Customer Service Center GmbH
Europaplatz 3, 69115 Heidelberg, Germany

Printed by Libri Plureos GmbH
in Hamburg, Germany